Lecture Notes in Logistics

For further volumes:
http://www.springer.com/series/11220

Michael Essig, Michael Hülsmann, Eva-Maria Kern,
and Stephan Klein-Schmeink (Eds.)

Supply Chain Safety Management

Security and Robustness in Logistics

Springer

Editors
Prof. Dr. Michael Essig
Bundeswehr University Munich
Neubiberg
Germany

Prof. Dr. Eva-Maria Kern
Bundeswehr University Munich
Neubiberg
Germany

Prof. Dr. Michael Hülsmann
Jacobs University Bremen
Bremen
Germany

Dr. Stephan Klein-Schmeink
Gesellschaft für Entwicklung,
Beschaffung und Betrieb mbH
Köln
Germany

ISSN 2194-8917
e-ISSN 2194-8925
ISBN 978-3-642-32020-0
e-ISBN 978-3-642-32021-7
DOI 10.1007/978-3-642-32021-7
Springer Heidelberg New York Dordrecht London

Library of Congress Control Number: 2012945329

Printed on acid-free paper

Springer is part of Springer Science+Business Media (www.springer.com)

Preface

Almost daily, newspapers are full of reports about insolvency, product recalls, and production stops. Such disruptive factors that pose a threat to a company's existence are not new, but have always been directly related to business operations. Moreover, due to the increase in labor division which, in turn, results from outsourcing activities – it is estimated that companies buy approx. 50% to 80% of their value added from outside – companies are increasingly facing disruptive factors that are beyond their control. Consequently, to meet these challenges, they have to adopt adequate, supply chain-oriented approaches and problem solving strategies. For this purpose, a high diversity of approaches – among them Supply Chain Risk Management, Supply Chain Security, and Crisis Management – has been provided by both academics and practitioners for many years now. The trend towards new and further developments is still continuing!

What are the salient characteristics of Supply Chain Safety Management? And what is our motivation for the comprehensive exploration of this concept in the course of this volume? We would like to put forward some arguments:

The current approaches offered for managing risk and uncertainty factors are characterized by their huge number. The issue of their diversity and individual specifics, however, has remained largely unanswered so far. In other words: Are all existing approaches equally suitable for companies that seek to manage their supply chains from a risk and uncertainty-related point of view? Or do distinguishing features exist, which may demonstrate that under certain conditions one approach is to be favored over another. This information is vital for a supply chain's functionality and a company's survivability. After all, hastily adopting an approach poorly suited to a company's or a supply chain's specific situation hardly helps them to manage underlying risks and uncertainties. Consequently, a systematized and structured analysis of the approaches offered is to be considered the basic prerequisite for getting an overview of the specific contributions and limitations of each approach.

Apart from this, according to the analysis of *Christopher/Holweg* (2011), the approaches currently offered lack sufficient depth. This results from the fact that their development came at a time and was based on conditions that do no longer exist. To put it more precisely: Since companies now act in an environment that is currently characterized by its turbulences and its high volatility, striving for stability is being replaced by striving for structural flexibility. As a consequence, the usefulness of the existing approaches has to be critically questioned.

Against this background, by developing Supply Chain Safety Management (SCSM), it is our ambition firstly, to bring together the existing preliminary works and the corresponding findings in a single integrated approach, and secondly, to be capable of meeting the currently prevailing conditions. The approach of SCSM includes various interrelated elements, which, in the course of managing risk and uncertainty factors, should generally be considered. These elements are part of a structured process, which is carried out in several phases and allows companies to make individual adjustments according to their specific situation. Hereby, companies shall be enabled to achieve an increased continuity of supply of their supply chains while also taking into consideration the economic goal of profitability. Thus, despite the existence of disruptive factors that supply chains are faced with, the aim is to efficiently and effectively provide customers with their requested goods and services.

The high variety of subscriptions within this book significantly contributes to emphasizing the integrative character of this approach. On the basis of the conceptual development of Supply Chain Safety Management, its underlying elements are analyzed by nationally and internationally recognized experts from science and practice. At the same time, feedback mechanisms between theoretical knowledge and practical experiences allow to sustainably strengthen our approach.

We hope that both scientists and practitioners will benefit from the book. Should this be the case, we would first like to express our deepest gratitude to the top-class selection of authors for their contributions. Thanks to them, we were able to launch this book. Lastly, we would like to thank Springer for the admission into its book program and for the excellent cooperation. However, Sandra Tandler has made a particularly significant contribution to the book: she was entrusted with the task of coordinating and editing, and consistently accompanied and pushed forward the whole process from the concept until its realization – many thanks for her excellent and dedicated work!

Munich, Cologne, Bremen
Summer 2012

Michael Essig
Eva-Maria Kern
Michael Hülsmann
Stephan Klein-Schmeink

Contents

1 Supply Chain Safety Management – The Concept

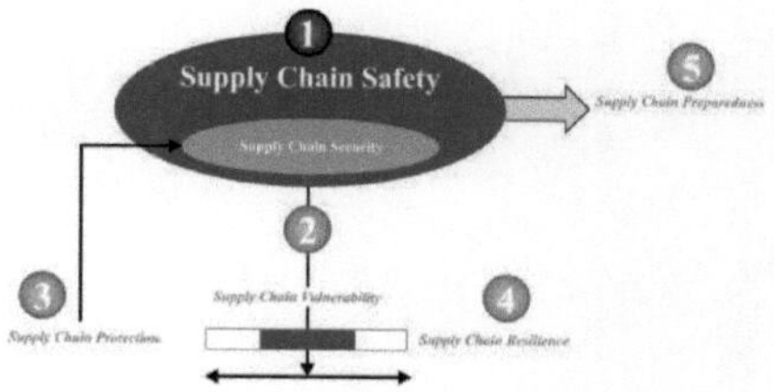

Conceptual Framework of Supply Chain Safety

Sandra Tandler and Michael Essig

Targets and Components of Supply Chain Safety Management: Structure of the Book

Eva-Maria Kern, Michael Hülsmann, Stephan Klein-Schmeink, and Michael Essig

Conceptual Framework of Supply Chain Safety

Sandra Tandler[1] and Michael Essig[2]

[1] Bundeswehr University Munich,
Research Assistant at the Chair of Materials Management and Distribution,
Werner-Heisenberg-Weg 39,
85577 Neubiberg,
Germany
sandra.tandler@unibw.de

[2] Bundeswehr University Munich,
Chair of Materials Management and Distribution,
Werner-Heisenberg-Weg 39,
85577 Neubiberg,
Germany
michael.essig@unibw.de

1 Introduction

In today's world, companies cannot afford to act on their own.[1] According to a study conducted by AMR Research in 2006 the average company has 36 contract manufacturers.[2] Out of this, 42 % of these companies report that more than 25 % of manufacturing output is produced by third-party contract manufacturers. Thus, it is evident from the study that outsourcing-related activities (predominantly concerning the functional areas of information technology, production, and logistics)[3] not only result in a higher number of companies involved in the supply chain – this in turn explains the fact that the length and the depth of a supply chain have increased manifold. In addition, companies manage today more than five different supply chains[4] because of the requirement to produce multiple products for multiple markets – supply chains can thus be characterized as networks. By consequence, companies experience a stronger mutual dependence on each other – the whole supply chain becomes more vulnerable to disruptions.

Due to conditions referring to drivers such as the number of customers, products, and variants, demand fluctuations, strategic suppliers, production sites, and distribution centres this trend is expected to be intensified over the next years. The Business Continuity Institute (2010) comes to the same conclusion:[5] Their study,

[1] See Roland Berger (2010), p. 4.

[2] The study is based on a survey of 455 companies in Europe and North America. See as follows AMR Research (2006).

[3] The same conclusion is found in a study conducted by FriceWaterhouseCoopers (2009), p. 8.

[4] This refers to the number of supply chains that, from a focal company's view, can be managed. In general, companies are part of even much more supply chains.

[5] See Business Continuity Institute (2010).

M. Essig et al. (Eds.): Supply Chain Safety Management, LNL, pp. 3–40.
springerlink.com

covering 35 countries, shows that 72% of the companies recorded (with an average of five) at least one and no more than 52 supply chain disruption/s in 2010. From the companies' perspective, the resulting consequences are: damages to brands or reputation, loss of productivity, and huge financial burdens. In contrast, from the customers' perspective, supply chain disruptions have a detrimental impact on the continuity of supply.

To take some examples: Nikon, a japanese camera and lens manufacturer suffered losses of production and, by consequence, supply shortages, that resulted from a flood damaging a manufacturing subsidiary located in the Rojana Industrial Park in Ayutthaya. In a statement of December 2011, Nikon estimated to push down net sales by 65 billion yen and operational income by 25 billion yen due to sales opportunity loss. Moreover, Nikon expects to restore its normal operations by the end of March 2012. Due to a recall of 18 million toys, Mattel sustained an estimated 30 million US dollars worth of damage in 2007. This kind of disruption was mainly attributed to a Chinese supplier who ignored Mattel's guidelines not to deploy toxic chemicals for the toy production. Additionally, Mattel took indirect financial damage in 2009 since it was imposed (by the US American Consumer Product Safety Commission) to pay a record fine of 2.3 million US dollars. In 2005, variations in quality of diesel pumps at Bosch resulted in significant losses of production at nearly all German automotive suppliers – among them were Daimler Chrysler AG, BMW AG, and Audi. The disruption had its origin in a purchase part of one of Bosch's suppliers that had been modified slightly. The outbrake of foot and mouth disease in Great Britain in 2001 forced car manufacturers like Volvo and Ford to stop production for certain car models. Due to the slaughtering of six million animals, leather to finish car interiors was not available – the estimated damage was around 12.6 million euros.

The named examples demonstrate not only the variety of disruptions of supply chains but also that risk and uncertainty factors take on a whole new dimension with the increasing integration of a company's activities as part of supply chain management. Beside new internal risks resulting from their own creation of goods and services, companies have to face up to a whole new set of supply chain- and environmental risks in the sense of a networking paradigm.[6] Thus, the adoption of a holistic concept is required that takes into consideration all supply chain risk and uncertainty factors and that develops appropriate action measures to maintain and enhance, the continuity of supply.

In principle, an extensive variety of concepts can already be found in the literature – what they have in common is their claim to make a crucial contribution to an increased continuity of supply in companies, respectively supply chains. However, a detailed analysis and comparison of the prevailing concepts is missing so far pointing out the individual achievements as well as the distinguishing features of each concept. Rather, statements can be found in the literature indicating that each concept can be characterized by its specific focus with regard to criteria, such as the underlying type of disruption, the object to be protected, or the management approach.

---

6 See Christopher/Peck (2004), pp. 4ff.

Based on this, it is the aim of this chapter to introduce a comprehensive overall concept called Supply Chain Safety Management (SCSM) which, at the same time, lays the foundation for the structuring of the book. Therefore, a systematic overview is given on the concepts that currently prevail in the literature. By doing this, similarities and differences existing between the concepts are emphasized. Finally, the results of the analyzed concepts are brought together and integrated in the concept of SCSM.

2 Preliminary Considerations

Our overview on existing, supply chain safety-related concepts is based on a literature analysis. In the course of this article, the literature analysis aims at identifying concepts for the management of risk and uncertainty factors. Here, a two-step procedure is adopted consisting of a general literature survey within monographs,[7] supplemented by a detailed analysis within 13 selected journals each having a particular focus on purchasing, logistics, and/or supply chain management[8].[9]

By combining both techniques, 16 concepts – including Risk Management, Supply Chain Risk Management, Business Continuity Management, Disaster Management, Emergency Management, Crisis Management, Supply Chain Event Management, Supply Chain Security und Supply Risk Management – have been identified. Beside this, six distinguishing features have been emerged which lay the foundation for conducting a consistent and stringent analysis of all identified concepts. These are as follows:

Type of disruption

Disruptions reflect conditions that result in a partial or entire interruption of an original plan of the operating processes.[10] In the literature, depending on the individual concept these conditions are described by specific types of disruptions which are presumed to differ not only in terms of terminology, but also in terms of content.

---

[7] Here, the so-called snowball technique is applied which is a heuristic technique to identify references which in turn lead to further references. See Ebster/Stalzer (2008), p. 45. Relevant sources are databases, such as Online Public Access Catalogue (OPAC) or websites, such as books.google.de and books.google.com. The selected search terms were as follows: Risiko (Risk) and Risikomanagement (Risk Management) as well as Supply Chain Risiko (Supply Chain Risk) and Supply Chain Risikomanagement (Supply Chain Risk Management).

[8] These include the International Journal of Logistics Management, International Journal of Logistics: Research and Applications, International Journal of Operations and Production Management, International Journal of Production Economics, International Journal of Production Research, Journal of Business Logistics, Journal of Physical Distribution and Logistics Management, Journal of Purchasing and Supply Management, Journal of Supply Chain Management, Production and Operations Management, Production Planning and Control, Supply Chain Management. An International Journal and Supply Chain Management Review.

[9] For a detailed discussion see Tandler/Essig (2011).

[10] See Czaja (2009), p. 222.

Management approach

Due to the character of risk and uncertainty factors possibly resulting in associated disruptions, an appropriate management approach has to be implemented. Here, different types can be distinguished, which differ among the individual concepts.

Process

(Strategic) management is executed in several (strategic) steps.[11] With regard to the management of risk and uncertainty factors, a process-oriented view is recommended as well. However, it is assumed that these steps differ among the individual concepts in terms of scope and level of detail.

Objective

Each of the concepts has a focus on managing risk and uncertainty factors – however, a focus on specific objects to be protected is to be supposed. This particularly refers to the protection of individuals, companies, parts of companies (e.g., functional areas), supply chains or parts of supply chains.[12]

Scientific discipline

Each concept has its origin in one or more specific scientific discipline(s). Thereby, conclusions can be drawn about the (horizontal) "proximity" of the individual concepts among one another as well as about the (vertical) "proximity" of each concept in relation to a holistic, all-embracing concept for the management of risk and uncertainty factors in supply chains.

Theoretical approach

By taking this feature into account, theoretical approaches can be captured that help explaining, respectively designing (specific facets of) a concept for the management of risk and uncertainty factors. At the same time, the theoretical approaches can be identified that prevail in the literature.

Table 1 provides an overview of the emerging grid which lays the foundation for analyzing and delineating each concept from one another.

Table 1 Grid for analyzing and delineating each concept from one another

Distinguishing feature	**Concept A**	**Concept B**	**Concept C**
Type of disruption			
Management approach			
Process			
Objective			
Scientific discipline			
Theoretical approach			

11 See a.o. Trux/Müller/Kirsch (1989), p. 6.

12 The differentiation of objects to be protected is based on Kaufmann (1973), p. 50ff., who emphasizes that the term is mainly used within the context of protecting individuals, objects, and conditions.

3 Analysis of the Identified Concepts

3.1 Risk Management

Risk Management (RM) can be considered the most popular concept for the management of risk and uncertainty factors. Thus, RM is the starting point for the identification of further concepts as well as for its further development at different levels.[13] Within the economic context the concept of RM has gained attention in the 1950s and 1960s in the US insurance industry for the first time.[14]

The concept focuses on the type of disruption called *risk*, whose strongly debated etymological origins are the Italian *risicare*,[15] the Arabic *risq*[16] or the Greek *rhiza*.[17] Risk is discussed within various scientific disciplines: Apart from engineering, mathematics, medicine and psychology, business administration, in particular – comprising areas such as insurance industry, decision sciences, finance, marketing, (strategic or international) management, accounting, operations and sales – have to be emphasized.[18] In consequence, various definitions of risk exist in the literature.[19] There is, however, an extensive agreement about the elements of the construct "risk": these are probability of loss and significance of loss.[20] Normally, risk is associated with negative deviations which can be expressed in terms of a threat, damage, loss, injury or other undesirable consequences.[21]

RM is predominantly adopted at the company-level[22] and aims at ensuring a company's survival, future success and at minimizing risk-related costs.[23] The latter shows that companies accept that a 100 % safety-level is not achievable – rather, their focus is on realizing an optimum cost-/benefit-ratio.[24] This results from the fact that the marginal benefit of the undertaken action measures is reduced by the increase of the safety level.

In the literature, the design of the RM-process differs in terms of level of detail (with respect to the number of steps and to the allocation of the associated activities). Goal setting reflecting the first step of the RM-process is comparatively

---

[13] The literature of the remaining concepts is often characterized by its references to the concept of RM and to the underlying risk construct, respectively.

[14] See Wolf/Runzheimer (2009), p. 30; Khan/Burnes (2007), p. 197.

[15] See Khan/Burnes (2007), p. 198; Bernstein (1996), p. 8.

[16] See Norrman/Lindroth (2004), p. 17f.

[17] See Romeike/Hager (2009), p. 31.

[18] For a detailed literature review see Rao/Goldsby (2009), p. 99; Manuj/Mentzer (2008), p. 134f.; Khan/Burnes (2007), p. 198; Wagner/Bode (2006), p. 303; Zsidisin (2003b), p. 217.

[19] See a.o. Manuj/Mentzer (2008), p. 134f.; Junginger (2005), p. 101.

[20] See Trkman/McCormack (2009), p. 249; Mitchell (1995), p. 115; The Royal Society (1992), p. 4.

[21] See Harland/Brenchley/Walker (2003), p. 52.

[22] See Kajüter (2007), p. 16.

[23] See Wolf/Runzheimer (2009), p. 31; Romeike (2003), p. 150; Mikus (2001), p. 11; Sauerwein/Thurner (1998), p. 35f.

[24] See as follows Romeike (2003), p. 253f.

seldom.[25] Normally, the goals of RM derive from the main objectives of the individual company and its risk attitude.[26] Rather, what the most RM-processes have in common is a structure based on the steps risk identification, analysis, assessment, management and control.[27] To manage the risk itself, in general both preventive and reactive action measures can be taken into consideration.[28] Preventive measures aim at eliminating the source of risk[29] and are therefore addressed to the probability of occurrence of a risk factor (the first element of the risk-construct). By contrast, reactive measures aim at minimizing the detrimental impact resulting from a risk that has occurred[30] and are therefore addressed to the significance of a risk factor (the second element of the risk-construct). However, conclusions for the management of risk can be drawn from the capital market theory[31] whose main characteristic is the distinction between unsystematic and systematic risks. Based on this view the adoption of risk-minimizing action measures is exclusively recommended for unsystematic risks as these can be influenced, whereas systematic risks cannot be influenced.

3.2 Supply Chain Risk Management

With the increasing tendency of co-operations between companies, concepts were developed that took the management of risks on the supply chain level into consideration. In this context, Supply Chain Risk Management (SCRM) is the most frequently represented concept in the SCM-literature. SCRM results from the intersection of the two areas SCM and RM[32] and, therefore, has its origin in business administration.

The concept focuses on the type of disruption called *supply chain risk*, which – in contrast to risks – has its origin no longer just in (company-related) internal and (company-, respectively supply chain-related) external, but also in supply chain-specific factors.[33] Consequently, supply chain risks occurring within or outside a supply chain (i.e., from raw supplier up to end customer) disrupt the

---

25 For a detailed overview see a.o. Moder (2008), p. 19 and Koppelmann (2004), p. 405ff.; Romeike (2003), p. 153; Mikus (2001), p. 13.

26 See Diederichs (2004), p. 12; Hahn/Hungenberg (2001), p. 38.

27 See a.o. Czaja (2009), p. 89; Trkman/McCormack (2009), p. 249; Junginger (2005), p. 194; Hallikas et al. (2004), p. 52; Pfohl (2002), p. 8; Brühwiler (2001), p. 80; White (1995), p. 36; Mensch (1991), p. 14.

28 See a.o. Czaja (2009), p. 89; Koppelmann (2004), p. 405ff.; Romeike (2003), p. 153; Rogler (2002), p. 29ff.; Brühwiler (2001), p. 80; Mikus (2001), p. 13.

29 See a.o. Götze/Mikus (2007), p. 46; Rogler (2002), p. 22f.

30 See a.o. Götze/Mikus (2007), p. 46; Rogler (2002), p. 22f.

31 See as follows Kajüter (2007), p. 17ff. and the references cited there.

32 See Paulsson (2007), p. 46; Jüttner/Peck/Christopher (2003), p. 198.

33 See Götze/Mikus (2007), p. 29; Jüttner (2005), p. 122; Jüttner/Peck/Christopher (2003), p. 201f. See also Kajüter (2003), p. 111, who indicates that the risks of a supply chain generally not correspond with the sum of all risks of the actors participating in the supply chain.

material-, information-, and/or financial flow and affect directly a company's ability to continue its operations respectively to deliver products and/or services to its (end) customer.[34] In this context, a neutral understanding is found at *Jüttner/Peck/Christopher* (2003).[35] However, supply chain risks are mainly associated with negative deviations.[36]

SCRM aims at reducing a supply chain's vulnerability as a whole[37] and can be divided into several process-oriented steps.[38] In this regard, conclusions can be drawn from various theoretical approaches. Agency theory, for instance, deals with relationships between principals and agents (i.e., actors who have individual motivations) which also exist in supply chains and which may cause (supply chain) risks. For this purpose, agency theory offers various approaches that give support to the management of supply chain risks.[39] Closely linked to this is the so-called transaction cost theory (TCT) which also focuses on relationships between actors and, by consequence, offers corresponding approaches.[40] However, the impact of proactive and reactive action measures, respectively, on the increased resilience of supply chains is emphasized by high reliability theory and normal accident theory.[41] Finally, portfolio theory can be considered a theoretical approach which specifies supply chain risks by distinguishing between two components (see also capital market theory).[42] Based on this, risks external to the supply chain that cannot be influenced correspond to systematic supply chain risk. By contrast, internal risks are subject to the control of a supply chain and thus, correspond to unsystematic supply chain risk. The distinction of these two components is particularly relevant for risk assessment and management. In this respect, approaches that allow the management of supply chain risks can be of preventive or of reactive[43] (but only rarely of proactive)[44] character. Inferences may be drawn from real options theory[45] which differentiates between different options (i.e., action measures) that can be exercised at different points in time. The process of SCRM is thus – except for the underlying type of disruption – quite similar to the process of RM.

34 See Jüttner (2005), p. 121f.; Jüttner/Peck/Christopher (2003), p. 200.
35 See Jüttner/Peck/Christopher (2003), p. 200.
36 See Kajüter (2003), p. 110.
37 See Jüttner (2005), p. 124; Jüttner/Peck/Christopher (2003), p. 201.
38 See for instance the process model of Manuj/Mentzer (2008), p. 137, which has been derived from a literature analysis.
39 See Ritchie/Brindley (2007), p. 304, p. 317; Khan/Burnes (2007), p. 208.
40 See Khan/Burnes (2007), p. 205 and the references cited there.
41 See Kleindorfer/Saad (2005), p. 56 and the references cited there.
42 See as follows Ritchie/Brindley (2007), p. 307f. in conjunction with Kleindorfer/Saad (2005), p. 55.
43 See Götze/Mikus (2007), p. 46; Jüttner/Peck/Christopher (2003), p. 206ff.
44 See Kleindorfer/Saad (2005), p. 56 and the references cited there.
45 See Cohen/Kunreuther (2007), p. 532, p. 536.

3.3 Supply Risk Management

Another concept to be discussed here is Supply Risk Management (SRM) which can be considered as a specific type of SCRM. Accordingly, the literature is filled with discussions surrounding SRM. The distinguishing feature is that SRM refers to dyadic supply chains – this means that only those risks are analyzed that exist at the interface between a purchasing company and its supplier(s).[46] Since these actors view each other as principal (purchasing company) and agent (supplier) respectively, agency theory[47] again, is appropriate for the analysis and management of such an interface.[48] Due to uncertainty from the principal's perspective – this follows from factors such as output uncertainty, goal conflict, length of relationship, adverse selection or moral hazard[49] – procurement risks need to be managed. Output-oriented and behavior-based contracts can be deemed to be appropriate solutions.[50] However, the concept focuses on the type of disruption called *supply risk* which occur at the interface between purchasing companies and its suppliers[51] and have detrimental effects.[52] The supply risk-construct is, analogously to the risk-construct, based on the two components: probability of occurrence and significance of impact.[53] According to *Zsidisin* (2003b), who heads the scientific discussion surrounding SRM, supply risk can be defined as "[…] the probability of an incident associated with inbound supply from individual supplier failures or the supply market occurring, in which its outcomes result in the inability of the purchasing firm to meet customer demand or cause threats to customer life and safety".[54] SRM therefore aims at ensuring the continuity of supply of the end customer. However, considering the design of the SRM-process, there is little evidence in the literature. *Harland/Brenchley/Walker* (2003) are the only authors who discuss a cyclic process including the steps *map supply network*, *identify risk and its current location*, *assess risk*, *manage risk*, *form collaborative supply network risk strategy* und *implement supply network risk strategy* comprising elements of risk identification, analysis and management.[55] For the purpose of conducting risk analysis, conclusions can be drawn from system theory.[56] Its aim is to identify and analyze input-output-systems which, in turn, make contribution

---

46 See Zsidisin (2003a), p. 20.

47 Agency theory is concerned with analyzing problems that may occur at the interface between a principal delegating tasks to an agent whose behaviour cannot be verified. See Eisenhardt (1989), p. 58f.

48 See as follows Zsidisin/Ellram (2003) and the references cited there.

49 For a literature review see Zsidisin et al. (2004), p. 399f.

50 See Eisenhardt (1989), p. 59ff.

51 See Zsidisin et al. (2004), p. 399.

52 See Zsidisin (2003a), p. 14.

53 See Zsidisin et al. (2004), p. 397.

54 Zsidisin (2003b), p. 222.

55 See Harland/Brenchley/Walker (2003), p. 56.

56 See Hallikas et al. (2004), p. 56.

to pointing out those critical interfaces that largely influence a supply chain with respect to its vulnerability. In this context, reactive as well as preventive action measures can be undertaken to manage supply risk.[57]

3.4 Crisis Management

Another concept that has to be considered here is Crisis Management (CM). The term CM has its origin in political science and is often associated with the Cuba crisis in 1962. In business administration however, the term is discussed for the first time with the beginning of the 1970s.[58] According to the literature analysis of *Natarajarathinam/Capar/Narayanan* (2009), CM is principally rather discussed at an economic than at a social level nowadays.[59]

The concept of CM focuses on the type of disruption called *crisis* which results from the Greek *krisis* and refers to German *Entscheidung* or *Wendepunkt*.[60] In this context, the Chinese synonym *wei-chi* highlights an important feature of a crisis: its ambivalence.[61] This means that a crisis can have both positive and negative impacts.[62] The term crisis is subject to various scientific disciplines – these include political and human sciences, psychology, theology, history, philosophy, medicine, laws, social sciences and economics.[63] Accordingly, the literature is filled with definitions that refer to the term crisis.[64] *Krystek* (1987) – one of the most popular authors in crisis research – describes crises as "[...] Bruch einer bis dahin kontinuierlichen Entwicklung und im engeren Sinne eine Entscheidungssituation, die den Wendepunkt bzw. Höhepunkt einer gefährlichen Entwicklung markiert".[65] A similar, but more concrete definition can be found at *Müller* (1985) who defines a crisis as "[...] an unwanted event which always seriously threatens the continued existence of the firm."[66] CM is thus characterized by its focus on specific risks – those that have the potential to substantially threaten the company's survival.[67] According to *Turner* (1994, 1976) who is deemed to be the founder of modern crisis research, crises are less the result of normal operations but more the result of a long incubation period.[68] Their occurrence is further

---

57 See Zsidisin et al. (2004), p. 397; Zsidisin/Ellram (2003), p. 15f.
58 See Fürst/Sattelberger/Heil (2007), p. 29; Schulten (1995), p. 3; Krystek (1987), p. 89.
59 See Natarajarathinam/Capar/Narayanan (2009), p. 546.
60 See Krummenacher (1981), p. 3.
61 See Krystek (1987), p. 3.
62 See Fürst/Sattelberger/Heil (2007), p. 11f.; Trauboth (2002), p. 13; Saynisch (1994), p. 52.
63 See Fürst/Sattelberger/Heil (2007), p. 12; Hülsmann (2005), p. 35; Burger (1988), p. 5.
64 See Krystek (1987), p. 3 in conjunction with Gabele (1981), p. 151; Hülsmann (2005), p. 36f. and the references cited there.
65 Krystek (1987), p. 3.
66 Müller (1985), p. 39.
67 See Moore/Lakha (2006), p. 86; Krystek (1987), p. 6f.; Müller (1982), p. 1. See also Mitroff/Shrivastava/Udwadia (1987), p. 283.
68 See as follows Turner (1994), p. 215ff.; Turner (1976), p. 389f.

intensified by sloppy management and problems that relate to information processing. In this context, conclusions explaining the causes of crises can be drawn from information economics.

As already indicated CM which is located at a company-level[69] aims at ensuring the company's survival.[70] Accordingly, ensuring solvency, realizing minimum profit and maintenance of potential for success can be deemed relevant factors.[71] While in recent years CM – in terms of *Supply Chain Crisis Management* (SCCr) – is discussed at a supply chain-level as well,[72] it still lacks a specification of the underlying type of disruption. Rather, supply chain crisis is seen as a general disruption resulting from the activities of one or more companies of the supply chain;[73] here, the characteristic feature of a crisis is not emphasized so far. Apart from this, the literature is filled with recommendations for the process design of CM.[74] However, typically[75] the process is divided into the steps of *signal detection, preparation/prevention, containment/damage limitation, recovery* and *learning*, which all embrace elements of risk identification, management and control.[76] What is remarkable (as they are not explicitly mentioned) is the exclusion of the steps of risk analysis and assessment. A possible reason for this is the fact that crises already reflect specific risks that do need no further prioritization. Instead, the process comprises the step of *preparation/prevention* which is based on the assumption that not every crisis can be eliminated. In consequence, action measures have to be undertaken that address to the occurrence of a crisis. Despite this proactive element, CM is primarily marked by its reactive character.[77]

3.5 Business Continuity Management

Business Continuity Management (BCM) is a comparatively new concept which deals with the management of risk and uncertainty factors and which is less discussed from a scientist's perspective and more from a practitioner's perspective at

---

69 See Kajüter (2007), p. 16; Krystek (2006), p. 45; Krystek (1987), p. 4ff.; Mitroff/Shrivastava/Udwadia (1987), p. 283ff., p. 291; Müller (1985), p. 38; Pearson/Mitroff (1993), p. 48.

70 See Müller (1985), p. 40, p. 42.

71 See Krystek (1987), p. 6f.; Müller (1982), p. 1.

72 See Natarajarathinam/Capar/Narayanan (2009), p. 535.

73 See Natarajarathinam/Capar/Narayanan (2009), p. 537.

74 For a detailed overview see e.g. Natarajarathinam/Capar/Narayanan (2009), p. 546. See also Krystek (1987), p. 92; Müller (1982), p. 48, which, in terms of the chosen terminology, strongly refer to the steps of the RM process.

75 See Natarajarathinam/Capar/Narayanan (2009), p. 546, who come to the same results in their literature analysis.

76 See Mitroff/Shrivastava/Udwadia (1987), p. 284f. and Pearson/Mitroff (1993), p. 52ff., the steps of which can be also found at Proff (2009), p. 209f.; Töpfer (2009), p. 185; Fürst/Sattelberger/Heil (2007), p. 29ff.

77 For a detailed discussion of CM in terms of its content and its focus see Rüsen (2009), p. 197f. See also Fürst (2004), p. 59.

that moment.[78] The term BCM has been mentioned for the first time in 1987 in the Disaster Recovery Journal.[79] As the journal addresses issues that relate to the area of (the US) information technology, this is also the origin of the concept. In detail, BCM is associated with the assumption that the disruption of information systems leads to a loss of billions of US-dollars.[80] This prognosis is widely known as the so-called *year 2000 problem* (Y2K) resulting in a historically unique "catch-up race" to safeguard for as many information systems as possible.[81] In this context, Y2K is also deemed to be a turning point for the concept of BCM.[82] This is expressed by the fact that BCM no longer applies to the area of information technology[83] (in this case, the concept is called Disaster recovery[84]), but to the whole company.[85]

The concept of BCM focuses on *continuity* which, from the company's perspective, can be considered the desirable condition.[86] Contrary to the remaining concepts, BCM lacks a focus on a concrete type of disruption. Rather, some authors refer to disruptions (of normal operations)[87] that are neither identifiable nor quantifiable[88] and use terms, such as *das Unbekannte* and *the unknown*[89] respectively, *the unexpected*[90] or *surprises*[91]. Nonetheless, BCM focuses on that type of disruptions that have a low probability of occurrence and a high significance of impact.[92]

---

78 See Zsidisin/Melnyk/Ragatz (2005a), p. 3402. Instead, the current literature is highly characterized by providing a guideline to companies which enables them to implement the concept. See a.o. the literature of Wieczorek/Naujoks/Bartlett (2010); Hiles (2004); Doughty (2001).

79 See von Rössing (2005), p. 26.

80 See Elliott/Swartz/Herbane (2010), p. 3; Heath (2008), p. 48; Moore/Lakha (2006), p. 3; Engel (2005), p. 41; von Rössing (2005), p. 26.

81 See von Rössing (2005), p. 28f.; see also Moore/Lakha (2006), p. 3.

82 See Peck (2006), p. 136.

83 See Herbane/Elliott/Swartz (2004), p. 439.

84 See Wieczorek/Naujoks/Bartlett (2010), p. X; Herbane/Elliott/Swartz (2004), p. 438. Apart from this, some authors also refer to crisis management and contingency management being further precursors of BCM. See Herbane/Elliott/Swartz (2004), p. 438; Norrman/Jansson (2004), p. 439 in conjunction with Chartered Management Institute (2002).

85 See Norrman/Jansson (2004), p. 439.

86 See Waters (2007), p. 218; Gibb/Buchanan (2006), p. 129; Moore/Lakha (2006), p. 5f. and the literature cited there; Engel (2005), p. 41, p. 45; von Rössing (2005), p. 33; Herbane/Elliott/Swartz (2004), p. 436f.; Norrman/Jansson (2004), p. 437f.

87 See Business Continuity Institute (2007), p. 5; Graham/Kaye (2006), p. 11, p. 15.

88 See Waters (2007), p. 214ff.

89 See Graham/Kaye (2006), p. 304; McKee/Guthridge (2006), p. 33; von Rössing (2005), p. 33, p. 35.

90 See Sikich (2008), p. 127; Elliott/Swartz/Herbane (2010), p. 126.

91 See Waters (2007), p. 214.

92 See Waters (2007), p. 231; Moore/Lakha (2006), p. 3; Engel (2005), p. 42; Norrman/Jansson (2004), p. 437f.

Currently, the concept primarily applies at the company-level;[93] besides this, an increased discussion surrounding *Supply Chain Continuity* (SCC) (Management) – here the terms Business and Supply (chain) continuity planning[94] are used – can be found in the literature. However, at a supply chain-level, SCC is rather deemed to be an instrument (e.g., for the management of supply risks) than a holistic concept.[95] Apart from this, BCM generally aims at restoring or improving operations or ensuring the continuity of operations after a disruption has occurred.[96] The objectives show that BCM is explicitly based on the assumption that disruptions cannot be eliminated as a whole; thus, the resulting detrimental impacts on a company's (or a supply chain's) operations have to be minimized.[97] BCM is routed through a process including several successive steps.[98] *Elliott/Swartz/Herbane* (2010) who rank among those authors that discuss the concept from a more scientific perspective, distinguish between the steps of *initiation and redefinition*, *planning for business continuity*, *implementation* and *operational management*. A characteristic feature is that these steps include elements of a holistic process.[99] However, Business continuity planning (here: *Planning for business continuity*) is deemed to be the core of the whole concept.[100] Its focus is on developing plans[101] which enable companies to respond to disruptions in a timely manner and thus, to proactively increase their *Preparedness*.[102] In this context, parallels can be drawn with the concept of CM which is seen as one of the precursors of BCM.[103] In contrast, BCM – as already indicated by the underlying objective – is mainly of proactive character.[104] The significance of Business continuity planning (e.g., with respect to the management of supply risks) is also highlighted by (open) systems theory and by institutional economics.[105] Due to their characteristic

---

93 See Gibb/Buchanan (2006), p. 129; Herbane/Elliott/Swartz (2004), p. 435, p. 438; Elliott/Swartz/Herbane (2010), p. 48.

94 See Zsidisin/Melnyk/Ragatz (2005a), p. 3402, p. 3412; Zsidisin/Ragatz/Melnyk (2005b), p. 46.

95 See Zsidisin/Melnyk/Ragatz (2005a), p. 3402; Elliott/Swartz/Herbane (2010), p. 158ff.

96 See Waters (2007), p. 218; Gibb/Buchanan (2006), p. 129; Moore/Lakha (2006), p. 5f. and the references cited there; Engel (2005), p. 41, p. 45; von Rössing (2005), p. 33; Herbane/Elliott/Swartz (2004), p. 436f.; Norrman/Jansson (2004), p. 437f.

97 See Gibb/Buchanan (2006), p. 129; Norrman/Jansson (2004), p. 437f.

98 For a detailed overview of different approaches see e.g .Gibb/Buchanan (2006), p. 129 and Zsidisin/Melnyk/Ragatz (2005a), p. 3405. See also Waters (2007), p. 224ff.; Moore/Lakha (2006), p. 9ff.; von Rössing (2005), p. 45ff.; Cornish (2004), p. 107; Elliott/Swartz/Herbane (2010), p. 5.

99 See below Elliott/Swartz/Herbane (2010), p. 5, p. 94ff.

100 See implicit Barnes (2011), p. 166; Business Continuity Institute (2007), p. 7.

101 See Elliott/Swartz/Herbane (2010), p. 137ff.; Norrman/Jansson (2004), p. 439; Rice/Caniato (2003), p. 27.

102 See Rice/Caniato (2003), p. 27, which refer to the term *resilience* instead of *preparedness*. However, conducting a business impact analysis which consists of the analysis of the company and its environment is deemed to be the most important element of the entire process.

103 See Elliott/Swartz/Herbane (2010), p. 4.

104 See Gibb/Buchanan (2006), p. 133ff.; Moore/Lakha (2006), p. 6f.; Elliott/Swartz/Herbane (2010), p. 101.

105 See as follows Zsidisin/Melnyk/Ragatz (2005a), p. 3402ff.

feature as open systems, companies are facing environmental conditions threatening the company's survival which require the adoption of appropriate instruments and strategies. Considerations of institutional economics show that these are almost identical for all companies. This in turn results from the fact that companies are subject to similar regulatory and legal circumstances.

3.6 Safety Management

Safety Management (SM) is a further concept which is rarely debated from the scientist's perspective at that moment. In the literature, SM is less discussed within the area of business administration[106] and more discussed within the area of information technology[107] which is assumed to be the origin of the concept.[108]

The concept of SM focuses on *Sicherheit* which is rooted in the Italian *securitas* and indicates a desirable condition instead of a type of disruption.[109] Apart from information technology, the term is also subject to further scientific disciplines including political, social and cultural sciences, laws, ecology, business administration and psychology.[110] In consequence, numerous definitions exist in the literature.[111] In the Anglo-American region, three types of safety can be distinguished that all fit to the business administration-context:[112] *Certainty* is grounded in knowledge; *safety* is interpreted as a kind of protection from unintended events – these include human and technical failure as well as forces of nature; *security* is interpreted as a kind of protection from intended events (i.e., attacks) – these include active and passive attacks as well as criminal acts. Safety research is filled with approaches from occupational psychology, organizational sociology and engineering sciences that are concerned with analyzing various types of safety (occupational vs. process safety), system components (human, human-machine-interaction, entire system) and the resulting recommendations.[113]

---

106 See a.o. Gundel/Mülli (2009); Müller (2005).

107 See the corresponding literature.

108 See Burns/McDermid/Dobson (1992), p. 3, who find out that the terms *safety* and *security* are often used to characterize computer systems.

109 See Kaufmann (1973), p. 52f., p. 70.

110 See a.o. Bridi (2008); Witt (2006); Brands (2005); Gärtner (2005), p. 127; Müller (2005); Faust (2002), p. 85f. and the references cited there; Kaufmann (1973), p. 55ff.

111 For a detailed discussion surrounding the term safety see Künzler (2001), p. 8ff. According to *United Nations Development Programme* (UNDP) (1994), (human) security can be subdivided into seven categories (which correspond to various scientific disciplines): economic security, food security, health security, environmental security, personal security, community security, and political security. See United Nations Development Programme (1994), p. 27.

112 See as follows Sheffi (2006), p. 317; Witt (2006), p. 66f.; Müller (2005), p. 17 in conjunction with p. 28f. See also Gundel/Mülli (2009), p. 4ff. and Brühwiler (2001), p. 7.

113 See Künzler (2001), p. 58 and for a detailed overview p. 28ff. See also Grote/Künzler (1996), p. 6ff.

Within the business administration-context, SM is located at the company-level[114] and aims at achieving a certain safety level which is individual for each company.[115] This is associated with the awareness that disruptions occur,[116] which in turn require the adoption of both preventive and reactive action measures. For this purpose, SM is divided into several steps[117] which correspond not only to the process of RM but also include elements of goal setting.[118] Consequently, the formulation of (protection-related) objectives and of the desired safety level respectively, is deemed to be the core of the process and the distinguishing feature of SM to RM.

3.7 Supply Chain Security

As it is the desired condition of Supply Chain Security (SCS) Management, *security* – a special type of safety – is recently discussed intensively at a supply chain-level as well. Often, the emergence of the concept is associated with the events of September 11, 2001[119] which resulted in an increased awareness on adopting action measures that make contribution to a better protection of supply chains, of the companies involved and the whole society. According to *Department of Homeland Security* (2007), SCS is of high relevance for ensuring national security.[120] Legal regulations– as also shown in institutional economics[121] – are thus considered to be the main drivers forcing the practical implementation of SCS.[122]

Contrary to the concept of SM discussed before, SCS specifically focuses on *security* and is rooted within business administration, especially within the areas of SCM and logistics.[123] Due to the events of September 11, 2001, security in terms of protecting assets during their distribution along the supply chain has undergone a substantive realignment:[124] The focus is no longer on protecting assets from leaving the supply chain (e.g., via theft) alone, but also on protecting assets from unauthorized entries into the supply chain (e.g., contraband, people or weapons of mass destruction).[125]

---

114 See Gundel/Mülli (2009), p. 3.
115 See Gundel/Mülli (2009), p. 4; Müller (2010), p. 26.
116 See Gundel/Mülli (2009), p. 4.
117 See Müller (2010), p. 26; Gundel/Mülli (2009), p. 7, p. 10ff.
118 See as follows Gundel/Mülli (2009), p. 7, p. 10ff. See also Müller (2005), p. 55.
119 See a.o. Lee/Whang (2005), p. 289 and as follows Williams et al. (2009), p. 595f.
120 See Department of Homeland Security (2007), p. 10.
121 See Williams et al. (2009), p. 596 and the references cited there.
122 See Williams/Lueg/LeMay (2008), p. 255. For a detailed overview on various initiatives founded by public institutions see Department of Homeland Security (2007), p. 64ff.
123 See Autry/Bobbitt (2008), p. 43; Williams/Lueg/LeMay (2008), p. 259; Hale/Moberg (2005), p. 196; Closs/McGarrell (2004), p. 17.
124 See Williams/Lueg/LeMay (2008), p. 255 and the references cited there.
125 See Closs/McGarrell (2004), p. 7.

Based on this, *Closs/McGarrell* (2004) define SCS as "[t]he application of policies, procedures, and technology to protect supply chain assets (product, facilities, equipment, information, and personnel) from theft, damage, or terrorism and to prevent the introduction or unauthorized contraband, people or weapons of mass destruction into the supply chain […]."[126] Although the significance of "[…] developing and maintaining an organizational culture that understands the importance and processes of security […]" is emphasized by *Williams/Lueg/LeMay* (2008),[127] there is no evidence found in the literature so far that contributes to the design of such a process.[128] However, further characteristic features surrounding the concept of SCS can be derived from the definition:

- The main objective is protecting a supply chain and its assets.[129]
- The concept focuses on those types of disruptions that refer to any attacks which are intended by human acts.[130]
- The focus is more on protecting the material flow and less on protecting the information flow.
- For the purpose of managing security, the adoption of preventive instead of reactive action measures is recommended.[131]

3.8 Uncertainty Management

Uncertainty Management (UM) is a concept which is discussed both at a company-[132] and a supply chain-level[133]. Similar to SM, there is only sporadic consideration for UM. A possible reason for this is: UM focuses on the type of disruption called *uncertainty* which is, implicitly and/or explicitly, subsumed by other concepts (e.g., BCM) as well. The term is distinguished by two characteristic features: Firstly, in contrast to risk, uncertainty has no negative connotation.[134] Secondly, a significant correlation between the terms uncertainty and information can be found.[135] In general, the latter emphasizes the absence of information[136] which can take the form of missing, incomprehensive or undifferentiated information.[137] In

126 Closs/McGarrell (2004), p. 8.
127 See Williams/Lueg/LeMay (2008), p. 262.
128 Instead, processes of other concepts or approaches, such as Emergency management or Six sigma are adopted. See Lee/Whang (2005), p. 293; Helferich/Cook (2002), p. 52ff.
129 See Williams et al. (2009), p. 596.
130 See ähnlich Williams et al. (2009), p. 599.
131 See Closs/McGarrell (2004), p. 8. See also Williams et al. (2009), p. 607.
132 See Grote (2009), p. 11.
133 See Prater (2005), p. 524.
134 See Grote (2009), p. 47.
135 See Grote (2009), p. 12 and Prater (2005), p. 524f., who discuss the significance of information in terms of information systems contributing to the management of uncertainty.
136 See Grote (2009), p. 12.
137 See Lipshitz/Strauss (1997), p. 151.

this context, information economics can be deemed to be an appropriate approach as it analyzes the existing relationship between uncertainty and information. Beside this, conclusions can be drawn from (open) systems theory which gives support to the explanation of the causes of uncertainty. Here, uncertainty results from a company's dependence from its environment and therefore calls for the adoption of appropriate strategies to minimize or to respond to uncertainty factors.[138] However, resource dependence theory makes a contribution to the design of UM:[139] Due to restricted options for control, it is recommended to adopt only strategies for those (selected) uncertainty factors that threaten critical resources and in consequence, the company's survival. Chaos theory which refers to complex, non-linear dynamic systems is based on a similar principle:[140] Based on the differentiation between non-deterministic (i.e., non-controllable) and deterministic (i.e., controllable) types of chaos, it can be shown that only the latter can be managed by using both preventive and reactive action measures.[141] Accordingly, activities that refer to UM have to pass through a process.[142] However, a holistic process model is missing so far – rather, a process orientation can be found.

3.9 Failure Management

A further concept focusing on risk- and uncertainty factors can be seen in Failure Management (FM). Again, only a few publications exist highlighting an interesting feature: The concept of FM, within the scientific discipline of business administration, both located at a company-level[143] and at a specific part of the company (here: manufacturing).[144]

The concept of FM as well as its underlying type of disruption (*failure*) is subject to various scientific disciplines including engineering sciences, occupational sciences and psychology.[145] Accordingly, numerous definitions exist that describe the term failure which can be seen as a deviation from an optimum condition or procedure[146] or as a characteristic not fulfilling the underlying requirements[147]. Thus, failures are associated with neutral connotations,[148] which can take the form of both positive and negative deviations.[149] According to

138 See Grote (2009), p. 15f., p. 30f. and the references cited there.
139 See Grote (2009), p. 17 and the references cited there.
140 See Prater (2005), p. 530f. and the references cited there.
141 See Grote (2009), p. 30ff.
142 See Grote (2009), p. 44ff.
143 See Mistele (2007), p. 41ff.
144 See Tao et al. (2010); Dangoumau/Craye/Lorimier (2008).
145 See Mistele (2007), p. 40.
146 See Ertl-Wagner/Steinbrucker/Wagner (2009), p. 148.
147 See Ertl-Wagner/Steinbrucker/Wagner (2009), p. 148.
148 See Mistele (2007), p. 40.
149 See Reason (2008), p. 29.

failure research,[150] failures can have different causes.[151] From the perspective of human-based approaches, failures result from individuals that act in a careless, unmindful, insufficient motivated or sloppy manner. From the perspective of technical-oriented approaches, failures result from the insufficient adaptation of machines in general and of machines to the individuals' needs in particular. Finally, from the perspective of system-oriented approaches, failures result from the coincidence of several technical, organizational, structural and human factors.

In the literature, hints can be found indicating that the concept of FM is based on the assumption that not all failures can be eliminated. Thus, the focus is not only on preventive, but also on proactive and reactive action measures.[152] This aspect is also subject to various theoretical approaches which are characterized by their respective focus on deriving recommendations that refer to specific types of failure.[153] Apart from this, the development of action measures is one element of the entire FM-process which complies with the RM-process to a large extent.[154]

3.10 Disruption Management

Disruption Management (DiM) is another concept that refers to the management of risk and uncertainty factors. In general, the term *disruption* is subject to various scientific disciplines including engineering sciences and business administration.[155] The latter distinguishes between two conflicting options. One group discusses DiM as a concept focusing on the analysis of production systems.[156] Consequently, disruptions are viewed as unplanned deviations from planned processes or from a systemic condition that has to be achieved.[157] According to considerations from open system theory, their occurrence results from a system's dependence of its complex environment.[158] Due to the detrimental impacts of disruptions on the efficiency[159] (in terms of capacity-, time- and cost-related objectives[160]), an adequate management is needed. For this purpose, the concept comprises preventive, anticipative and reactive action measures.[161]

---

150 This is to be considered a specific discipline which is part of security, risk, and accident research. See Mistele (2007), p. 40.
151 See as follows Mistele (2007), p. 41ff. and the references cited there.
152 See Reason (2000), p. 769 in conjunction with Reason (1997), p. 125; Hoyos (1992), p. 14.
153 See Weingardt (2004), p. 147 and the references cited there.
154 See a.o. Ellouze (2010), p. 44ff.; Michell-Auli/Schwemmle (2008), p. 150f.
155 See Heil (1995), p. 15ff.
156 See Fischäder (2007), p. 27; Patig (1999), p. 1.
157 See Fischäder (2007), p. 27 and the references cited there. See also Heil (1995), p. 1.
158 See as follows Fischäder (2007), p. 20 and the references cited there.
159 See Heil (1995), p. 1.
160 See Patig (1999), p. 1.
161 See Fischäder/Schneider (2009), p. 295f. in conjunction with Patig (1999), p. 7 and Fischäder (2007), p. 31ff.; Patig/Thorhauer (2002), p. 355; Patig (1999), p. 1.

By contrast, another group discusses the term disruption (instead of the concept) at a supply chain-level.[162] However, a clear definition is missing so far – rather, disruptions are viewed as unplanned events resulting in a deviation from the plan,[163] as any kind of events[164] or as unpredictable, inevitable events.[165] What the majority of the definitions have in common is the negative character of disruptions.[166] Apart from this, disruptions can occur both internal and external to a supply chain[167] and in turn, result in Supply chain risks.[168] Accordingly, the management of disruptions is primarily part of other concepts, such as SCRM.[169]

3.11 Incident, Problem and Event Management

Incident Management (IM), Problem Management (PM) and Event Management (EM) are three other concepts that take the management of risk and uncertainty factors at an IT level into consideration and thus can be assigned to a specific corporate division. From a scientific view, the concept is only implicitly discussed. This means that the terms *incident* and *event* are, without being specified, generally used for describing disruptions or risks in supply chains. By contrast, from a practical view, it can be found that the concepts are extensively discussed within the context of a super ordinate approach.[170] Thus they do not meet the requirements of being holistic concepts.

The focus of IM is on the type of disruption called *incidents*[171] which describe outages or (programming) errors. Incidents cause unplanned interruptions or quality reductions of an IT service[172] and are either notified by end users or technical staff, respectively detected automatically by monitoring tools.[173] IM aims at rapidly restoring normal operations, at minimizing negative effects resulting from incidents[174] and at achieving the agreed service levels.[175] For this purpose, a

162 See Cauvin/Ferrarini/Tranvouez (2009), p. 430, p. 434.
163 See Cauvin/Ferrarini/Tranvouez (2009), p. 429.
164 See Kovács/Tatham (2009), p. 216.
165 See Kovács/Tatham (2009), p. 216; Skipper/Hanna (2009), p. 406.
166 See Kovács/Tatham (2009), p. 216; Skipper/Hanna (2009), p. 404f.; Wagner/Bode (2006), p. 301; Hendricks/Singhal (2005), p. 35. In contrast to this, a neutral view is taken by Jüttner/Peck/Christopher (2003), p. 200.
167 See Wagner/Bode (2006), p. 303.
168 See Wagner/Bode (2006), p. 303.
169 Only *Cauvin/Ferrarini/Tranvouez* (2009) deal with the management of disruptions with an own concept but touch the supply chain level at best implicit.
170 This is an ITIL-approach based on five process fields. Each field has different subgoals and –tasks and integrates adequate individual concepts and elements respectively. Among these, an intense interconnectedness can be observed emphasizing their complementary character. See a.o. van Bon (2008b), p. 38; Bächle/Kolb (2007), p. 15ff.; Zarnekow/Hochstein/Brenner (2005), p. 19f.
171 See van Bon (2008b), p. 144; Beims (2009), p. 139.
172 See van Bon (2008b), p. 144; Beims (2009), p. 139; Brunnstein (2006), p. 71.
173 See van Bon (2008b), p. 144; Beims (2009), p. 139f.
174 See Beims (2009), p. 139; See Victor/Günther (2005), p. 34.

standardized process including the steps of *incident detection and recording, initial classification and support, investigation and diagnosis, resolution and recovery, incident closure* and *ownership, monitoring, tracking* and *communication* is postulated[176] which largely coincides with the steps of the RM process. Hereby, it becomes evident that IM is primarily of reactive character.[177]

The focus of PM is on the type of disruption called *problems* which can be considered to be the unknown cause of one or several incidents.[178] The objective is to prevent incidents from occurring and to minimize negative effects resulting from it.[179] Similar to IM a standardized process is postulated for PM as well which largely corresponds to the process of RM.[180] In the literature, evidence can be found that the concept of PM also provides reactive action measures.[181] However, due to its focus on developing preventive action measures to minimize or to eliminate the probability of occurrence of incidents, PM is primarily of preventive character.[182]

Finally, the focus of EM is on the type of disruption called *events*. Generally, events can be defined "[…] as any detectable or discernible occurrence that has significance for the management of the IT infrastructure or the delivery of IT service and evaluation of the impact a deviation might cause to the services."[183] As EM enables companies to early detect incidents, it is closely connected with IM.[184] Thus, EM aims at identifying, analyzing and managing events[185] and is based on a largely standardized process including steps that correspond with the process of RM.[186] However, in contrast to PM, EM mainly refers to reactive action measures.

---

175 See Victor/Günther (2005), p. 34. In this regard distinct overlaps with the BCM-concept can be identified, since the latter is based on a similar but at a company level aggregated objective.

176 See Office of Government Commerce (2002), p. 73ff., and for the steps in detail see p. 36f. The steps of the IM-process are to a large extent congruent with the content of the steps of the RM-process – this also applies to the missing integration of the goal setting in a first step.

177 See a.o. Ebel (2008), p. 463; Victor/Günther (2005), p. 46.

178 See Beims (2009), p. 151; Ebel (2008), p. 484; Thejendra (2008), p. 82; Addy (2007), p. 163; Brenner (2007), p. 38.

179 See Beims (2009), p. 151; See also Brenner (2007), p. 38.

180 See Thejendra (2008), p. 84ff. as well as Ebel (2008), p. 490; Addy (2007), p. 166; Brenner (2007), p. 38.

181 See a.o. Beims (2009), p. 153; Thejendra (2008), p. 83.

182 See Addy (2007), p. 164; Brenner (2007), p. 38; Köhler (2006), p. 82. Among the IM and PM-concept temporal and textual interdependencies exist. However, while the reactive-driven IM aims at restoring business processes after an incident has occurred, the preventive-driven PM deals with the source(s) of potential incidents. See Brenner (2007), p. 38.

183 Office of Government Commerce (2007), p. 94.

184 See Office of Government Commerce (2007), p. 95.

185 See van Bon (2008a), p. 285.

186 See Office of Government Commerce (2007), p. 95 in connection with van Bon (2008a), p. 286ff.

3.12 Supply Chain Event Management

Supply Chain Event Management (SCEM) is a further concept for the management of risk and uncertainty factors which can be considered to be a specific type of EM. In the literature, it is also described as being a software solution and a software component respectively.[187] Thus, the focus of SCEM is on the type of disruption called *events*, too.[188] However, events can be seen here as milestones,[189] which show deviations between a target and an actual state[190] that have to be minimized.[191] As the concept aims at reducing complexity in terms of managing the flood of information, only specific events – those that are of high relevance and require the need of implementing action measures – are taken into consideration.[192]

SCEM can be considered to be a new concept integrating various approaches of different scientific disciplines[193] such as business administration[194] and information technology.[195] On the one hand, SCEM is influenced by the business management-related approaches of Management by exception (MbE)[196] and Event-driven planning.[197] Both approaches are characterized by their flexible and rapid reaction to environmental changes[198] and thus, explain the proactive[199] as well as reactive[200] management of (supply chain) events. On the other hand, SCEM is influenced by the information technology-related[201] approach of Tracking & Tracing (T&T) systems[202] which explains the objective of SCEM. Due to their integration of the material and information flow, T&T systems do not only contribute to the cross-company provision of information along the supply chain.[203] Moreover,

---

187 For a detailed article see Otto (2003). See also Karrer (2003), p. 189 and Steven/Krüger (2004), p. 184, according to which SCEM is both a concept and an information system.

188 See Bensel/Fürstenberg/Vogeler (2008), p. 5; Steven/Krüger (2004), p. 181.

189 See Otto (2003), p. 2f.

190 See Heusler/Stölzle/Bachmann (2006), p. 21; Stölzle (2008), p. 542.

191 See Otto (2003), p. 3.

192 See Heusler/Stölzle/Bachmann (2006), p. 20f.; Otto (2003), p. 1.

193 See Otto (2003), p. 1.

194 See Karrer (2003), p. 188.

195 See Bensel/Fürstenberg/Vogeler (2008), p. 3; Enarsson (2006), p. 322; Hunewald (2005), p. 9.

196 MbE is a management method which focuses on the reduction of a manager's monitoring and control activities, whose capabilities and attention are only required in case of extraordinary, unexpected incidents that occur during a process. See Hunewald (2005), p. 11 and the references cited there.

197 Event-driven planning is based on realizing a rescheduling in case of the occurrence of a critical classified event. This results in the variation of all previously valid limitations and target variables. See Hunewald (2005), p. 11 and the references cited there.

198 See Hunewald (2005), p. 11.

199 See Enarsson (2006), p. 321; Heusler/Stölzle/Bachmann (2006), p. 20; Hunewald (2005), p. 32; Steven/Krüger (2004), p. 179.

200 See Karrer (2003), p. 188.

201 See Heusler/Stölzle/Bachmann (2006), p. 20.

202 T&T systems deal with the connection of a transporting logistical unit (these are, for instance, containers, boxes, pallets, packed units, packaged goods) with an information system. See Stölzle (2008), p. 543 in conjunction with Hunewald (2005), p. 12.

203 See Bretzke et al. (2002), p. 29.

T&T systems enable supply chains to create cross-company transparency within the information flow.[204] To realize the objective, a process is recommended which includes the steps of *monitor, notify, simulate, control and measure*. However, the process of SCEM lacks elements of goal formulation, risk analysis and risk assessment.[205]

3.13 Supply Chain Resilience

Supply Chain Resilience (SCR) is a comparatively new concept emerging from recent events such as fuel protests in 2000 or foot-and-mouth-disease in 2001. In 2003, SCR was mentioned for the first time in a study on how British companies manage supply chain disruptions.[206] Similar to BCM, SCR lacks the focus on a specific type of disruption.[207] Rather, the concept gives attention to *resilience* which can be deemed to be the main objective of supply chains.

The term resilience is used in several scientific disciplines – among them are engineering, social sciences (in particular: developmental psychology, sociology and business administration) and ecology.[208] Accordingly, various definitions exist that describe the term.[209] Within business administration, SCR integrates aspects of both areas RM and SCM and thus, is located at a supply chain level.[210] However, scientific debates surrounding SCR suggest that it is less a holistic concept and more the objective of supply chains. For instance, according to *Christopher/Peck* (2004) "[T]he challenge to business today is to manage and mitigate [that] risk through creating more resilient supply chains."[211]

As already indicated, SCR is quite similar to BCM: the purpose of SCR is to enable a system which has been affected by a disruption to rapidly reestablish its original state or rather to achieve a better new state.[212] The objective of supply chains to create resilience can be explained with recourse to system theory.[213]

---

204 Transparency is the main objective of the SCEM-concept. See Heusler/Stölzle/Bachmann (2006), p. 19; Otto (2003), p. 4; Nissen (2002), p. 477.

205 See Enarsson (2006), p. 324; Heusler/Stölzle/Bachmann (2006), p. 22f.; Stölzle (2008), p. 543; Bretzke et al. (2002), p. 38f.

206 See Christopher/Peck (2004), p. 3 in conjunction with Cranfield University (2003).

207 See e.g. Ponomarov/Holcomb (2009), p. 125, Stewart/Kolluru/Smith (2009), p. 347 as well as Christopher/Peck (2004), p. 1, who address the concept to disruptions in general, catastrophes or supply chains risks.

208 See Pettit/Fiksel/Croxton (2010), p. 3f.; Ponomarov/Holcomb (2009), p. 124, p. 127; Stewart/Kolluru/Smith (2009), p. 347; Fiksel (2006), p. 16; Peck (2005), p. 211.

209 See Pettit/Fiksel/Croxton (2010), p. 3f. and the references cited there. For a detailed discussion of the term "resilience" within the respective science disciplines see also Ponomarov/Holcomb (2009), p. 125ff.

210 See Ponomarov/Holcomb (2009), p. 124f.

211 Christopher/Peck (2004), p. 1.

212 See Christopher/Peck (2004), p. 2, p. 7, p. 10. See also Ponomarov/Holcomb (2009), p. 124.

213 See as follows Ponomarov/Holcomb (2009), p. 128.

Systems, such as supply chains, are subject to dynamic environmental conditions and thus, aim at maintaining stability which can be achieved in two ways: inherently (by means of substituting or reallocating resources) or adaptive (by means of expanding resources). Apart from this, two conclusions can be drawn from the definition of SCR: SCR is based on the awareness that it is not possible to identify, avoid, control or eliminate all types of disruptions.[214] Thus, to enable supply chains to achieve higher degrees of flexibility and agility, the implementation of proactive action measures is required. Both flexibility and agility can be deemed to be the main levers of SCR emphasizing the proactive character of SCR.[215] The analysis of those factors that contribute to SCR is also subject of the resource-based view[216] which deals with the effects of a supply chain's resources (e.g., logistical capabilities) on its resilience.

3.14 Supply Chain Vulnerability

Supply Chain Vulnerability (SCV) can be considered to be another concept for the management of risk and uncertainty factors. As the events mentioned above highlighted the vulnerability of supply chains, the emergence of the concept is directly linked with SCR.[217]

The focus of SCV is on *vulnerability*, the term of which is discussed within various scientific disciplines including political and social sciences, geography and ecology.[218] Within business administration, SCV can be deemed to be a concept[219] which combines elements of both areas RM and SCM.[220] *Svensson* (2002a) defines vulnerability as "[…] a condition that affects a firm's goal accomplishment dependent upon the occurrence of negative consequences of disturbance. The degree of vulnerability for a given disturbance may be interpreted as being proportional to the chance of disturbance and the expected negative consequence of the disturbance, given that it has occurred."[221] Accordingly, the concept lacks any action measures to avoid or minimize the vulnerability of a supply chain. Rather, it aims at highlighting the degree of a supply chain's vulnerability. In this context, the increased interconnectedness of companies is to be considered to be one of the main drivers of vulnerability. From the system theory's view, this trend results from the character of companies as open systems:[222] here, a company's output is, at the same time, another company's input. These findings make an important

214 See as follows Peck (2006), p. 132. See also Pettit/Fiksel/Croxton (2010), p. 5; Ponomarov/Holcomb (2009), p. 129; Fiksel (2006), p. 16; Christopher/Peck (2004), p. 2.

215 See Pettit/Fiksel/Croxton (2010), p. 2; Christopher/Peck (2004), p. 7.

216 See Ponomarov/Holcomb (2009), p. 125, p. 133ff., who emphasize the absence of a consistent theory for the SCR concept.

217 See Peck (2005), p. 210, who implicitly refers to a study of the Cranfield University (2003).

218 See Dietz (2006), p. 13.

219 See Svensson (2000), p. 731f.

220 See Peck (2006), p. 127 and the references cited there.

221 Svensson (2002a), p. 112.

222 See as follows Peck (2005), p. 216ff.

contribution to the identification of the causes and drivers of the vulnerability of supply chains.[223] The interconnectedness of companies is also subject of normal accident theory which is based on the assumption that, within tightly coupled systems, accidents are inevitable and normal.[224] Here, the higher the degree of a supply chain's coupling the higher is its vulnerability to accidents and disruptions. The interconnectedness of companies is associated with the emergence of interdependencies which can be deemed to be a characteristic feature of supply chains. According to *Svensson* (2002a, 2002b), three types can be distinguished:[225] *Time dependence* refers to chronological or sequential dependencies existing within business-related activities of supply chains. *Relationship dependence* describes other relationship-specific factors that influence the collaboration of companies. *Functional dependence*, finally, results from the specialization of the respective companies and from the associated complementary character existing within supply chains.

3.15 Disaster Management

Disaster Management (DM) is another concept dealing with the management of risk and uncertainty factors. Since the beginning of the 1920's, DM is considered to be one of the main tasks of the public sector.[226] Accordingly, the concept has its origins in social sciences. Besides this, DM is subject to further scientific disciplines, such as laws, political and economic sciences.[227] With regard to business administration, the concept is increasingly discussed within the areas of SCM and logistics.[228] However, it is striking that in the literature, DM as a concept and disaster as the underlying type of disruption are neither conceptually debated nor sufficiently delineated from other concepts and type of disruptions, such as crisis or disaster/emergency preparedness, relief and response.[229] The latter can be considered to be some of the steps of the process of DM. According to *Helferich/Cook* (2002) who adopt the recommendations of FEMA (1993) the process is subdivided into the steps of *planning*, *mitigation*, *detection*, *response and recovery*.[230] Apart from the recognition that these steps generally include elements of the process of RM, the most relevant step is to be seen in planning (here: preparedness).[231]

---

223 See Peck (2005), p. 218ff.

224 See Wagner/Bode (2006), p. 302 and the references cited there.

225 See Svensson (2002a), p. 110f. and Svensson (2002b), p. 169 and the references cited there.

226 See Hale/Moberg (2005), p. 197.

227 See Canton (2007), p. XIV; Adam (2006), p. 60ff.; Müller et al. (1997), p. 21ff.

228 See Maon/Lindgreen/Vanhamme (2009), p. 149.

229 See i.a. Natarajarathinam/Capar/Narayanan (2009); Oloruntoba/Gray (2009); Perry (2007); Hale/Moberg (2005), p. 195f.; Pettit/Beresford (2005). Relief and response are deemed to be synonyms. See Perry (2007), p. 410.

230 See Hale/Moberg (2005), p. 200f. i.V.m. FEMA (1993) and Helferich/Cook (2002), p. 52ff. See also Kovács/Spens (2009), p. 510; Kovács/Tatham (2009), p. 217; Kovács/Spens (2007), p. 101ff.; Perry (2007), p. 416; Pettit/Beresford (2005), p. 316.

231 See Perry (2007), p. 411 and the references cited there.

DM focuses on the type of disruption called *disaster* which is, in the literature, predominantly described as being *large-scale.*[232] Concretely, a disaster is normally associated with both a comparatively low probability of occurrence and a high severity.[233] *United Nations* (1992) define a disaster as "[…] a serious disruption of the functioning of society, causing widespread human, material or environmental losses which exceed the ability of the affected people to cope using only its own resources."[234] Thus, disasters are of negative character and generally lead to humanitarian,[235] ecological and economic damages.[236]

In addition, the occurrence of a disaster generally exceeds the capabilities of an effected nation, region or society to cope with it by using its own resources.[237] Consequently, national and international assistance from third parties[238] is not only to be deemed a characteristic feature of DM. Moreover, a network-related view on the concept is suggested. Such a network consists of various actors, like aid organizations, non-governmental organizations, logistics providers, military, government, suppliers and donors[239] who all contribute different resources.[240] This gives rise to the main SCM- and logistics-related issues of DM that have to be solved – these include not only aspects of how to coordinate[241] and of how to configure resources[242] but also aspects of how to establish supply chains.[243] The reasons for the latter are as follows: firstly, it is neither possible nor efficient to establish supply chains before a disaster occurs.[244] Secondly, the occurrence of a disaster requires rapid assistance and thus, the rapid establishment of supply chains.

3.16 Emergency Management

A last concept to be mentioned here is Emergency Management (EM). This concept has its origins in social sciences[245] and here, in particular, within the areas of

---

232 See Kovács/Tatham (2009), p. 216; Oloruntoba/Gray (2009), p. 487 and the references cited there.
233 See Moore/Lakha (2006), p. 3.
234 United Nations (1992), p. 27.
235 See Plate/Merz (2001), p. 1.
236 See Sundar/Sezhiyan (2007), p. 9f.; Carr (1932), p. 211f.
237 See also Florida Division of Emergency Management (2011).
238 See Adam (2006), p. 97, p. 108.
239 See Oloruntoba/Gray (2009), p. 490; Kovács/Spens (2007), p. 106 as well as Perry (2007), p. 425; Pettit/Beresford (2005), p. 314.
240 See Kovács/Tatham (2009), p. 221f. and the references cited there.
241 See Perry (2007), p. 412f. and the references cited there; Canton (2007), p. XV. For detailed information about coordination problems that refer to catastrophes see Canton (2007), p. 57ff. See also Kovács/Spens (2007), p. 103.
242 Pettit/Beresford (2005), p. 314 implicitly refer to the significance of resource configurations, too.
243 See Kovács/Tatham (2009), p. 216; Oloruntoba/Gray (2009), p. 488; Kovács/Spens (2007), p. 103 and the references cited there.
244 See as follows and in detail the article of Kovács/Tatham (2009).
245 See Canton (2007), p. 35.

public administration and national security.[246] Besides this, EM is increasingly subject to business administration and here, in particular, to the areas of SCM and logistics. The literature surrounding EM, however, fails to provide a clear delineation from DM. Rather, both concepts DM and EM are highly interrelated by continually making references to each other. For instance, the definition of EM found at *Bumgarner* (2008) is based on the DM-related literature.[247] It is therefore not surprising that both concepts overlap each other.

The focus of EM is on the type of disruption called *emergency* which can be defined as "[...] an exceptional event that exceeds the capacity of normal resources and organization to cope with it."[248] In addition, an emergency is generally associated with a comparatively low probability of occurrence and a high severity[249] and therefore is similar to a disaster. However, the two terms differ from each other. An emergency is defined as "[...] a dangerous event that normally can be managed at the local level", while "Disasters are distinguished from emergencies by a greater level of response required."[250] *Bumgarner* (2008) extends the differentiation by precising the term: in contrast, an emergency is described as a routine event that cannot be coped with local resources and does not pose a threat to the stability of an affected region.[251] Thus, the concept of EM is more located at a local level and – as this is the case for the concept of DM – less located at a national and international level.[252] However, both concepts overlap with respect to the processual implementation. Thus, DM is based on a comparatively holistic process model including the preventive-oriented steps of *mitigation* and *preparedness* as well as the reactive-oriented steps of *response* und *recovery*.[253] In this context it gets clear, that *mitigation* is of special interest as this step differs from the other steps in terms of its underlying long-term orientation and its high cost intensity.

4 Interim Result: State of Research

Based on the analyzed concepts the assumption can be confirmed that a holistic, all-embracing concept that takes into consideration all types of risk and uncertainty factors of supply chains is missing so far. The reasons for this can be summed up as follows: several concepts (in the figure, these "concepts" are coloured grey) do not meet the requirements of a concept, since these either represent elements of

246 See Bumgarner (2008), p. 1. See also Christopher/Peck (2004), p. 3, who ascribe the functions of EM predominantly to public institutions.

247 See Bumgarner (2008), p. 1. It is somewhat similar with *Canton* (2007) and *Coppola* (2011) who ostensibly focus on the one (i.e., EM) or the other (i.e., DM) concept in their publications. However, both concepts are virtually treated equally without a stringent differentiation.

248 Alexander (2002), p. 1.

249 See Moore/Lakha (2006), p. 3.

250 See Florida Division of Emergency Management (2011).

251 See Bumgarner (2008), p. 13 as well as Canton (2007), p. 41.

252 See Canton (2007), p. 51.

253 See a.o. Bumgarner (2008), p. 17ff.; Canton (2007), p. 23, as follows Bumgarner (2008), p. 17ff.; Canton (2007), p. 158ff.

a higher-level approach (this applies to IM, EM, and PM) or can be understood as a supply chain's objective (this applies to SCR). Beside this, the majority of the concepts discussed here fails already since they are not located at a supply chain-level – what remains is the concept of SCRM alone. However, it seems worth to take a critical look at the character of this concept. For this purpose, the distinguishing features *type of disruption*, *process*, and *management approach* will be discussed in detail.

Generally, SCRM is characterized by its explicite focus on (supply chain) risks and, associated with this, by its renunciation of (supply chain) uncertainties. At this point, it should be emphasized that both terms differ in terms of the criteria of identifiability and quantifiability.[254] Taking this into account, SCRM focuses on identifiable and quantifiable (supply chain) risks by excluding non-identifiable and non-quantifiable (supply chain) uncertainties.[255] Against this background, *Waters* (2007) recommends to complement SCRM by integrating aspects of BCM. This is also shown in the respective management approaches: the concept of SCRM is based on the intention to management (supply chain) risks with recourse to preventive and reactive management approaches. In contrast, the concept of BCM is explicitly based on proactive management approaches. Besides this, evidence has been found in the literature that the step of goal formulation is missing – thus, the process of SCRM can be deemed to be insufficient.

Considering these findings it can be stated that SCRM meets the requirements of a holistic, all-embracing concept for the management of risk and uncertainty factors in supply chain only partially.

The illustration below sums up the findings generated so far by means of the three dimensions level (1), degree of comprehensiveness of the considered type of disruptions or risk and uncertainty factors, respectively (2), and management approach (3):

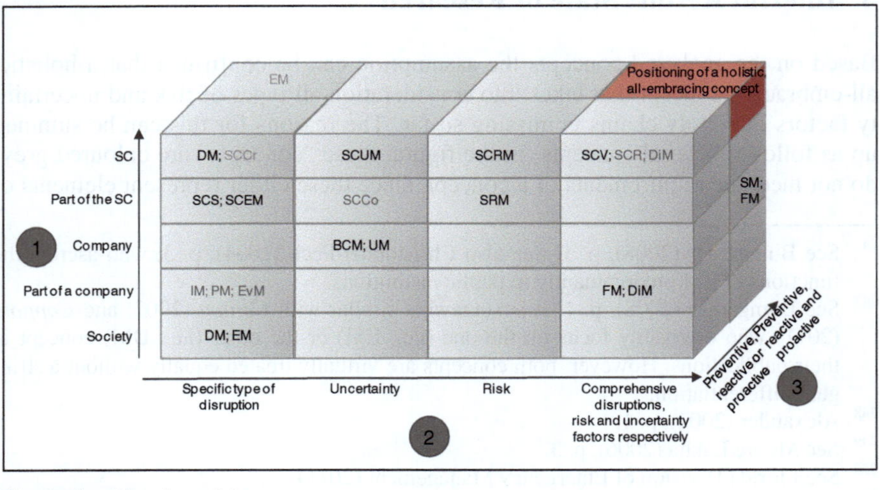

Fig. 1 Classification of the concepts

[254] See Waters (2007), p. 216, p. 231.
[255] See Waters (2007), p. 94.

Indeed an impressingly high number of concepts for the management of risk and uncertainty factors exist, it is generally apparent that some of the fields still remain unfulfilled. However, since it is the main objective to develop a holistic, all-embracing concept for the management of risk and uncertainty factors, the article is positioned within the red coloured field.

Against this background a new concept called Supply Chain Safety Management (SCSM) will be introduced now.

5 The Concept of SCSM

Before describing the concept in detail, the term safety has to be defined. This results from the fact that the English language offers two specifications – these are safety and security. Both terms have their origins in the area of IT – generally they are used to characterize computer systems. Due to the predominant inconsistent use, *Burns/McDermid/Dobson* (1992) have undertaken the attempt to define safety- and security-critical computer systems within their article 'On the meaning of safety and security'.[256] Despite this contribution focussing on the area of IT, the heterogeneous use of these terms can be identified within the context of supply chains, too. Thus, a more extensive analysis of the terminological foundations of supply chain safety is recommended.[257]

On the one hand, supply chain safety embraces the term security, which in turn can be subdivided into physical and digital security.[258] Physical security refers to protecting all tangible goods from any intended attacks that aim to damage the supply chain. By contrast, digital security refers to protecting information (systems) from any intended attacks that aim to damage the supply chain. Thus, security can be interpreted as the protection of the material and immaterial elements of a supply chain against intended attacks in the form of organised crime and international terrorism.[259]

On the other hand, supply chain safety embraces the term safety, which is being interpreted as a kind of protection from unintended hazards. The focus here is on the protection from random events, such as natural catastrophes or carelessness and negligence.[260] In the literature only a few approaches can be identified that refer to safety as defined here.[261] However, according to the authors, safety will be used as the comprehensive term of the concept of SCSM and therefore should be

---

256 See Burns/McDermid/Dobson (1992), p. 3.

257 See Lange (2005), p. 28; Egger (1992), p. 51f.

258 See Rice et al. (2003), p. 28.

259 See Lange (2005), p. 38 in connection with Matschke/Ick (1998), p. 14.

260 See Lange (2005), p. 38; Matschke/Ick (1998), p. 14.

261 The term food safety is to be considered one exception. See Grunert (2005), p. 369; Henson/Traill (1993), p. 158.

understood to include measures taken to achieve security. Based on this, SCSM aims at ensuring the continuity of supply while also taking the economic goal of profitability into consideration.[262] This condition is based on the fact that the achievement of absolute safety is neither possible nor economic. Hence, the goal is to achieve relative safety by means of taking adequate action alternatives and downsizing potential risks on a tolerable degree.[263] This understanding goes in line with the realization that "[f]irms need to understand the value of prevention and not merely the reaction to security risks".[264]

Considering the primary goal of ensuring the continuity of supply, all safety-related action measures should be aligned with the minimisation of disruptions that supply chains are faced with.[265] These action measures can be distinguished by their point of intervention and, thus, categorized into supply chain protection and supply chain resilience. Supply chain protection includes preventive action measures, which are intended to avoid disruptions or interruptions of the supply chain.[266] Inspections, data back-ups, introduction of (international) standards and certifications as well as enhancing safety precautions are deemed to be such preventive action measures.[267] By contrast, supply chain resilience includes reactive action alternatives that aim at enabling a supply chain to swiftly react to unexpected events.[268] Generally, the resilience of a supply chain is determined by both its flexibility and its redundancy. Flexibility includes the creation of capabilities and infrastructure, which are used during normal operations.[269] Flexibility-oriented action measures are postponement, the use of alternative modes and routes of transport, and parallelizing processes. By contrast, redundancy includes the creation of capacities that are needed in case a disruption occurs.[270] Redundancy-oriented action measures are multiple sourcing, enhancing safety stocks, and redundant resources.

In the general framework of the concept of SCSM, supply chain preparedness is the stated goal which aims at enabling a supply chain to continue or to rapidly restore its operations in case a disruption occurs.[271] The results up to now can be summarised in the following figure:

[262] See Large (2006), p. 48f.
[263] See Steven/Tengler (2005), p. 346.
[264] See Giunipero/Eltantawy (2004), p. 698.
[265] See Craighead et al. (2007), p. 132.
[266] See Sheffi et al. (2004), p. 3; Lee/Wolfe (2003), p. 25.
[267] See Deutch (2002), p. 2; Sheffi (2002), p. 7.
[268] See Rice/Caniato (2003), p. 25.
[269] See Rice et al. (2003), p. 32.
[270] See Sheffi/Rice (2005), p. 44.
[271] See Waters (2007), S. 218; Herbane/Elliott/Swartz (2004), S. 436f.

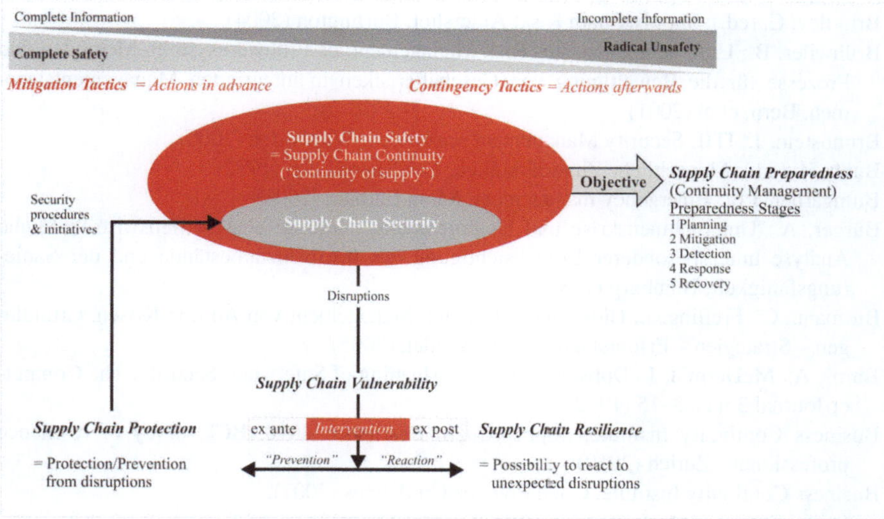

Fig. 2 Overview of the concept of SCSM

Reference List

Adam, V.: Hochwasser-Katastrophenmanagement. Wirkungsprüfung der Hochwasservorsorge und -bewältigung österreichischer Gemeinden, Wiesbaden (2006)

Addy, R.: Effective IT Service Management. To ITIL and Beyond!, Berlin, Heidelberg, New York (2007)

Alexander, D.: Principles of Emergency Planning and Management, New York (2002)

AMR Research, AMR Research Report on Managing Supply Chain Risk, Inc. (2006)

Autry, C.W., Bobbitt, L.M.: Supply chain security orientation: conceptual development and a proposed framework. The International Journal of Logistics Management 19(1), 42–64 (2008)

Bächle, M., Kolb, A.: Einführung in die Wirtschaftsinformatik, München (2007)

Barnes, P.: Business Impact Analysis, Hiles, A., Barnes, P. (eds.), pp. 166–182 (2011)

Beims, M.: IT Service Management in der Praxis mit ITIL® 3. Zielfindung. Methoden. Realisierung, München (2009)

Bensel, P., Fürstenberg, F., Vogeler, S.: Supply Chain Event Management. Entwicklung eines SCEM-Frameworks, Berlin (2008)

Bernstein, P.: Against the Gods: The Remarkable Story of Risk, Chichester (1996)

Bogaschewsky, R., Essig, M., Lasch, R., Stölzle, W. (eds.): Supply Management Research. Aktuelle Forschungsergebnisse 2011, Wiesbaden (2011)

Brands, G.: IT-Sicherheitsmanagement, Berlin, Heidelberg, New York (2005)

Brenner, M.: Werkzeugunterstützung für ITIL-orientiertes Dienstmanagement. Ein modellbasierter Ansatz, Norderstedt (2007)

Bretzke, W.-R., Stölzle, W., Karrer, M., Ploenes, P.: Vom Tracking & Tracing zum Supply Chain Event Management. Aktueller Stand und Trends, Studie der KPMG Consulting AG, Düsseldorf (2002)

Bridi, A.: IT Sicherheitsmanagement. Ihr Praxis-Leitfaden, Norderstedt (2008)

Brindley, C. (ed.): Supply Chain Risk, Aldershot, Burlington (2004)
Brühwiler, B.: Unternehmensweites Risk Management als Frühwarnsystem. Methoden und Prozesse für die Bewältigung von Geschäftsrisiken in integrierten Managementsystemen, Bern, et al (2001)
Brunnstein, J.: ITIL Security Management realisieren, Wiesbaden (2006)
Bubb, H. (ed.): Menschliche Zuverlässigkeit, Landsberg/Lech (1992)
Bumgarner, J.B.: Emergency management, Santa Barbara (2008)
Burger, A.: Unternehmenskrise und Unternehmenssanierung: eine betriebswirtschaftliche Analyse unter besonderer Berücksichtigung der Insolvenztatbestände und der Sanierungsfähigkeit, Hamburg (1988)
Burmann, C., Freiling, J., Hülsmann, M. (eds.): Management von Ad-hoc-Krisen. Grundlagen – Strategien – Erfolgsfaktoren, Wiesbaden (2005)
Burns, A., McDermid, J., Dobson, J.: On the Meaning of Safety and Security. The Computer Journal 35(1), 3–15 (1992)
Business Continuity Institute, Supply Chain Resilience 2010. BCI survey of resilience professionals, Zurich (2010)
Business Continuity Institute, Good Practice Guidelines (2007), `http://www.thebci.org/CHAPTER1BCIGPG071.pdf` (retrieved April 12, 2011)
Canton, L.G.: Emergency Management. Concepts and Strategies for Effective Programs, Hoboken (2007)
Carr, L.J.: Disaster and the Sequence-Pattern Concept of Social Change. American Journal of Sociology 38(2), 207–218 (1932)
Cauvin, A.C.A., Ferrarini, A.F.A., Tranvouez, E.T.E.: Disruption management in distributed enterprises: A multi-agent modelling and simulation of cooperative recovery behaviours. International Journal of Production Economics 122(1), 429–439 (2009)
Chartered Management Institute, Business Continuity and Supply Chain Management (2002), `http://www.thebci.org/2809-01%20Bus%20Continuity%20Summ.pdf` (retrieved May 24, 2011)
Christopher, M., Peck, H.: Building the Resilient Supply Chain. The International Journal of Logistics Management 15(2), 1–13 (2004)
Closs, D.J., McGarrell, E.F.: Enhancing Security Throughout the Supply Chain, Washington (2004)
Cohen, M.A., Kunreuther, H.: Operations Risk Management: Overview of Paul Kleindorfer's Contribution. Production & Operations Management 16(5), 525–541 (2007)
Coppola, D.P.: Introduction to International Disaster Management, 2nd edn., Burlington (2011)
Cornish, M.: Business Continuity Management Methodology, Hiles, A., Barnes, P. (eds., 2011), pp. 121–136 (2004)
Craighead, C.W., Blackhurst, J., Rungtusanatham, M.J., Handfield, R.B.: The Severity of Supply Chain Disruptions: Design Characteristics and Mitigation Capabilities. Decision Sciences 38(1), 131–156 (2007)
Cranfield University, Creating Resilient Supply Chains. A Practical Guide, Cranfield University, Cranfield (2003)
Czaja, L.: Qualitätsfrühwarnsysteme in der Automobilindustrie, Wiesbaden (2009)
Dangelmaier, W., Gajewski, T., Kösters, C. (eds.): Innovationen im E-Business, Paderborn (2003)

Dangoumau, N., Craye, E., Lorimier, L.: Co-operative functional/hybrid behavioural models for dynamical failure assessment. International Journal of Production Research 46(19), 5313–5336 (2008)

Department of Homeland Security, Strategy to enhance International Supply Chain Security (2007), `http://www.dhs.gov/xlibrary/assets/plcy-international supplychainsecuritystrategy.pdf` (retrieved May 18, 2011)

Deutch, J.: Assessment of the Security Threat, Rice, J.B (ed.), pp. 1–2 (2002)

Dickmann, P. (ed.): Schlanker Materialfluss mit Lean Production, Kanban und Innovationen, 2nd edn., Berlin, Heidelberg (2009)

Diederichs, M.: Risikomanagement und Risikocontrolling, München (2004)

Dietz, K.: Vulnerabilität und Anpassung gegenüber Klimawandel aus sozial-ökologischer Perspektive. Aktuelle Tendenzen und Herausforderungen in der internationalen Klima- und Entwicklungspolitik, Berlin (2006)

Doughty, K.: Business Continuity Planning: Protecting Your Organization's Life, Boca Raton, et al (2001)

Ebel, N.: ITIL® V3 Basis-Zertifizierung. Grundlagenwissen und Zertifizierungsvorbereitung für die ITIL Foundation-Prüfung, München (2008)

Ebster, C., Stalzer, L.: Wissenschaftliches Arbeiten für Wirtschafts- und Sozialwissenschaftler, 3rd edn., Wien (2008)

Egger, E.: Datenschutz und Datensicherheit – Grundlagen oder Spezialgebiet? Datenschutz und Datensicherung 16(10), 512–516 (1992)

Eisenhardt, K.: Agency theory: an assessment and review. Academy of Management Review 14(1), 57–74 (1989)

Elliott, D., Swartz, E., Herbane, B.: Business Continuity Management. A Crisis Management Approach, 2nd edn., New York, Abington (2010)

Ellouze, W.: Entwicklung eines Modells für ein ganzheitliches Fehlermanagement. Ein prozessorientiertes Referenzmodell zum effizienten Fehlermanagement, Aachen (2010)

Enarsson, L.: Future Logistics Challenges, Kopenhagen (2006)

Engel, H.: Gesprengte Ketten – Absicherung der Supply Chain durch ein unternehmensweites Business Continuity Management. Risk News 2(5), 38–45 (2005)

Ertl-Wagner, B., Steinbrucker, S., Wagner, B.: Qualitätsmanagement & Zertifizierung. Praktische Umsetzung in Krankenhäusern Reha-Kliniken, stationären Pflegeeinrichtungen, Heidelberg (2009)

Faust, D.A.: Effektive Sicherheit, Wiesbaden (2002)

FEMA, Emergency Management Guide for Business and Industry. A Step-by-Step Approach to Emergency Planning, Response and Recovery for Companies of All Sizes, Washington, Rockville (1993)

Fiksel, J.: Sustainability and Resilience: Toward a Systems Approach. Sustainability: Science, Practice, & Policy 2(2), 1–8 (2006)

Fischäder, H.: Störungsmanagement in netzwerkförmigen Produktionssystemen, Wiesbaden (2007)

Fischäder, H., Schneider, H.M.: Störparameter im Materialfluss und in Produktionssystemen, Dickmann, P. (ed.) pp. 294–297 (2009)

Florida Division of Emergency Management (2011), `http://www.floridadisaster.org/EMIT/introductionem.htm` (retrieved May 24, 2011)

Fürst, R.A.: Preiswettbewerb in Krisen. Auswirkungen der Terror-Attentate des 11. September 2001 auf die Luftfahrtbranche, Wiesbaden (2004)

Fürst, R.A., Sattelberger, T., Heil, O.P.: 3D-Krisenmanagement. Bewältigung von Krisen in Krisen, München (2007)
Gabele, E.: Ansatzpunkte für ein betriebswirtschaftliches Krisenmanagement. Zeitschrift für Organisation 50(3), 150–158 (1981)
Gareis, R. (ed.): Erfolgsfaktor Krise: Konstruktionen, Methoden, Fallstudien zum Krisenmanagement, Wien (1994)
Gärtner, H.: Internationale Sicherheit: Definitionen von A-Z, Baden-Baden (2005)
Gibb, F., Buchanan, S.: A framework for business continuity management. International Journal of Information Management 26(2), 128–141 (2006)
Götze, U., Henselmann, K., Mikus, B. (eds.): Risikomanagement, Heidelberg (2001)
Götze, U., Mikus, B.: Der Prozess des Risikomanagements in Supply Chains, Vahrenkamp, R., Siepermann, C. (eds.), pp. 29–58 (2007)
Graham, J., Kaye, D.: A Risk Management Approach to Business Continuity: Aligning Business Continuity with Corporate Governance, Brookfield (2006)
Grote, G.: Management of Uncertainty. Theory and Application in the Design of Systems and Organizations, London (2009)
Grote, G., Künzler, C.: Sicherheitskultur, Arbeitsorganisation und Technikeinsatz, Zürich (1996)
Grunert, K.G.: Food Quality and Food Safety: Consumer Perception and Demand. European Review of Agricultural Economics 32(3), 369–391 (2005)
Gundel, S., Mülli, L.: Unternehmenssicherheit, München (2009)
Hahn, D., Hungenberg, H.: PuK. Planung und Kontrolle. Planungs- und Kontrollsysteme. Planungs- und Kontrollrechnung. Wertorientierte Controllingkonzepte, 6th edn., Wiesbaden (2001)
Hale, T., Moberg, C.R.: Improving supply chain disaster preparedness: A decision process for secure site location. International Journal of Physical Distribution & Logistics Management 35(3), 195–207 (2005)
Hallikas, J., Karvonen, I., Pulkkinen, U., Virolainen, W.-M., Tuominen, M.: Risk management processes in supplier networks. International Journal of Production Economics 90(1), 47–58 (2004)
Harland, C., Brenchley, R., Walker, H.: Risk in supply networks. Journal of Purchasing & Supply Management 9(2), 51–62 (2003)
Heath, R.: A crisis management perspective of business continuity, Hiles, A. (ed.), pp. 47–58 (2008)
Heil, M.: Entstörung betrieblicher Abläufe, Wiesbaden (1995)
Helferich, O.K., Cook, R.L.: Securing the Supply Chain, Oak Brook (2002)
Hendricks, K.B., Singhal, V.R.: An Empirical Analysis of the Effect of Supply Chain Disruptions on Long-Run Stock Price Performance and Equity Risk of the Firm. Production & Operations Management 14(1), 35–52 (2005)
Henson, S., Traill, B.: The Demand for Food Safety: Market Imperfections and the Role of Government. Food Policy 18(2), 152–162 (1993)
Herbane, B., Elliott, D., Swartz, E.: Business Continuity Management: time for a strategic role? Long Range Planning 37(5), 435–457 (2004)
Heusler, K.F., Stölzle, W., Bachmann, H.: Supply Chain Event Management. Grundlagen, Funktionen und potentielle Akteure. WiSt 35(1), 19–24 (2006)
Hiles, A.: Business Continuity: Best Practices – World-Class Business Continuity Management, 2nd edn., Brookfield (2004)
Hiles, A., Barnes, P. (eds.): The Definitive Handbook of Business Continuity Management, 2nd edn., West Sussex (2011)

Hinterhuber, H., Sauerwein, E., Fohler-Norek, C. (eds.): Betriebliches Risikomanagement, Wien (1998)

Hoyos, C.G.: Die Zuverlässigkeit des Menschen, Bubb, H. (ed.), S.11–S.15 (1992)

Hülsmann, M.: Ad-hoc-Krise – eine begriffliche Annäherung, Burmann, C., Freiling, J., Hülsmann, M. (eds.), pp. 33–57 (2005)

Hunewald, C.: Supply Chain Event Management. Anforderungen und Potentiale am Beispiel der Automobilindustrie, Wiesbaden (2005)

Hutzschenreuter, T., Griess-Nega, T. (eds.): Krisenmanagement. Grundlagen, Strategien, Instrumente, Wiesbaden (2006)

Junginger, M.: Wertorientierte Steuerung von Risiken im Informationsmanagement, Wiesbaden (2005)

Jüttner, U.: Supply chain risk management. Understanding the business requirements from a practitioner perspective. International Journal of Logistics Management 16(1), 120–141 (2005)

Jüttner, U., Peck, H., Christopher, M.: Supply chain risk management. Outlining an agenda for future research. International Journal of Logistics 6(4), 197–210 (2003)

Kajüter, P.: Risikomanagement in der Supply Chain: Ökonomische, regulatorische und konzeptionelle Grundlagen, Vahrenkamp, R., Siepermann, C. (eds.), pp. 13–27 (2007)

Kajüter, P.: Instrumente zum Risikomanagement in der Supply Chain, Stölzle, W., Otto, A. (eds.), pp. 107–135 (2003)

Karrer, M.: Supply Chain Event Management. Impulse zur ereignisorientierten Steuerung von Supply Chains, Dangelmaier, W., Gajewski, T., Kösters, C. (eds.), pp. 187–198 (2003)

Kaufmann, F.-X.: Sicherheit als soziologisches und sozialpolitisches Problem, 2nd edn., Stuttgart (1973)

Khan, O., Burnes, B.: Risk and supply chain management: creating a research agenda. The International Journal of Logistics Management 18(2), 197–216 (2007)

Klaus, P., Krieger, W. (eds.): Gabler Lexikon Logistik. Management logistischer Netzwerke und Flüsse, 3rd edn., Wiesbaden (2008)

Kleindorfer, P.R., Saad, G.H.: Managing Disruption Risks in Supply Chains. Production & Operations Management 14(1), 53–68 (2005)

Köhler, P.T.: ITIL, 2nd edn., Berlin, Heidelberg, New York (2006)

Koppelmann, U.: Beschaffungsmarketing, 4th edn., Berlin, et al (2004)

Kovács, G., Spens, K.M.: Identifying challenges in humanitarian logistics. International Journal of Physical Distribution and Logistics Management 39(6), 506–528 (2009)

Kovács, G., Spens, K.M.: Humanitarian logistics in disaster relief operations. International Journal of Physical Logistics & Distribution Management 37(2), 99–114 (2007)

Kovács, G., Tatham, P.H.: Responding to Disruptions in the Supply Network – From Dormant to Action. Journal of Business Logistics 30(2), 215–229 (2009)

Krummenacher, A.: Krisenmanagement. Leitfaden zum Verhindern und Bewältigen von Unternehmenskrisen, Zürich (1981)

Krystek, U.: Krisenarten und Krisenursachen, Hutzschenreuter, T., Griess-Nega, T. (eds.), pp. 41–66 (2006)

Krystek, U.: Unternehmenskrisen. Beschreibung, Vermeidung und Bewältigung überlebenskritischer Prozesse in Unternehmen, Wiesbaden (1987)

Künzler, C.: Kompetenzförderliche Sicherheitskultur. Ein Ansatz zur ganzheitlichen Gestaltung risikoreicher Arbeitssysteme, Zürich (2001)

Lange, J.A.: Sicherheit und Datenschutz als notwendige Eigenschaften von computergestützten Informationssystemen: Ein integrierender Gestaltungsansatz für vertrauenswürdige computergestützte Informationssysteme, Wiesbaden (2005)

Large, R.: Strategisches Beschaffungsmanagement: Eine praxisorientierte Einführung, 3rd edn., Wiesbaden (2006)

Lee, H.L., Whang, S.: Higher supply chain security with lower cost: Lessons from total quality management. International Journal of Production Economics 96(3), 289–300 (2005)

Lee, H.L., Wolfe, M.: Supply chain security without tears. Supply Chain Management Review 7(1), 12–20 (2003)

Lipshitz, R., Strauss, O.: Coping with uncertainty: a naturalistic decision-making analysis. Organizational Behavior and Human Decision Making Processes 69(2), 149–163 (1997)

Manuj, I., Mentzer, J.T.: Global supply chain risk management. Journal of Business Logistics 29(1), 133–156 (2008)

Maon, F., Lindgreen, A., Vanhamme, J.: Developing supply chains in disaster relief operations through cross-sector socially oriented collaborations: a theoretical model. Supply Chain Management: An International Journal 14(2), 149–164 (2009)

Matschke, K.-D., Ick, R.: Security Quality Management Handbuch: Grundsätze und Verfahren für umfassende Unternehmenssicherheit, Ingelheim (1998)

McKee, K., Guthridge, L.: Leading People Through Disasters. An Action Guide. Planning for and Dealing with the Human Side of Crises, San Francisco (2006)

Mensch, G.: Risiko und Unternehmensführung. Eine systemorientierte Konzeption zum Risikomanagement, Frankfurt am Main (1991)

Michell-Auli, P., Schwemmle, M.: Integriertes Management mit der Balanced Scorecard. Ein Praxisleitfaden für Sozialunternehmen, Stuttgart (2008)

Mikus, B.: Risiken und Risikomanagement – ein Überblick. Götze, U., Henselmann, K., Mikus, B (eds.), pp. 3–28 (2001)

Mistele, P.: Faktoren des verlässlichen Handelns. Leistungspotentiale von Organisationen in Hochrisikoumwelten, Wiesbaden (2007)

Mitchell, V.W.: Organisational risk perception and reduction: a literature review. British Journal of Management 6(2), 115–133 (1995)

Mitroff, I.I., Shrivastava, P., Udwadia, F.E.: Effective crisis management. The Academy of Management Executive 1(3), 283–292 (1987)

Moder, M.: Supply Frühwarnsysteme. Die Identifikation und Analyse von Risiken in Einkauf und Supply Management, Wiesbaden (2008)

Moore, T., Lakha, R.: Tolley's Handbook of Disaster and Emergency Management. Principles and Practice, Oxford, Burlington (2006)

Müller, K.-R.: Handbuch Unternehmenssicherheit. Umfassendes Sicherheits-, Kontinuitäts- und Risikomanagement mit System, 2nd edn., Wiesbaden (2010)

Müller, K.-R.: IT-Sicherheit mit System, 2nd edn., Wiesbaden (2005)

Müller, U., Zimmermann, W., Neuenschwander, P., Tobler, A., Wyss, S., Alder, R.: Katastrophen als Herausforderung für Verwaltung und Politik. Kontinuitäten und Diskontinuitäten, Zürich (1997)

Müller, R.: Corporate Crisis Management. Long Range Planning 18(5), 38–48 (1985)

Müller, R.: Krisenmanagement in der Unternehmung. Ein Beitrag zur organisatorischen Gestaltung des Prozesses der Krisenbewältigung, Frankfurt am Main, Bern (1982)

Natarajarathinam, M., Capar, I., Narayanan, A.: Managing supply chains in times of crisis: a review of literature and insights. International Journal of Physical Distribution & Logistics Management 39(7), 535–573 (2009)

Nissen, V.: Supply Chain Event Management. Wirtschaftsinformatik 44(5), 477–480 (2002)

Norrman, A., Jansson, U.: Ericsson's proactive supply chain risk management approach after a serious sub-supplier accident. International Journal of Physical Distribution & Logistics Management 34(5), 434–456 (2004)

Norrman, A., Lindroth, R.: Categorization of Supply Chain Risk and Risk Management, Brindley, C. (ed.), pp. 14–27 (2004)

Office of Government Commerce, The Official Introduction to the ITIL Service Lifecycle, Edinburgh (2007)

Office of Government Commerce, Best Practice for Service Delivery. ITIL®. The key to Managing IT Services, Birmingham, et al (2002)

Oloruntoba, R., Gray, R.: Customer service in emergency relief chains. International Journal of Physical Distribution & Logistics Management 39(6), 486–505 (2009)

Otto, A.: Supply Chain Event Management: Three Perspectives. The International Journal of Logistics Management 14(2), 1–13 (2003)

Patig, S.: Ansatzpunkte und Rechnerunterstützung des produktionsorientierten Störungsmanagements: Ergebnisse einer Literaturanalyse (1999), `http://www-wi.cs.uni-magdeburg.de/~patig/publikationen/preprint_nr6_1999.pdf` (retrieved July 28, 2011)

Patig, S., Thorhauer, S.: Ein Planungsansatz zum Umgang mit Störungen bei der Produktion: Die flexible Produktionsfeinplanung mithilfe von Planungsschritten. Wirtschaftsinformatik 44(4), 355–366 (2002)

Paulsson, U.: On Managing Disruption Risks in the Supply Chain. The DRISC Model, Lund (2007)

Pearson, C., Mitroff, I.: From Crisis-Prone to Crisis-Prepared: A Framework for Crisis Management. Academy of Management Executive 7(1), 48–59 (1993)

Peck, H.: Reconciling supply chain vulnerability, risk and supply chain management. International Journal of Logistics: Research and Applications 9(2), 127–142 (2006)

Peck, H.: Drivers of supply chain vulnerability: an integrated framework. International Journal of Physical Distribution and Logistics Management 35(4), 210–232 (2005)

Perry, M.: Natural disaster management planning: A study of logistics managers responding to the tsunami. International Journal of Physical Distribution & Logistics Management 37(5), 409–433 (2007)

Pettit, S., Beresford, A.: Emergency relief logistics: an evaluation of military, non-military and composite response models. International Journal of Logistics: Research and Applications 8(4), 313–331 (2005)

Pettit, T.J., Fiksel, J., Croxton, K.L.: Ensuring Supply Chain Resilience: Development of a Conceptual Framework. Journal of Business Logistics 31(1), 1–21 (2010)

Pfohl, H.-C.: Risiken und Chancen: Strategische Analyse in der Supply Chain, Pfohl, H.-C. (ed.), pp. 1–56 (2002)

Pfohl, H.-C. (ed.): Risiko- und Chancenmanagement in der Supply Chain. Proaktiv – ganzheitlich – nachhaltig, Berlin (2002)

Plate, E.J., Merz, B.: Naturkatastrophen. Ursachen – Auswirkungen – Vorsorge, Stuttgart (2001)

Ponomarov, S.Y., Holcomb, M.C.: Understanding the concept of supply chain resilience. The International Journal of Logistics Management 20(1), 124–143 (2009)

Prater, E.: A framework for understanding the interaction of uncertainty and information systems on supply chains. International Journal of Physical Distribution & Logistics Management 35(7), 524–539 (2005)

PriceWaterhouseCoopers, Global Sourcing. Outsourcing comes of age: The rise of collaborative partnering (2009), http://www.pwc.com/en_GX/gx/operations-consulting-services/pdf/outsourcingcomesofage.pdf (retrieved July 28, 2011)

Proff, H.: Systematische Krisenbewältigung. WiSt 38(4), 209–212 (2009)

Rao, S., Goldsby, T.J.: Supply chain risks: a review and typology. International Journal of Logistics Management 20(1), 97–123 (2009)

Reason, J.T.: The Human Contribution. Unsafe Acts, Accidents, and Heroic Recoveries, Farnham, Burlington (2008)

Reason, J.: Human error: models and management. British Medical Journal 320(7237), 768–770 (2000)

Reason, J.: Managing the Risk of Organizational Accidents, Ashgate (1997)

Rice, J.B. (ed.): Supply Chain Response to Terrorism – Planning for the Unexpected, http://web.mit.edu/supplychain/repository/scresponse_120502v1.pdf (retrieved August 1, 2011)

Rice, J.B., Caniato, F.: Building a Secure and Resilient Supply Network. Supply Chain Management Review 7(5), 22–30 (2003)

Rice, J.B., Caniato, F., Fleck, J., Disraelly, D., Lowtan, D., Lensing, R., Pickett, C.: Supply Chains Response to Terrorism: Creating Resilient and Secure Supply Chains", Supply Chain Response to Terrorism Project, Interim Report of Progress and Learning, MIT Center for Transportation and Logistics, Massachusetts (2003)

Ritchie, B., Brindley, C.: Supply chain risk management and performance: A guiding framework for future development. International Journal of Operations & Production Management 27(3), 303–322 (2007)

Rogler, S.: Risikomanagement im Industriebetrieb: Analyse von Beschaffungs-, Produktionsrisiken, Wiesbaden (2002)

Roland Berger Strategy Consultants, Von der Kostenbremse zum Gewinntreiber. Supply Chain Management ist eine Königsdisziplin des Managements. Und ein strategischer Wettbewerbsvorteil dazu. Aber nur, wenn auch der Supply Chain Fit gelingt, Hamburg (2010)

Romeike, F.: Der Prozess des strategischen und operativen Risikomanagements. Romeike, F., Finke, R.B. (eds.), pp. 147–161 (2003)

Romeike, F., Finke, R.B. (eds.): Erfolgsfaktor Risiko-Management: Chance für Industrie und Handel. Methoden, Beispiele, Checklisten, Wiesbaden (2003)

Romeike, F., Hager, P.: Erfolgsfaktor Risiko-Management 2.0. Methoden, Beispiele, Checklisten. Praxisbuch für Industrie und Handel, 2nd edn., Wiesbaden (2009)

Rüsen, T.A.: Krisen und Krisenmanagement in Familienunternehmen, Wiesbaden (2009)

Sauerwein, E., Thurner, M.: Der Risikomanagement-Prozess im Überblick, Hinterhuber, H., Sauerwein, E., Fohler-Norek, C. (eds.), pp. 19–39 (1998)

Saynisch, M.: Krisenmanagement - Chancennutzung oder Risikoabsicherung, Gareis, R. (ed.), pp. 49–73 (1994)

Schulten, M.F.: Krisenmanagement, Berlin (1995)

Sheffi, Y.: Worst-Case-Szenario. Wie Sie Ihr Unternehmen auf Krisen vorbereiten und Ausfallrisiken minimieren, Landsberg am Lech (2006)

Sheffi, Y.: Supply Chain Security, Rice, J.B. (ed.), pp. 6–7 (2002)

Sheffi, Y., Rice, J.B.: A Supply Chain View of the Resilient Enterprise. MIT Sloan Management Review 47(1), 41–48 (2005)

Sheffi, Y., Rice, J.B., Fleck, J.M., Caniato, F.: Supply Chain Response to Global Terrorism: A Situation Scan, Conference Proceedings, University of Milan, Supply Chain Conference, Milan (June 2004)

Sikich, G.W.: Protecting Your Business in a Pandemic. Plans, Tools, and Advice for Maintaining Business Continuity, Westport (2008)

Skipper, J.B., Hanna, J.B.: Minimizing supply chain disruption risk through enhanced flexibility. International Journal of Physical Distribution & Logistics Management 39(5), 404–427 (2009)

Spengler, T., Voss, S., Kopfer, H. (eds.): Logistik Management. Prozesse, Systeme, Ausbildung, Heidelberg (2004)

Steven, M., Krüger, R.: Supply Chain Event Management für globale Logistikprozesse. Charakteristika, konzeptionelle Bestandteile und deren Umsetzung in Informationssysteme, Spengler, T., Voss, S., Kopfer, H. (eds.), pp. 179–195 (2004)

Steven, M., Tengler, S.: Informationssicherheit im Supply Chain Management. Wirtschaftswissenschaftliches Studium 25(6), 345–348 (2005)

Stewart, G.T., Kolluru, R., Smith, M.: Leveraging public-private partnerships to improve community resilience in times of disaster. International Journal of Physical Distribution & Logistics Management 39(5), 343–364 (2009)

Stölzle, W.: Supply Chain Event Management, Klaus, P., Krieger, W. (eds.), pp. 541–546 (2008)

Stölzle, W., Otto, A. (eds.): Supply Chain Controlling in Theorie und Praxis, Wiesbaden (2003)

Sundar, I., Sezhiyan, T.: Disaster Management, New Delhi (2007)

Svensson, G.: A conceptual framework of vulnerability in firms' inbound and outbound logistics flows. International Journal of Physical Logistics & Distribution Management 32(2), 110–134 (2002a)

Svensson, G.: A typology of vulnerability scenarios towards suppliers and customers in supply chains based upon perceived time and relationship dependencies. International Journal of Physical Logistics & Distribution Management 32(3), 168–187 (2002b)

Svensson, G.: A Conceptual Framework for the Analysis of Vulnerability in Supply Chains. International Journal of Physical Distribution & Logistics Management 30(9), 731–749 (2000)

Tandler, S., Essig, M.: Supply Chain Safety Management: Konzeption und Gestaltungsempfehlungen, Bogaschewsky, R., Essig, M., Lasch, R., Stölzle, W. (eds.), pp. 57–92 (2011)

Tao, F., Hu, Y., Zhao, D., Zhou, Z.: Study of failure detection and recovery in manufacturing grid resource service scheduling. International Journal of Production Research 48(1), 69–94 (2010)

Thejendra, B.S.: Practical IT Service Management, Ely (2008)

The Royal Society, Risk: Analysis, Perception and Management - Report of a Royal Society Study Group, London (1992)

Töpfer, K.: Krisenmanagement. Verlauf, Bewältigung und Prävention von Krisen. WiSt 38(4), 180–187 (2009)

Trauboth, J.H.: Krisenmanagement bei Unternehmensbedrohungen. Präventions- und Bewältigungsstrategien, Stuttgart (2002)

Trkman, P., McCormack, K.: Supply chain risk in turbulent environments – A conceptual model for managing supply chain network risk. International Journal of Production Economics 119(2), 247–258 (2009)

Trux, W., Müller, G., Kirsch, W.: Das Management strategischer Programme, München (1989)
Turner, B.A.: Causes of Disaster: Sloppy Management. British Journal of Management 5(3), 215–219 (1994)
Turner, B.A.: The Organizational and Interorganizational Development of Disasters. Administrative Science Quarterly 21(3), 378–397 (1976)
United Nations, Internationally agreed glossary of basic terms related to Disaster Management, Genf (1992)
Vahrenkamp, R., Siepermann, C. (eds.): Risikomanagement in Supply Chains. Gefahren abwehren, Chancen nutzen, Erfolg generieren, Berlin (2007)
Van Bon, J.: Foundations of It Service Management basierend auf Itil® V3, 3rd edn., Zaltbommel (2008a)
Van Bon, J.: IT Service basierend auf ITIL V3. Das Taschenbuch, Zaltbommel (2008b)
Victor, F., Günther, H.: Optimiertes IT-Management mit ITIL, 2nd edn., Wiesbaden (2005)
Von Rössing, R.: Betriebliches Kontinuitätsmanagement, Bonn (2005)
Wagner, S.M., Bode, C.: An empirical investigation into supply chain vulnerability. Journal of Purchasing & Supply Management 12(6), 301–312 (2006)
Waters, D.: Supply Chain Risk Management. Vulnerability and Resilience in Logistics, London, Philadelphia (2007)
Weingardt, M.: Fehler zeichnen uns aus. Transdisziplinäre Grundlagen zur Theorie und Produktivität des Fehlers in Schule und Arbeitswelt, Kempten (2004)
White, D.: Application of system thinking to risk management: a review of the literature. Management Decision 33(10), 35–45 (1995)
Wieczorek, M., Naujoks, U., Bartlett, B. (eds.): Business continuity: it risk management for international corporations, Berlin, Heidelberg, New York (2010)
Williams, Z., Lueg, J.E., Taylor, R.D., Cook, R.L.: Why all the changes? An institutional theory approach to exploring the drivers of supply chain security (SCS). International Journal of Physical Distribution & Logistics Management 39(7), 595–618 (2009)
Williams, Z., Lueg, J.E., LeMay, S.A.: Supply chain security: an overview and research agenda. The International Journal of Logistics Management 19(2), 254–281 (2008)
Witt, B.C.: IT-Sicherheit kompakt und verständlich. Eine praxisorientierte Einführung, Wiesbaden (2006)
Wolf, K., Runzheimer, B.: Risikomanagement und KonTraG. Konzeption und Implementierung, 5th edn., Wiesbaden (2009)
Zarnekow, R., Hochstein, A., Brenner, W.: Serviceorientiertes IT-Management. ITIL-Best-Practices und -Fallstudien, Berlin, Heidelberg, New York (2005)
Zsidisin, G.A.: Managerial Perceptions of Supply Risk. Journal of Supply Chain Management 39(1), 14–26 (2003a)
Zsidisin, G.A.: A grounded definition of supply risk. Journal of Purchasing & Supply Management 9(5-6), 217–224 (2003b)
Zsidisin, G.A., Ellram, L.M.: An Agency Theory Investigation of Supply Risk Management. Journal of Supply Chain Management 39(3), 15–27 (2003)
Zsidisin, G.A., Ellram, L.M., Carter, J.R., Cavinato, J.L.: An analysis of supply risk assessment techniques. International Journal of Physical Distribution & Logistics Management 34(5), 397–413 (2004)
Zsidisin, G.A., Melnyk, S.A., Ragatz, G.L.: An institutional theory of business continuity planning for purchasing and supply chain management. International Journal of Production Research 43(16), 3401–3420 (2005a)
Zsidisin, G.A., Melnyk, S.A., Ragatz, G.L.: The Dark Side of Supply Chain Management. Supply Chain Management Review 9(2), 46–52 (2005b)

Targets and Components of Supply Chain Safety Management: Structure of the Book

Eva-Maria Kern[1], Michael Hülsmann[2], Stephan Klein-Schmeink[3], and Michael Essig[4]

[1] Bundeswehr University Munich,
Chair of Knowledge Management and Business Process Design,
Werner-Heisenberg-Weg 39, 85577 Neubiberg,
Germany
eva-maria.kern@unibw.de
[2] Jacobs University Bremen,
Associate Professor of Systems Management, International Logistics – School of Engineering and Science Campus Ring 1, 28759 Bremen,
Germany
m.huelsmann@jacobs-university.de
[3] Gesellschaft für Entwicklung, Beschaffung und Betrieb (g.e.b.b.) mbH,
Business Unit Manager, Ferdinand-Porsche-Str. 1a, 51149 Köln,
Germany
stephan.klein-schmeink@gebb.de
[4] Bundeswehr University Munich, Chair of Materials Management and Distribution,
Werner-Heisenberg-Weg 39, 85577 Neubiberg,
Germany
michael.essig@unibw.de

Based on the introductory chapter, five elements can be distinguished that are all part of the concept of SCSM. First at all (1), since various concepts already exist in the literature, the necessity of developing a new concept for the management of risk and uncertainty factors in company spanning supply chains has to be emphasized. Then (2), risk and uncertainty factors threatening a supply chain's safety have to be identified and analyzed. Following this, adequate action measures have to be taken: Preventive action measures aim at eliminating the source of risk and uncertainty factors and therefore make contribution to a supply chain's protection (3). Reactive action measures, by contrast, aim at minimizing the detrimental impact resulting from risk and uncertainty factors that already have occurred and therefore make contribution to a supply chain's resilience (4). Finally, a management process has to be implemented enabling a supply chain to improve it's overall preparedness (5).

In accordance with the concept of SCSM and its elements, the intended book will be subdivided into five chapters. Before describing the chapters in detail, the following figure gives an overview of SCSM and the structure of the book:

M. Essig et al. (Eds.): Supply Chain Safety Management, LNL, pp. 41–48.
springerlink.com

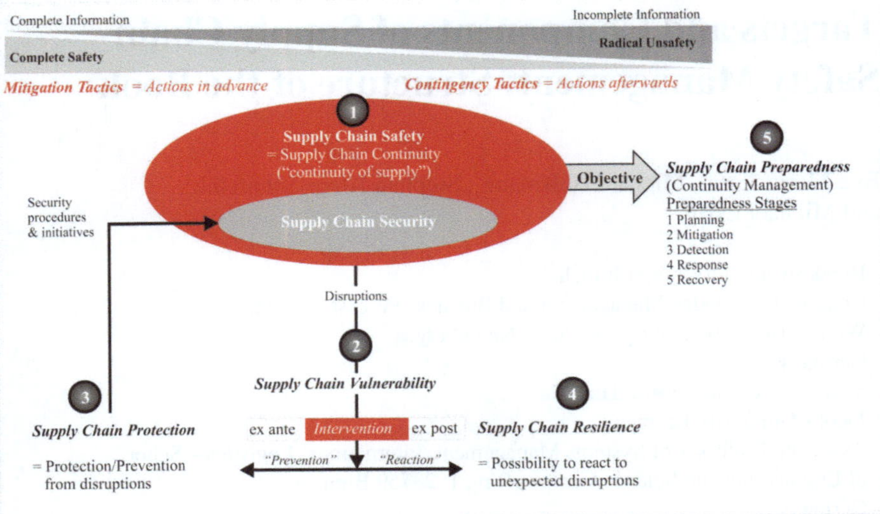

Chapter 1 included the introduction of SCSM being characterized as a conceptually funded, integrated concept which aims at ensuring the continuity of company spanning supply chains.

Sandra Tandler and Michael Essig therefore discussed the state of research and outlined the need for a new concept. Based on an extensive literature analysis, they identified 16 concepts for the management of risk and uncertainty factors that currently prevail in the literature. Moreover, six distinguishing features emerged along with it which laid the foundation for conducting a consistent and stringent analysis of all identified concepts and for emphasizing similarities and differences existing between the concepts. The results provided evidence to the lack of an existing, holistic concept for the management of risk and uncertainty factors in supply chains. Against this background, Sandra Tandler and Michael Essig introduced the concept of SCSM and its underlying elements.

Based on the explanation of the conceptual foundations, the relevance and omnipresence of the given research field will be highlighted. Therefore, ***chapter 2*** discusses various (potential) risk and uncertainty factors, which result in the increased vulnerability of supply chains and, by consequence, disruptions.

Supply chains strongly depend on the availability of scarce natural resources to ensure their continuity of supply. While the continuity of supply is – with the exception of China – primarily viewed as a task for the industry, governments are responsible for ensuring market access and fair competition. However, since countries differ with respect to their national wealth in raw materials and to their industrial structure, they prioritize varying key aspects in their raw material strategies. Based on this, Stormy-Annika Mildner, Gitta Lauster, and Lukas Boeckelmann give, in their article *Scarce Metals and Minerals as Factors of Risk: How to Handle Criticality*, an overview of studies of scarce natural resources and highlight that the two predominant risks on resource markets are price and supply risks.

They then discuss national raw material strategies by taking into consideration the different sets of countries and conclude by stating that despite these differences, all countries considered in their study face one common challenge: securing metals and minerals at sustainable prices.

At the latest since 2001, September 11, highly developed industrial nations must accept that potential state and often also non-state adversaries are spying out their weak points – the hub of all power – and preparing to attack them. However, there is still a lack of imagination with regard to "new" forms of threats and attacks and to the possibilities that can result from the combination of military and non-military instruments. In their article *Hybrid Threats and Supply Chain Safety Management*, Marc Oprach and Boris Bovekamp therefore take a new view on risk factors threatening the continuity of supply since their focus is on hybrid threats comprising all state and non-state adversaries who, in a conflict, use the full spectrum of conventional, criminal, terrorist and irregular measures. Their aim is both to sensitize industrial nations about the topic by highlighting the options that hybrid adversaries have and to shed light on how industrial nations can successfully respond to those hybrid threats.

The political environment is becoming more and more volatile and demand driven, therefore unforeseeable for supply chains operating within it. Thus, since the political environment can have significant impacts on their activities, it poses a significant risk factor for supply chains. However, speaking about political environment and risk requires a clear-cut definition to cut through the confusion existing in the contemporary body of scholarly literature. To overcome this, Carlo Masala aims in his article *Political Environment as a Factor of Risk* at contributing to a conceptual clarification of two concepts political environment and risk. Therefore, he firstly comes up with discriminating definitions of both terms and discusses their connections afterwards. Then, he discusses some empirical examples to illustrate how the political environment might be considered as a risk factor. Finally, some theoretical considerations are presented regarding potential strategies for managing risks emanating from specific political environments.

Besides legally required, information on the risk situation is mainly demanded by shareholders, potential investors, and other stakeholders such as employees to access and appraise the future performance of the company. Therefore, risk disclosures in annual reports become the main type of risk communication between a company and its stakeholders. As companies nowadays normally act within supply chains, Christoph Bode, René Kemmerling, and Stephan M. Wagner aim, with their article *Internal versus External Supply Chain Risks: A Risk Disclosure Analysis,* at providing a tool that allows them to systematically identify and analyze their supply chain risks. Based on this, they propose a simple classification system for the analysis of supply chain risks. This two-level system distinguishes on the top level between internal-driven and external-driven supply chain risks and on the second level between five risk categories. Finally, the analysis of the annual 10-K report of 219 companies between 2007 and 2009 shows that the importance of internal-driven supply chain risks has increased in the last years.

For the benefit of a higher protection of supply chains, ***chapter 3*** deals with the identification of preventive action measures. In this context, particular emphasis will be placed on the discussion of security procedures and initiatives which specifically aim at strengthening logistics chains being part of company spanning supply chains.

Aviation security is of high significance since it enables various process chains to achieve a higher safety level. However, due to its often time-consuming scheduling processes, aviation security makes it increasingly difficult to achieve just-in-time delivery. Against this background, Gerhard Wirth sheds, in his article *The secure process chain in aviation security,* light on the topic of what is meant by aviation security. By examining the processes that relate to the arrival and, respectively, to the departure of passengers, he highlights the complexity surrounding aviation security resulting from its great sensitivity to changes induced by laws and regulations. He concludes by stating that appropriate management systems are needed to ensure the "just-in-time" positioning of personnel and equipment to handle passengers and luggage, at the right time and in the right location, and thus ensuring a functional process chain.

The physical transportation of goods requires the construction of infrastructures, such as highways, bridges, and others. Those built infrastructures are key elements of physical supply chains. However, recent events comprising multiple threats at the same time, show that physical supply chains are highly vulnerable and that, due to their strong interdependencies, even tiny causes can have the potential to disrupt physical supply chains. Given this background, Norbert Gebbeken deals in his article *Protection of Buildings* with the safety of critical built infrastructures against multiple threats due to natural disasters, technical disasters or terrorist attacks. He highlights today's possibilities to assess and to design critical built infrastructures and concludes by stating that numerical simulations not only help to study threat scenarios or to assess already existing infrastructures but also help to design new buildings avoiding or reducing physical tests that are usually time consuming and expensive or even impossible to carry out in the 1:1 scale.

In recent years, cargo theft in European road freight transport chains has increased and has become very carefully organized. Due to its negative effects on the partners in supply chains, Irene Sudy, Sebastian Kummer, and Ellis Lehner develop in their article *Risk response measures for the management of the risk of theft and organized crime in road freight transport chains* a set of risk response measures for the management of cargo theft in road transportation and categorize them according to their abilities to eliminate, reduce, transfer, or accept the risk of theft. In order to get a deep practical insight and to ensure the practicability of measures proposed, the proposed risk response measures are based on a thorough literature review with personal expert interviews from internationally operating logistics service providers as well as insurance companies in Austria and Germany. This approach allows the alignment of risk response measures that can be found in the risk management literature with the measures applied in practice.

Since they are responsible for the transportation of goods, logistics service providers play an important role for and within supply chains. They are therefore required to set-up and operate appropriate global supply chains. Thereby, logistics service providers are facing two challenges that have to be met: on the one hand, customers ask them to create cost-efficient and high-performance supply chains; on the other hand, since there can be observed a rise of threats, such as terrorism, or extreme weather conditions, logistics service providers are obliged to secure the goods being transported across the supply chain. Karl Engelhard and Christian Böhm, employees at Hellmann Worldwide Logistics, take in their article *Security of Supply Chains from a Service Provider's Perspective* a practical view on the requirements logistics service providers are facing today by describing preventive measures that have been successfully implemented to achieve supply chain security.

Social networks, online banking, e-health, e-marketplaces – these exemplarily listed trends highlight that today's society is moving towards a networked society. However, since ubiquitous connectivity means also widespread vulnerability, such trends not only include potential advantages but also potential risks with respect to security and privacy. The task to protect relevant assets in the digital world is becoming even more demanding. Based on this, Gabi Dreo Rodosek and Mario Golling describe in their article *Cyber Security: Challenges and Application Areas* various threats of cyber security and give an overview of possible countermeasures. They conclude by stating that the approaches existing so far do not sufficiently face the current threats of cyber security. Research in developing new approaches is therefore of mandatory importance.

Today, it is not unusual when a product is developed in the USA, produced in Asia and sold in Europe. The world-wide flow of good and global supply chains already allow a society to profit from the positive effects of globalization. However, the failure of a system – whether transportation, communication or energy supply – shocks almost all areas of life and production and triggers various domino effects. Individual nations and international organizations have therefore the difficult task of ensuring public security as efficiently and effectively as possible. In doing so they receive support from the logistics sector that helps create public security with its solutions by supporting governmental administrations and organizations in various ways. Based on this, Matthias Witt emphasizes in his article *How logistics can create and support public security* how – with recourse to his practical experiences at LOG mbH – the logistics sector can contribute to the maintenance of public security by examining two selected examples in detail.

For the benefit of a higher resilience of supply chains, ***chapter 4*** deals with the identification of reactive action measures.

The Panama Canal extension project is deemed to be perhaps the most important transportation project in the world today. The 5.5 billion US dollar project will enable the Canal to handle up to 12,600 TEU, Post-Panamax vessels, instead of the current maximum of 4,400 TEU, Panamax Vessels. It allows most Post-Panamax vessels to lower their shipping costs by using the canal and is likely to change transportation flow patterns throughout North and South America, as well as port loads and transportation flows inland in the Americas. Since the Panama Canal expansion will impact cargo throughout the Americas, Liliana Rivera and Yossi Sheffi give in their article *Panama Canal Update* a short update on the status quo of the project by presenting new opportunities, leading to new transportation routes, new distribution patterns and new logistics hubs formation.

There is an increasing amount of influences that can endanger logistics systems to carry out their functions. Hence, the ability to resist against them is increasingly important for logistics companies as well as whole logistics networks to gain and maintain competitiveness through offering and ensuring a high reliability of logistics services. Thus, the robustness of logistic systems, i.e., the ability to restore their operational reliability after being damaged, becomes an increasingly important topic. Based on this, Philip Cordes and Michael Hülsmann introduce in their article *Self-Healing Supply Networks – A Complex Adaptive Systems Perspective* the concept of self-healing processes that allow logistics systems to take decisions on their own. For this purpose, the authors focus on analysing the technological and organizational pre-conditions for self-healing processes in modern supply networks and how they contribute to a logistics system's robustness. Finally, the associated developed hypotheses show that there are both potentials to increase the robustness of logistics systems and potential limitations that have to be taken into consideration.

Critical infrastructures subsuming organizations and institutions of great importance for the national community are of specific character as nobody recognizes the value of them as long as they work. However, if there is a break-down which can result in long-term interruption of supplies, critical disruption of public safety and other dramatic results, one realizes, how important critical infrastructures really are. Supply chains are dependent on critical infrastructures' safety as well as, for example, energy is needed to produce goods. Given this background, Albrecht Broemme discusses in his article *Supply Chains – How to Support Critical Infrastructures' Safety, Protection, Preparedness, and Resilience* a process model including several steps that have to be taken into consideration to achieve a higher safety level of critical infrastructure. Being the president of the *Technisches Hilfswerk*, he presents this governmental organisation and shows, by providing some examples, how the THW gives support to restore critical infrastructures after a break-down.

Ensuring product availability outbound to customers has become a major customer requirement over the past decades. Consequently, distributors, traders, and original equipment manufacturers (OEM) across the supply chain are forced to optimize their operations and to achieve continuity of supply. Companies therefore require hands-on management concepts that warn them about and trigger them to respond to potential product shortages as early as possible. Yet, the high complexity of today's supply chains makes this task very demanding. Based on this, Joerg S. Hofstetter and Wolfgang Stölzle introduce in their article *Supply Chain Event Management – concept and use in business practice* the concept of Supply Chain Event Management (SCEM). They address the existing heterogeneity of SCEM understandings and SCEM approaches found in business practice and academia today and the missing hands-on measurement for SCEM use. Finally, they shed light on the darkness about the current use of SCEM in business practice.

Ensuring agility along with disruption resistance and resilience are crucial issues in supply chain (SC) planning. To overcome this challenge, Dmitry Ivanov, Boris Sokolov, and Joachim Käschel develop in their article *Adaption-based supply chain resilience* an adaptation-based supply chain resilience framework which is provided to companies to achieve maximal economic performance and stability in their supply chains. Therefore, they propose a detailed analysis of supply chain resilience based on a mutual classification of flexibility and reliability elements. Subsequently, they present an algorithm of decision-making on supply chain planning which contributes to both supply chain reliability and flexibility. The resulting supply chain resilience framework and tools make it possible to take into account individual risk perceptions of managers, different strategies with regard to risk management, and to consider not only supply chain economic performance but also supply chain stability.

Finally, ***chapter 5*** is concerned with the implementation of the elements of Supply Chain Safety Management discussed above within the underlying management process.

Increasing economic globalization and rising competitive pressure represent a challenge to companies, which have to ensure cost-effective purchasing in global sourcing and offer products on various international sales markets. End-to-end monitoring of supply and production chains as well as the certification of products and processes are therefore critical for ensuring that the requirements of different target markets are complied with and that the expectations of target groups from different cultural backgrounds are fulfilled. By outsourcing this task to special service providers, companies can focus on their core competencies, minimize the risks of global sourcing and exploit the opportunities offered by new markets. Against this background, Axel Stepken describes – with recourse to four companies TÜV Süd had accompanied – in his article *Monitoring and Certification of Supply Chain Safety* how testing and inspection along the global supply and production chains as well as the role of product and process certification work in practice.

Global supply chains – and some business sectors (e.g., the military), in particular – are increasingly forced with the requirement of complying with complex laws, regulations and standards. Ensuring end-to-end supply chain compliance has therefore become a critical success factor for all elements and relations, material movements and information flows within and along global supply chains. Due to its significant impact on the risk and chance dimension of a supply chain's performance, compliance affairs have to be managed. Against this background, Josef Mauermair shows in his article *Compliance and Supply Chain Safety* how, from a theoretical and a supplier's perspective, a compliance system can be developed, implemented and run. He therefore refers to a life cycle model of rules which aims at to ensuring supply chain preparedness and outlines the possibilities and limitations of planning, running and controlling rules as the core objects of the compliance system.

Innovation is a precondition for economic development, as it opens up opportunities for competitive advantages and long-term success. However, innovation also means risk, as it takes place under conditions of uncertainty. The current literature surrounding the concept of SCRM is characterized by its prevailing disregard of interrelations with opportunities and innovation, and of definition of objectives (which is a basic requirement to identify both positive and negative deviations). Against this background, the article *Supply Chain Innovation and Risk Assessment (SCIRA) Model* provided by Stephan Klein-Schmeink and Thomas Peisl represents an advanced approach of SCRM located at a strategic level, since it sheds light on how objectives for supply chains can be defined, and how the tension between innovation, opportunities, and risks can be successfully managed.

In accordance with the frequently cited business wisdom 'You can't manage, what you don't measure' supply chain managers need support in quantifying and thus mitigating supply chain risks. A look into the relevant literature shows, however, that only few publications can be identified which choose simulations and mathematical models as an alternative to large-scale surveys. Thus, Andreas Brieden, Peter Gritzmann, and Michael Öllinger introduce in their article *Supply Chain Safety: A Diversification Model based on Clustering* a novel quantitative algorithm that provides a multiple covering of the commodity graph via constrained clustering. It can be used to measure the risk of the status quo of the supply chain of a production by calculating the (conditional) probability of failure. This risk can be judged in comparison to best assignments of suppliers to different supply chain components of the same size. By using this quantitative approach, it is demonstrated that the risk of failure can be significantly reduced.

Auctions are a common tool which not only allows to optimize ex-factory prices but also to allocate the production capacity in exactly these markets (internationally) and to those dealers (intra-nationally) that can reach these maximum prices with their customers. The control effort is minimized as the optimal distribution is defined as a self-adjusting mechanism and sales channel risks are lowered. In the automotive industry, auctions are implemented in various areas so far. However, in sales of new automobiles there is no comprehensive application of auctions in place yet. Thomas Ruhnau and Thomas Peisl therefore propose in their article *Risk Management through Flexible Capacity Allocation and Price Control – Auctions in the New Car Sales Process* a model for a forward auction with one supplier and a large number of bidders for the indirect sales of new vehicles aiming at risk reduction in a manufacturer's sales channel. They define and explain the auction model by a process flow and conclude by discussing challenges and suggesting solutions.

2 Supply Chain Vulnerability

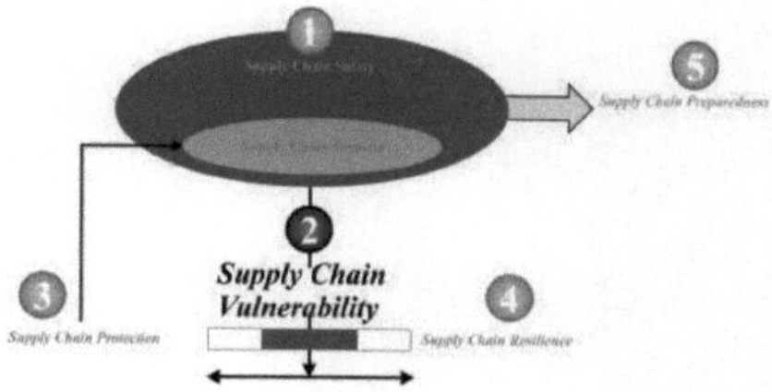

Scarce Metals and Minerals as Factors of Risk: How to Handle Criticality*

Stormy-Annika Mildner[1], Gitta Lauster[2], and Lukas Boeckelmann[3,**]

[1] Stiftung Wissenschaft und Politik, Deutsches Institut für Internationale Politik und Sicherheit, Mitglied der Institutsleitung, Ludwigkirchplatz 3-4, 10719 Berlin, Germany
stormy-annika.mildner@swp-berlin.org
[2] Stiftung Wissenschaft und Politik, Deutsches Institut für Internationale Politik und Sicherheit, Forschungsassistentin in der Institutsleitung, Ludwigkirchplatz 3-4, 10719 Berlin, Germany
gitta.lauster@swp-berlin.org
[3] Student für Volkswirtschaftslehre an der University of Bristol, Great Britain
lukas.boeckelmann@gmail.com

Scarce natural resources are increasingly perceived as a risk by governments and industries alike. The turn of the century brought starkly rising and fluctuating prices. High demand due to technological innovation and fast growth in emerging economies, speculation, and increasing government interventions in commodity markets are only a few of the factors that contribute to this development, fuelling worries about sufficient and affordable supply of raw materials in many countries. Governments around the world are now looking at what needs to be done to secure supply of critical materials; many of them, such as Germany and Japan, have passed integrated resource strategies. Resources also rank high on this year's G20 agenda under the French presidency. Apart from the introduction and conclusion, our paper is divided into two parts. We start with an overview of studies on critical metals and minerals, comparing their methodology and findings, as well as highlighting the major risks on commodity markets. Despite differences in their framework and methodology, a number of metals and minerals is identified as critical by most of them (e.g. rare earths). The two predominant risks on commodity markets highlighted in the studies are price and supply risks. We then discuss national raw materials strategies of two sets of countries: countries/regions that are heavily dependent on raw materials imports (Germany, EU, and Japan) and resource-rich countries (the United States, Canada, China). With the exception of China, supply security is commonly viewed as a primary task for industry, not government. The government's primary chore is to ensure market access and fair competition. However, depending on the national wealth in raw materials and

* This article has been composed in spring 2011 and is thus based on facts and figures as of June 2011.

** Dr. Stormy-Annika Mildner (M.Sc.) is a member of the Executive Board of the Stiftung Wissenschaft und Politik (SWP); Gitta Lauster, M.A., M.A., is a Phd. student in resource governance at the SWP, Lukas Boeckelmann interned at SWP in spring 2011. We also thank our intern Paruvana Ludszuweit for her assistance in writing this article.

M. Essig et al. (Eds.): Supply Chain Safety Management, LNL, pp. 51–88.
springerlink.com

on a country's industrial structure, the countries prioritize varying key aspects in their raw materials strategies.

1 Introduction

Until recently, only specialists knew what the term *rare earths* stands for. Yet for almost a year, rare earth elements such as neodymium, terbium and cer have made quite a career in media, science and politics as their prices are skyrocketing and supply struggles to keep pace with demand. Many countries have now declared rare earths a critical resource and try to find alternatives to escape supply shortages.

Rare earths are not the only metals experiencing steep price hikes. Although prices for metals and minerals dropped considerably during the financial and economic crisis in 2008/2009, they are, following the global economic recovery, already on the rise again. There is no ground to fear metal ore reserves running out any time soon. However, prices will likely continue to rise and experience noticeable fluctuations. The most important driver for this development is strong and rising demand from China (China effect). A second decisive structural factor is technical innovation in electronic and environmental products which has pushed up demand for many metals. A more cyclical but nonetheless important factor is speculation on the markets. Fuelled by the current dollar weakness, low interest rates and inflation fears as well as uncertainties about economic developments in both the European Union and the United States, metals are becoming attractive investment alternatives (Hilpert et al. 2011).

Growing competition as well as increasing prices and price volatility have raised concerns about future access to key natural resources at sustainable prices in many countries. The qualms are fuelled by highly uneven geographical distributions of many metals across the globe. This unequal distribution could be considered a naturally-given opportunity for an international division of labor – but only if markets function well, which is hardly the case with regard to resources (Bardt 2010). Resource markets are anything but transparent, and strategic government intervention in the markets is on the rise, including, most notably, export restrictions such as tariffs and quotas. High and fluctuating prices, geographical and market concentrations together with strategic interventions pose considerable risks to the security of supply. To reduce these risks, many countries have recently developed national resource strategies. Resources also rank high on this year's G20 agenda under the French presidency.

In the light of these developments, we ask two sets of questions: 1. How can we measure criticality? Which metals are critical? And what are the main risks on the resource markets? 2. How are countries reacting to these risks and are their strategies adequate? Apart from the introduction and conclusion, the paper is divided into two sections: First, we will give an overview of how criticality and risks are measured by reviewing the most groundbreaking studies on this issue and highlighting the metals which are perceived as critical. In the second part of our paper, we will take a closer look at resource strategies, dividing the selected countries into two groups: resource-poor and import-dependent countries (or regions) (Germany, EU and Japan) and resource-rich countries (the United States, Canada, and China).

2 Critical Metals and Minerals

2.1 Measuring Criticality

In recent years, a multitude of studies has been published which analyze the markets for metals and identify critical elements. The methodologies applied range from simple (U.S. National Academy of Science 2007, U.S. Department of Energy 2010) and more elaborate semi-quantitative criticality/risk matrixes (Institut der deutschen Wirtschaft Köln, IW Consult 2009) to qualitative factor-based (American Physical Society Panel 2011, Elsner 2011, Öko-Institut/UNEP 2009, Behrendt et al. 2007) and quantitative (Angerer et al. 2009) studies. The majority of the frameworks apply only short-term perspectives, although mid-term (U.S. DOE, BGR) and long-term (Öko-Institut/UNEP; Angerer et al.) analyses can also be found.

While these studies differ with regard to their methodology (e.g. the selection and weight of criteria) and the specific elements under consideration, they have many common features. The starting point of all studies is increasing demand for metals and minerals, driven in particular by developments in green technology (e.g. wind turbines, solar technologies, electric mobility) and the information technology sector. It is therefore not surprising that a majority of the reviewed studies focuses on elements that are required for production in these sectors. These are often elements which, in the past, have not been widely extracted, traded or utilized and have not been in the center of well-established markets.

Furthermore, the studies use similar determinants for criticality: supply risks and economic importance. Indicators for supply risks are: (1) geological and market concentration, (2) country risks (e.g. political instability, market interventions), (3) degree of substitutability and recyclability, (4) response times in production and utilization, and (5) environmental and social considerations.

Despite differences in the methodology, the studies' results are surprisingly homogenous for certain elements: rare earths, platinum group metals, indium, niobium, lithium, germanium and terbium are assessed as critical metals by a number of studies. However, other materials receive mixed results, sometimes listed as critical and sometimes as non-critical (e.g. cobalt). Diverging results can be explained by different regional (e.g. local resource endowments) and sectoral foci (IT sector, green tech sector, economy as a whole). Furthermore, they can be attributed to diverging frameworks for assessing criticality. Thus, the inclusion of factors such as environmental concerns, recyclability or response times in production and utilization influences the risk estimations. This heterogeneity in frameworks and results highlights the need for normed assessment standards. The following section gives a short overview of the most important studies.

2.1.1 The North American Perspective

In *Minerals, Critical Minerals, and the U.S. Economy* (NAS 2007), the National Academy of Science notes the lack of an established methodology for identifying critical minerals in the United States. Therefore, the report offers a new methodology for the assessment of a mineral's degree of criticality – the criticality matrix.

The matrix consists of two dimensions: 'the importance in use' and the 'likelihood of supply restrictions'. As a number of materials are placed on the matrix in respect to their ratings, the matrix allows for a comparison and ranking amongst them.

The study defines an element as critical if it is both important in use and potentially subject to supply restrictions. The level of importance results from the demand for a respective mineral from different sectors of the U.S. economy and the availability of alternatives (substitutability): "The greater the difficulty, expense, or time to find a suitable substitute for a given mineral, the greater will be the impact of a restriction in the mineral's supply" (NAS 2007, p. 2). Over the long term (more than about ten years) five factors are identified which affect the availability of metals and minerals ('likelihood of supply restrictions'): "geologic (does the mineral resource exist?), technical (can we extract and process it?), environmental and social (can we produce it in environmentally and socially accepted ways?), political (how do governments influence availability through their policies and actions?), and economic (can we produce it at a cost users are willing and able to pay?)" (NAS 2007, p. 5). Over the short- and medium term the study identifies further factors which may result in the element's physical unavailability or in higher prices: a significant and unexpected increase in demand, relatively thin (or small) markets, high production concentration in a small number of mines, companies or producing countries, risky or fragile byproduct production of a mineral and low levels of recyclability.

The report introduces a criticality matrix using 11 minerals or mineral groups: copper, gallium, indium, lithium, manganese, niobium, platinum group metals, rare earth elements, tantalum, titanium and vanadium. However, the authors make clear that this is not a comprehensive coverage of minerals, but rather an incomprehensive selection for the demonstration of the matrix. Platinum group metals, rare earths, indium, manganese and niobium were identified as most critical among the assessed elements, caused foremost by high demand, a lack of substitutes and high risk in supply markets.

In *Critical Materials Strategy* (U.S. DOE 2010), the U.S. Department of Energy (DOE) examines the role of rare earth metals and other metals and minerals in the clean energy economy. To assess critical resources in the light of energy innovations, the report applies an adaptation of the NAS methodology. The *Critical Materials Strategy* approach utilizes a two-dimensional matrix, which allows arranging, ordering and comparing minerals in respect to their criticality in both the short and the long term. Scores for 'supply risk' and the 'importance to clean energy' are based on weighted attributes. For each attribute, the selected materials receive factor scores from 1 (least critical) to 4 (most critical). The variable 'supply risk' is determined by basic availability, competing technology demand, political, regulatory and social factors, co-dependence on other markets and producer diversity. The variable 'importance to clean energy' consists of the attributes clean energy demand and substitutability. The matrix defines three areas: critical (red), near-critical (yellow), and non-critical (green).

In congruency with the NAS report, the study lists minerals as critical if they receive high scores for both, 'importance' and 'supply risk' (U.S. DOE 2010, p. 95). As the DOE report focuses on 14 elements used in clean energy technologies, the results are not directly comparable with the NAS study. Indium as well as the rare earths elements dysprosium, europium, neodymium, terbium, and yttrium are considered critical in the short-term. The minerals cerium, lanthanum and tellurium are classified as near-critical while gallium, cobalt, lithium, praseodymium, and samarium are not critical. For some metals and minerals, the importance to clean energy and the supply risk change from the short- to the medium term perspective. The only material whose criticality increases is lithium (from non-critical to near-critical). This change is due to the anticipated importance of lithium to clean energy as market penetration of vehicles using lithium-ion batteries is projected to increase. For all other minerals the criticality level is expected to decrease over the medium term. Decreasing criticality levels can be explained by a combination of "expanded supply and increased alternatives for substitution at different levels of the supply chain" (U.S. DOE 2010, p. 99).

The most recent study focusing on metals and minerals in the green technology sector is *Energy Critical Elements—Securing Materials for Emerging Technologies* published by the American Physical Society Panel on Public Affairs and the Material Research Society (APS 2011). In analogy to the DOE report, the authors expect demand to rise for a number of metals and minerals employed in green technologies. As a shortage of these elements could confine the American ability to compete internationally, such materials are labeled energy-critical elements (ECEs). According to the report, ECEs are those that feature a "potential for major impact on energy systems" and for which "a significantly increased demand might strain supply, causing price increases or unavailability, thereby discouraging the use of some new technologies" (APS 2011, p. 4).

The APS report, which builds on the NAS study, adopts a qualitative framework for the assessment of critical elements. It introduces five indicators for the availability of energy-critical elements: (1) crustal abundance, concentration and distribution of ECEs, (2) geopolitical risks, (3) risk of joint production, (4) environmental and social concerns, and (5) response times in production and utilization. The report finds that ECEs are only to be found in 0.1 percent of the Earth's crust by weight. Therefore, the concentration of ECEs is usually not high enough to be obtained as a primary product itself given today's prices. Instead, ECEs are commonly extracted as by-products from the mining, production and refining of primary ores, such as lead, zinc and copper. Although the co-production of several raw materials comprise big advantages, e.g. economies of scale (as the number of materials produced increases, the average total cost of production per material decreases), it bares certain risks. For example, the obtainability of ECEs is constrained by the amount of primary ores produced. Furthermore, the extraction of by-products creates a complex and difficult environment for investment in these elements. Concerning geopolitical risks, the report finds that these are particularly prevalent when the number of producing countries, companies or mines is small. In the case of poorer developing countries with rather weak forms of governance, the report highlights risks arising due to political instability and conflict. In the case of emerging economies, the report warns against strategic interventions in markets and price manipulation. Last but not least, the report identifies the response time in production and utilization as an important factor. Notably, it often takes 5 to 15 years to render a new mine operative. Moreover, developing new technologies to extract certain elements from ores is often a long and uncertain process.

Based on the qualitative framework the report identifies a number of elements as ECEs: gallium, germanium, indium, selenium, silver and tellurium, for example, are essential components of advanced photovoltaic solar cells. Dysprosium, neodymium, praseodymium, samarium and cobalt are employed in particular in high-strength permanent magnets. Most rare earths, known for their unusual magnetic or optical properties, as well as lithium and lanthanum are required for high performance batteries. Helium is used, for example, in advanced nuclear reactor designs; platinum, palladium, cerium and other platinum group elements are required as catalysts in fuel cells. Finally, rhenium is a crucial component of high performance alloys for advanced turbines.

2.1.2 The German Perspective

Seltene Metalle: Maßnahmen und Konzepte zur Lösung des Problems konfliktverschärfender Rohstoffausbeutung am Beispiel Coltan, published by the Umwelt Bundes Amt (Behrendt et al. 2007) focuses on critical materials in the information and communication industry. The study explicitly concentrates on scarce metals. It offers three indicators for scarcity: (1) prices, (2) dynamic and static range of stocks, and (3) concentration of production. Elements with high scarcity ratings are antimony and indium. Cobalt, gold, iridium, palladium, platinum, rhenium, rhodium, ruthenium, zinc and tin are rated as medium scarce, while rare earths (among other elements) were not classified as scarce metals owing to the applied criteria.

Another study focusing on future technologies is *Rohstoffe für Zukunftstechnologien: Einfluss des branchenspezifischen Rohstoffbedarfs in rohstoffintensiven Zukunftstechnologien auf die zukünftige Rohstoffnachfrage* published by the Fraunhofer Institute for Systems and Innovation Research ISI and the Institute for Future Studies and Technology Assessment (Angerer et al. 2009). At the center of their analysis stands the 'vulnerability' of certain metals and minerals needed in the high-tech industry. An element is classified as vulnerable if it is particularly important to the national economy, concentrated on a few countries and produced in politically unstable regions. A further determinant is its prospective functional and quantitative importance for the development of future technologies. The study assesses the prospective demand for 22 metals and minerals for certain high-technologies in Germany in 2030. The report concludes that gallium will experience the strongest increase in demand. Due to technical innovations, demand of gallium in 2030 will be around six times as high as in 2006. The demand of neodymium is likely to increase by a factor of 3.8. Demand is also expected to grow for indium (factor 3.3), germanium (factor 2.4), scandium (factor 2.2), and platinum (factor of 1.6). Smaller demand increases are expected for tantalum (factor 1), for silver and tin (both factor 0.8), for cobalt (factor 0.4), for titanium (factor 0.3), for copper (factor 0.2) and for selenium (factor 0.3) (rounded).

In *Rohstoffsituation Bayern: Keine Zukunft ohne Rohstoffe* (IW Consult 2009), the Institut der deutschen Wirtschaft Köln, contracted by the Bavarian business association, analyzes the risk adherent to commodity supply in Bavaria. While the report zooms in on one specific region, it is interesting with regard to the methodology it applies to determine short-term and long-term risks: the 'commodity-risk index'. The index identifies seven indicators which influence the supply with raw metals and minerals. Among these indicators are four of quantitative and three of qualitative nature. (1) The static life index measures the period currently known for which reserves will last at current annual consumption rates. (2) Country risks estimate the political stability in a country, based on the Global Political Risk Index. (3) Country concentration denotes the proportion of the world production of a mineral/metal accounted for by the three largest producing countries. (4) Market concentration is measured by the percentage of world production accounted for by the three largest companies. (5) Economic importance is determined by the prominence of a mineral/metal for future technologies. (6) Risk of strategic employment analyzes the risk of political or strategic intervention. (7) Finally, the last indicator is substitutability.

Based on these seven indicators, the study assesses the supply risk of 37 metals and minerals. Points are awarded to each element, and a commodity-risk index is calculated. Contrary to many other studies presented (except Angerer et al. 2009), the framework allows a comparison and accurate ranking according to criticality. The authors divide the materials into a red, yellow and green box according to their level of supply risk. The red box contains 14 elements which are likely to suffer from prospective supply shortage (in order of criticality): yttrium, neodymium, cobalt, scandium, wolfram, phosphate, niobium, selenium, germanium, platinum group, lithium, chrome, indium, and molybdenum. Three of the four most risky elements belong to the rare earths group. The orange box contains 14 elements exhibiting a medium risk: fluorite, graphite, magnesium, manganese, tin, gallium, silver, tantalum, copper, titanium, gold, zinc, aluminum, and barite. Low-risk materials are categorized in the third, green box: nickel, lead, steel, potassium salt, bentonite, mica, feldspar, kaolin, gypsum, and anhydrite.

In *Kritische Versorgungslage mit schweren Seltenen Erden – Entwicklung "Grüner Technologien" gefährdet?* (Elsner 2011) published by the German Federal Institute for Geosciences and Natural Resources (BGR) Harald Elsner returns to the supply risk for green technology components, zooming in on one specific metal group – heavy rare earths, a sub-group of rare earths. These are (in the order of atomic weight): yttrium, samarium, europium, gadolinium, terbium, dysprosium, holmium, erbium, thulium, ytterbium and lutetium. In contrast to the studies discussed above, Elsner does not present a criticality framework. While he does not explicitly define criticality, implicitly he uses high demand, lack of substitutes, high prices and an insufficient supply as defining characteristics. Elsner finds a high level of production concentration for these elements: China is the only producer of heavy rare earths. He concludes that despite the expected increase in production of heavy rare earths by other countries and advancements in recycling technology, Chinese production cutbacks combines with a strong increase in demand will lead to a gap in supply of europium, dysprosium and terbium by 2015. Manufacturers and consumers of illuminants and permanent magnets will be affected particularly strongly by this development.

2.1.3 The European and Global Perspective

In *Critical Raw Materials for the EU* (EU 2010) the ad-hoc Working Group on defining critical raw materials (a sub-group of the Raw Materials Supply Group, chaired by the European Commission) shifts the regional focus of critical resource assessment to the European Union, the first such analysis conducted for the EU as an entity. The report focuses on 41 non-energy minerals and metals and on a time frame of ten years. The framework applies a relative concept of criticality: raw materials are labeled critical "when the risks of supply shortage and their impact on the economy are higher than for most of the other raw materials" (EU 2010, p. 23). The assessment of criticality is based on three aggregated indicators: the first two indicators 'economic importance' and 'supply risks' resemble the NAS approach. However, the authors expand the framework with the factor 'environmental country risk'. The 'economic importance' of a raw materials is measured by "breaking down its main uses and attributing to each of them the value added of the economic sector that has this raw materials as input" (EU 2010, p. 24). 'Supply risk' denotes the risk of supply shortage due to several factors: (1) the concentration of worldwide production (measured by the Herfindahl-Hirschman Index); (2) the political and economic stability of the producing countries (using the Worldwide Governance Indicators by the World Bank); (3) the potential to substitute the raw materials (based on expert opinion by the Fraunhofer ISI); (4) the recycling rate of the metal (defined by the Recycled Content (RC) rate). 'Environmental country risk' can arise due to government measures to protect the environment (measured with the Environmental Performance Indexes).

The analysis follows a two-step approach. First, the materials are placed in a two-dimensional matrix with the x-axis denoting economic importance and the y-axis designating supply risk. While this step resembles the methodology of the NAS and DOE framework the second step is comparably new: materials exceeding a certain threshold of economic importance are reassessed with regard to environmental country risk. Overall, materials are classified as critical if they lie above the threshold in economic importance and feature a high assessment in either supply risk or environmental risk (EU 2010, p. 32).

The report identifies three clusters. (1) 14 metals are classified as critical: antimony, indium, beryllium, magnesium, cobalt, niobium, fluorspar, platinum group metals, gallium, rare earths, germanium, tantalum, graphite, and tungsten. These elements do not only feature high degrees of production concentration (in China, Brazil, the Democratic Republic of the Congo and Russia) but also low substitutability and recycling rates. (2) The second sub-cluster consists of elements that are characterized by high economic importance but rather low supply risks: aluminum, bauxite, chromium, iron, magnesite, manganese, molybdenum, nickel, rhenium, tellurium, vanadium, and zinc. (3) Materials featuring relatively low economic importance and supply risk are grouped in the third sub-cluster.

Contrary to the national/regional focus of the preceding studies, *Critical Metals for Future Sustainable Technologies and their Recycling Potential* published by the Institute for Applied Ecology (Öko-Institut e.V.), commissioned by the United Nations Environment Programme (Öko-Institut/UNEP 2009) applies a global assessment. The report analyzes the availability of critical resources and their recycling potential, proposing possible courses of action. The study focuses on the following metals/metal groups: indium, germanium, tantalum, tellurium, cobalt, lithium, gallium, rare earths and platinum group metals such as ruthenium, platinum, and palladium. These elements are commonly labeled 'green minor metals' as they are crucial for certain future sustainable technologies (FST). Such technologies include electrical and electronics equipment, photovoltaic technologies, battery technologies and catalysts. The Öko-Insitut/UNEP determine those 'green minor metals' as critical that are likely to become short in supply and increasingly costly due to a demand increase.

The report bases its analysis on recent studies such as Minerals, Critical Minerals and the U.S. Economy (NAS 2007) but considers additionally "underestimated but very important issues like minor-product phenomena on the demand side and special challenges like postconsumer recycling of metals in dissipative usages". The report uses three indicators to measure criticality of 'green minor metals': demand growth, supply risk, and recycling restrictions. Supply risks emerge due to regional concentration of mining (i.e. the major three countries account for a share of more than 90 percent of global mining), physical scarcity, temporary scarcity and/or structural and technical scarcity. While supply risks and growing demand are thoroughly discussed in other reports as well, the Öko-Institut/UNEP's assessment places greater emphasis on restrictions to recycling. The report differentiates between pre-consumer and post-consumer recycling. The former denotes the recycling of production scrap from manufacturing processes. The latter identifies the recycling of critical metals from old scrap.

The indicators are used to identify critical 'green minor metals' in the short-term (5 years), medium term (until 2020) and long term (until 2050). In the short term, which is characterized by rapid growth of demand, serious supply risks and moderate recycling restrictions, tellurium, indium and gallium are classified as critical. For the medium term, the authors apply two scenarios: In a scenario of rapid demand growth and serious recycling restrictions, rare earths, lithium and tantalum will become critical. Under a second medium-term scenario (moderate supply risks and moderate recycling restrictions), palladium, platinum and ruthenium will be critical. The long-term perspective foresees a moderate demand growth, moderate supply risks and moderate recycling restrictions. In this setting, germanium and cobalt will be critical.

2.2 *Summary: Price and Supply Risks*

The discussion of criticality allows us to derive two predominant risks on the resource markets: price risks and supply risks.

Price Risks: The turn of the century brought sharply rising prices in all major commodity markets, including metals and minerals. While prices dropped temporarily during the financial and economic crisis in 2008-2009, they have spiked to new heights in the recent months (see figure 1). Strong demand from the emerging economies (especially China) is largely responsible for the long-term surge in demand since the turn of the century, along with changes in demand structure through the growth of particular sectors such as information (IT) and environmental technologies. Other long-term trends that fuel demand are demographic developments (with the global population set to grow by one third by 2050), urbanisation (and associated increases in material use), and changing patterns of mobility caused by rising incomes (especially in developing countries) (Hilpert et al. 2011). Given these long-term structural factors, it is a safe bet that prices will continue to rise (Bardt 2011).

In the short-run, prices have revealed high volatility as recent episodes like the steep drop in silver and gold prices in early May 2011 illustrate. These are driven by numerous political, financial and physical factors. In the light of high metal prices, the current dollar weakness, low interest rates and inflation fears as well as uncertainties about economic developments in both the EU and the U.S., metals are becoming attractive investment alternatives. The markets, however, are very sensitive and show a tendency to overreact. Prices are also influenced by temporary production outages, for example due to weather phenomena or political conflicts (U.S. DOE 2010). Overall, price fluctuations pose a considerable risk to companies as they create uncertainties in the planning process.

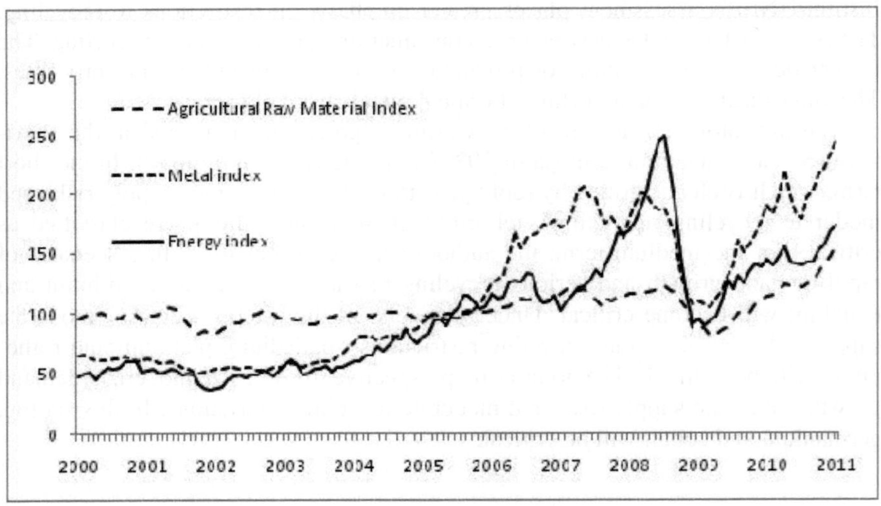

Source: IWF, Primary Commodity Prices, <http://www.imf.org/external/np/res/commod/index.asp> (last accessed on 25.5.2011).

Fig. 1 Prices for Agricultural commodities, energy and metals 2000–2010 (2005=100)

Supply Risks: High geographical and market concentrations of many metals and minerals render supply vulnerable to disturbances such as weather phenomena and political interventions (see figure 2). For example, 76 percent of the global lithium reserves are concentrated in Chile (Argentina: 8 percent, Australia: 6 percent). Of the global indium reserves, 73 percent are found in China. Likewise, more than half of the global reserves of cobalt, namely 52 percent, are located within the Democratic Republic of the Congo (U.S. DOE 2010, p. 28).

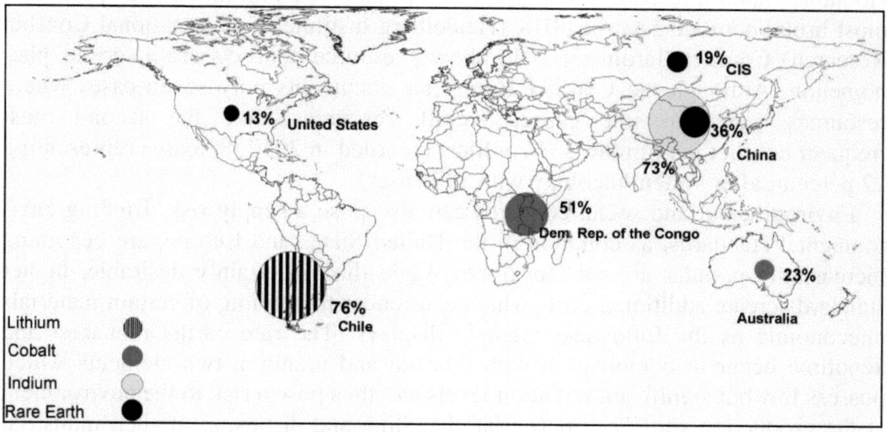

Based on: U.S. Department of Energy, *Critical Materials Strategy*, Washington D.C., December 2010, p. 28.

Fig. 2 Geographical Concentration of Reserves

A significant political risk derives from strategic interventions in the resource markets (such as export restrictions) and political conflicts in production countries. Export restrictions can take many different forms such as taxes, quotas and export bans, mandatory minimum export prices, reductions of value added tax (VAT) rebates on exports, and stringent export licensing requirements. China, now facing a trade dispute at the WTO, has applied export taxes and quotas on numerous metals and minerals, including rare earths. The list of countries using export restrictions is long: Russia, for example, has implemented an export tax on copper scrap of 50 percent; India charges an export tax of 15 percent on iron ore. Motivations for export restrictions are manifold: to nurture infant industries, to underpin social policy and income distribution, to buttress government revenues, to protect the environment and to preserve natural resources (BDI 2011).

Conflicts within producing countries constitute another considerable political risk. Mineral extraction is the most important source of national income for many developing countries and emerging economies. Control of mining and participation in its profits is therefore always a question of power, one which is sometimes contested violently. For decades minerals and metals have played a role in civil war economies as conflict resources, where the profits from rich reserves are used to fund warfare (resource curse). Most conflict resources are easy to extract (lootable) with a high value/volume ratio making them easy to transport and smuggle (for example gold). One of the best-known examples is coltan mining in the Democratic Republic of the Congo. Coltan contains the very rare metal tantalum which is used in metallurgy and in electrolytic capacitors (found in mobile phones, computers and digital cameras); substituting it involves increased cost or loss of quality (Cunningham 1998, pp. 143-45). The lucrative trade in coltan and other "lootable" natural resources like gold has for many years inflamed one of Africa's most brutal conflicts. As the HIIK (Heidelberg Institute for International Conflict Research) Conflict Barometer 2010 shows, resource conflicts are a serious phenomenon. Although the Conflict Barometer documents only seven cases where resources were the sole cause, overall, resources were the second most frequent conflict item in the 363 conflicts recorded in 2010 (80 cases representing 22 percent; after system/ideology with 117 cases).

Environmental and social concerns can also pose a supply risk. Binding environmental standards, as common in the United States and Europe, are becoming increasingly popular around the globe. While this is certainly desirable, higher standards create additional costs which can render the mining of certain materials uneconomic as the following example displays. The rare earths monazite and xenotime occur in combination with thorium and uranium, two elements which possess low but significant radiation levels and thus pose a risk to the environment at the production site. The appropriate handling and disposing of such industrial waste is expensive, wherefore the mining of monazite and xenotime is widely regarded as uneconomic (APS 2007).

Table 1 Determinants of Criticality

Indicator	*Factors*
Importance to an industry, sector, economy as a whole	• Demand
Price risks	• Long-term and short term price developments, in particular price fluctuations
Supply risks	• Geological and market concentration • Political Risks o Political and economic stability of the producing countries o Strategic interventions o Environmental and social standards • Potentials for substitution • Potentials for recycling

Authors' compilation.

Without wanting to minimize the gravity of these risks, it should not be forgotten that international resource markets are in flux: a partially dominant position of the Chinese in specific raw materials sectors, for example, can rapidly fall back to the levels of other market players. After all, only a fraction of the existing raw materials potential is known. New deposits are constantly being discovered. Due to technological developments in the exploration, sinking mining costs and increasing prices for raw materials, previously uneconomical deposits may become economically profitable.

3 Tackling Resource Risks: Country Strategies

How are countries reacting to these risks? The next section of this paper analyzes the raw materials strategies of two different groups of countries and regions:: those that are poor in domestic resources and feature a high dependence on imports – Germany, the EU and Japan – and those, which are relatively well-endowed with natural resources – the U.S., Canada, and China. Several of these base their strategies on the reports discussed above.

3.1 High Demand, Little Domestic Supply

3.1.1 Germany

When it comes to natural resources in general, Germany is not exactly poor. But while the country possesses vast amounts of industrial raw materials like sands, gravel and stones, it depends heavily on imports of metals and minerals. The supply risk is amplified as many of such commodities have been imported from only very few countries for years (high import concentration). In 2009, for instance, Germany imported 54.8 percent of its rare earths from China, 82.5 percent of lithium-carbonate from Chile, 85.3 percent of chrome from South Africa and 73.3 percent of bauxite from Guinea (BGR 2010). Despite high dependence and vulnerability, supply security of metals and minerals has moved to the top of the political agenda only recently, cumulating in a new raw materials strategy, which the German government presented in October 2010. The *Rohstoffstrategie der Bundesregierung* builds on *Elemente einer Rohstoffstrategie der Bundesregierung* (2007) and an interim report, which was brought before the German Bundestag by the inter-ministerial committee for resources in 2009 (Deutscher Bundestag 2010, p. 3). The new strategy was developed in an interagency process, accompanied by three resource summits (Rohstoffgipfel, 2005, 2007, 2010) jointly held by the government and the German Federation of Industries (BDI).

Similar to previous strategies, the new document places the responsibility to secure supply into the hands of industry (exploration, extraction, transport, contracts, stockpiling) without direct interference by the government. However, compared with the 2007 report, the updated version pays more attention to the development of new technologies and its effect on demand for certain metals and minerals, the strategic orientation of commodity politics by developing and emerging countries, and the need for an integrated resource approach of Germany and the EU (Deutscher Bundestag 2010 p. 7). The new strategy is also heavily influenced by the experiences with the financial and economic crisis, increasing price volatility, and severe shortcomings in market transparency. Acknowledging increasing supply risks, the German government announced to assist industry by improving market transparency and information, promoting competitive and open markets, supporting foreign investment with guarantees and concluding bilateral resource partnerships. In order to support this strategy, a new institution was founded as an interface and information platform for industry and politics: the German Agency for Raw Materials.

The German raw materials strategy stands on four pillars: (1) enhancing research in resource and product efficiency, in the development of substitutes for critical elements and recycling technologies, (2) improving supply by utilizing domestic reserves and forming resource partnerships with resource-rich countries, (3) enforcing international market discipline, and (4) supporting measures that address bad governance and corruption in resource-rich countries.

Recycling is a key source for raw materials in Germany. For instance, 54 percent of copper is recycled in Germany, marking this the highest recycling rate worldwide (EU: 45%, USA: 41%, world: 13%). Some 35 percent of aluminum is recycled on average and 59 percent of lead; recycling of steel even hits the 90 percent margin (BMWi 2010). For some metals, such as the rare earths, the recycling rate is, however, still very low. One of the goals is to change this. The strategy further recommends a stronger exploitation of domestic reserves and the implementation of bilateral partnerships to improve access abroad. To promote these bilateral resource partnerships, the German government offers technical and development assistance to the partner countries to help modernize their raw materials sector, to improve skills and knowledge and to upgrade ecological and social standards. The Federal Institute for Geosciences and Natural Resources (BGR) has created a list of potential and attractive partner countries, following specific criteria including reserves, current production, the amount of raw materials critical for German production, and the amount and value of German raw materials imports from the respective country. The most important partners in Africa are South Africa, Zimbabwe, and the DR Congo; in Asia, these are China, India and Indonesia. Of the CIS countries, the BGR lists Russia, Kazakhstan and the Ukraine; and in South America, prospective partners are Brazil, Chile and Peru (Vasters et al. 2010). An agreement with Kazakhstan, a country with high endowments of rare earths, is underway, as both countries have signed a joint declaration of intent on May 24th 2011. Kazakhstan is expected to provide Germany with many much-needed resources while the latter will support Kazakhstan in its economic development (Reuters Deutschland 2011). Eighteen German corporations have expressed their interest in a resource partnership with Kazakhstan (Dow Jones Stahl Aktuell 2011).

Much more diversely debated is the establishment of a *German Association of Raw Materials*; the current working title of the group is "Deutsche Rohstoff-NewCo". German companies pulled out of the commodity market in the 1990s when resources could be cheaply obtained on international markets. There is very little involvement in foreign mining. While investment in mining abroad could reduce supply risks, German companies, particularly the 'Mittelstand' (small and medium-sized enterprises), are too small and companies' interests in minerals and metals too diverse for this strategy to really work. Under the Deutsche Rohstoff-NewCo, on the other hand, German companies could bundle their raw materials purchases and thereby unfold greater market dominance. So far, corporations like

ThyssenKrupp, Siemens, BASF and Evonik have endorsed the endeavor (Dohmen and Jung 2011). Even though the German Federal Ministry of Economics and Technology opposes a direct government involvement in such an association, the former Federal Minister of Economics, Rainer Brüderle, had considered supporting the industry with guarantees (Stratmann 2010).

Last but not least, Germany investigates alternative sources for raw materials: the ocean floors. In 2006, the BGR acquired an exploration license for the so-called "Manganese Nodules Belt" in the Pacific Ocean. The permit covers a total of 75,000 square kilometers and is issued by the International Seabed Authority (ISA). The area under the German license presumably contains metal resources of copper, nickel, cobalt and manganese that exceed the 1-, 10-, 50- and 30 – fold annual worldwide consumption. The Pacific manganese nodules could possibly satisfy the German demand for nickel for the next 100 years (Bünger and Unbehend 2009). However, their exploitation is still prohibitively expensive and not yet profitable, despite currently high prices.

3.1.2 The European Union

Like Germany, the European Union as a whole is vulnerable to price shocks on resource markets and disruptions of supply given its limited mineral and metal reserves. While the EU is well endowed with construction materials like gypsum and natural stone and with some industrial materials, it still has to import most metals and minerals. The import dependence of metallic minerals is particularly high in the EU as it accounts for only 3 percent of global production (European Commission 2008, p. 3f). The EU covers 100 percent of its consumption of antimony, cobalt, molybdenum, niobium, platinum, rare earth minerals, tantalum, titanium minerals, and vanadium with imports. More than 80 percent of currently consumed manganese ore, iron ore, bauxite and tin stem from abroad. The import-to-consumption ratio for zinc, chromium ores and copper ranges around 50 percent in the EU in 2008 (European Commission 2008a, p. 4). European production of refined metals is not sufficient when compared to the needs of the EU. In 2009, the EU imported non-ferrous metals worth €34 billion, while exports amounted to only €26 billion, causing a trade deficit of €8 billion (European Commission 2010).

As early as in 2006, the EU Commission included access to raw materials in its trade policy strategy *Global Europe. Competing in the World* (European Commission 2006, p. 7). In 2008, the European Commission adopted a raw materials initiative, *The Raw Materials Initiative – Meeting Our Critical Needs for Growth and Jobs in Europe* (2008), based on three pillars: (1) "ensure access to raw materials from international markets under the same conditions as other industrial competitors"; (2) "set the right framework conditions within the EU in order to foster sustainable supply of raw materials from European sources; (3) "boost overall resource efficiency and promote recycling to reduce the EU's consumption of primary raw materials and decrease the relative import dependence".

In 2009, the European Commission Directorate-General for Trade published its first *Raw Materials Policy 2009 Annual Report* (European Commission 2009). The Directorate-General for Trade highlighted three goals: 1. integrating trade disciplines relevant to raw materials in ongoing trade negotiations at the WTO and FTA negotiations; 2. enforcing the rules, tackling illegal trade barriers; and 3. reaching out to third countries to exchange views and analyses. DG Trade kept its promise: The EU, together with the U.S. and Mexico filed a complaint against China at the WTO in June 2009; in December 2009, a dispute settlement panel was established upon the request of the three complainants to deal with China's export restrictions. According to the complainants, China subjects the exportation of bauxite, coke, fluorspar, silicon carbide, and zinc to quantitative restrictions in the form of quotas and the export of most of these and other materials like magnesium and manganese with export duties (European Commission Directorate-General for Trade 2009, p. 17). China is the EU's second most important trading partner after the United States. Around 500,000 jobs and industries that represent four percent of the general industrial activity are affected by China's export restrictions (European Commission 2009). Two consecutive hearings were held by the WTO dispute panel in August/September and November 2010, and on April 1 2011, the panel circulated the confidential final report to the dispute parties (European Union 2011).

This was followed by the Commission's report on *Critical Raw Materials for the EU* in mid-2010 (see discussion above). The EU plans to update its list of critical raw materials every three years (EurActiv 2011). On February 2, 2011, the Commission published the EU's new resource strategy *Tackling the Challenges in Commodity Markets and on Raw Materials*. The strategy stands on three pillars: (1) securing fair and sustainable supply with raw materials from world markets through a new resource diplomacy; (2) fostering resource production within the EU through a National Minerals Policy that helps plan and monitor the member states' achievements and actions in the resource sector; (3) enhancing research and development for improved recycling and substitution technologies as well as resource and product efficiency. Each pillar includes a long list of individual measures. The overarching theme of the strategy is innovation, which is to be pursued along the entire value chain of raw materials. For the crucial rare earths and other strategic metals, stockpiling is also under discussion.

After the experiences with the global economic and financial crisis, stability of commodity derivatives, market regulation, integrity and transparency were considered further crucial points for the European raw materials strategy (Commission to the European Parliament 2011, p. 2f; 20). "In order to secure supply of raw materials for the European industry for coming years, we need to link this policy with our reforms of the regulatory framework for financial markets", Commission President José Manuel Barroso stressed the need to better understand the synergy between resources and financial markets (EurActiv 2011). France in particular pushed for curtailing financial market speculation and placed the issue high on the 2011 agenda of the G8 and G20, of which it held the presidency. In addition, environmental protection remains a core issue within resource politics of the EU: in 2010, the EU Commission formulated guidelines to regulate the extraction of metals and minerals in nature conservation areas.

Resource diplomacy is the new catchphrase in the European resource strategy. Apart from reducing trade barriers and enforcing international trade rules through the WTO and Free Trade Agreements, the EU intends to conclude strategic partnerships with resource-rich countries, especially with the African countries, including the African Union. "Europe needs to maintain and gain access to raw materials from third countries, but it aims at doing so in a way that is fair to both sides, and creates win-win situations. This can be achieved through [...] the promotion of good governance, human rights, conflict resolution, non-proliferation and regional stability in resource-rich countries (European Commission Enterprise and Industry 2011). First steps were undertaken in June 2010, when the EU and the African Union agreed to cooperate bilaterally on governance, infrastructure, geological knowledge and skills in the raw materials sector. The EU supports governance projects under the European Development Fund; geological surveys are financed by the EU-Africa Infrastructure Fund to enable resource-rich countries to better assess their mineral and metal endowments (European Commission Enterprise and Industry 2011). The EU further aims at improving governance, curtailing corruption and reducing conflict potential through transparency initiatives such as the Extractive Industries Transparency Initiative (EITI) (EurActiv 2011a).

3.1.3 Japan

The pressure to address supply security is particularly high in Japan. While the country has large reserves of industrial minerals (such as dolomite, iodine, limestone, pyrophyllite, silica sand, and silica stones) and owns gold, magnesium and silver deposits, which partially meet domestic demand, Japan is a resource-poor country and depends on metal and mineral imports such as bauxite, copper and iron ore to supply its manufacturing industry (Kuo 2010, p. 13.1). In 2009, Japan's Nippon Steel Corp. was the fourth-largest producer of steel worldwide after Luxembourg's ArcelorMittal, China's Baosteel Group, and the South Korean POSCO. Yet, the country depends almost entirely on imports of iron ore. Consequently, rising prices severely affect Japan's steel industry (MarketResearch.com 2010). The same applies to the automobile industry, in particular the production of electronic and hybrid cars (Hiranuma 2009). How vulnerable the country is towards supply shocks was underlined when China temporarily banned its exports of rare earths to Japan in 2010, following a collision between a Chinese fishing boat and a Japanese naval patrol vessel. The reason behind the export ban had little to do with rare earths, and everything with a festering territorial dispute over the Senkaku Islands and their oil- and gas-rich territorial waters.

The Ministry of Economy, Trade and Industry (METI), supported by the Geological Survey of Japan, presented its *Strategy for Ensuring Stable Supplies of Rare Metals* in July 2009 (METI 2009). METI identified 31 metals, including rare earth elements, as rare; the goal is to secure the industry's supply (Hiranuma 2009). Japan's strategy, not unlike the ones of Germany and the EU, has three components: (1) *Focused, Strategic Approach*: The government is to evaluate the supply situation and to determine the levels of priority for the different kinds of rare metals; (2) *Four Pillars for Securing Rare Metals*: securing overseas resources, recycling, development of alternative materials, stockpiling; (3) *Development of a Common Infrastructure toward Securing of Rare Metals*: human resources development in the resources sector, enhancing the technical capabilities of the resources sector, and integrated efforts (multi-stakeholder process). In July 2010, the METI and the foreign ministry presented an integrated "one-stop-system" to secure industrial resources. Central to this strategy is a dual approach that combines the support of foreign small ("junior") mining companies and to open up new sources for the extractives (Rehn 2010).

Unlike its German and EU counterparts the Japanese government, however, has not shied away from a stronger involvement in the resource sector. The government has supported its industry for years, promoting research and development as well as holding strategic reserves. The state-owned raw materials corporation JOGMEG (Japan Oil, Gas and Metals National Corp.) holds national stockpiles equivalent to 42 days of standard consumption in Japan of seven rare metal commodities: chromium, cobalt, manganese, molybdenum, nickel, tungsten, and vanadium. It has the authority to release reserves if necessary to stabilize the country's economy (JOGMEG 2007). Apart from managing the strategic reserves, JOGMEG is tasked with providing information, mineral exploration, financing of mine development, and R&D for recycling and substitutes.

Japanese companies are intensely searching for alternatives for expensive rare earths. Toshiba, for instance, successfully developed a samarium-cobalt magnet, which is to reduce dependence on the rare earth dysprosium, found mainly in Southern China. By means of nanotechnology, the University of Kyoto succeeded in developing an alloy very similar to the rare earth element palladium. In autumn 2010, the Hokkaido University announced its successful development of a ferrite iron-based motor for hybrid and electronic cars that does not require the performance-enhancing rare earth elements in the alloy (Goldinvest.de 2011; Kölling 2011). The Japanese government extensively supports these research projects.

To secure supply of critical elements, the government also assists domestic companies investing in mining projects and companies abroad. There are numerous examples of such partnerships. Sojitz, for instance, holds 25 percent of the Endako molybdenum mine in British Columbia, Canada. The Canadian company Thompson Creek Metals accounts for the remaining 75 percent. In Kazakhstan, Kazatomprom and Sumitomo have founded the joint venture Summit Atom Rare Earth Company (SARECO) in March 2010. The stake of the Japanese company amounts to 49 percent. The Nippon Export and Investment Insurance (NEXI) and the Japan Bank for International Cooperation support such programs through guarantees. Companies seeking to acquire foreign mining rights are also assisted by JOGMEG. Moreover, the Japanese government seeks to strengthen bilateral relationships with producer countries, for instance via technical support of geological studies, training, infrastructure and environmental projects paid for by overseas development assistance (U.S. DOE 2010; Rehn 2010).

Furthermore, Japan has intensified its cooperation with other resource-dependent countries: in November 2010, Japanese officials held a U.S.-Japanese roundtable on natural resources, continued by talks in December. In February

2011, a Japanese delegation followed an invitation by the European Parliament. Deputy Director General Keiichi Kawakami of METI said in Brussels, "All of the [rare earths-] consuming countries' problems need to be solved through cooperation". He suggested a 'triangular cooperation' network for Japan, the U.S. and the EU (EurActiv 2011b).

Finally, in early 2010 Japan announced plans to resume the search for resources in the seabed. About 340,000 square kilometers of Japan's Exclusive Economic Zone (EEZ) in the East China Sea and the Pacific Ocean will be explored for metals (NTV 2010). Huge deposits of high-tech metals and other materials are expected to be located near the south Japanese island province Okinawa and the peninsula Izu near Tokyo. JOGMEC currently supports the development of a remote-controlled robot to explore metals in depths of up to 2,000 meters. However, more than 10 years might pass until this project's technology will be ready for the market. Moreover, new disputes with China are on the horizon as the ownership of the resources in the East China Sea and especially around the Senkaku-/Diaoyu Islands has not been resolved. The dispute is currently addressed by the Commission on the Limits of the Continental Shelf (CLCS).

3.2 Resource-Rich Countries

3.2.1 The United States

In contrast to highly import-dependent countries such as Germany and Japan, the U.S. is much better endowed with metals and minerals. For instance, around 10 percent of global gold production, approximately 11 percent of the lead production and 14 percent of the global rhenium production can be found within the U.S. (USGS 2008, p. 237). Nonetheless, the U.S. is still highly dependent on imports of many strategic metals. In the case of indium, for instance, the U.S. covers 100 percent of domestic demand with imports; the same applies to gallium (99%) and germanium (90%) (USGS 2010a, p. 6). The import to consumption ratios for lithium is also above 50 percent (USGS 2010a, p. 92). Similar to the German case, most of supply sources of the U.S. are highly concentrated. The U.S. obtains 91 percent of its rare earths from China (Japan/ France: 3% each, Russia: 1%, 2005-2008) (USGS 2010a, p. 128). 40 percent of indium imports originate in China (Japan: 19%; Canada: 18%) (USGS 2010a, p. 74); lithium comes from mainly Chile (63%) and Argentina (35%) (USGS 2010a, p. 92); cobalt (import dependence rate: 75%) from Norway, Russia and China (19%; 17%; 12%) (USGS 2010a, p. 6; 46). It is not exclusively the enormous demand of the U.S. that accounts for its import dependence. Due to low world market prices for raw materials and strict environmental regulations in the 1990s, mining of certain metals such as rare earths was either abandoned or drastically reduced.

Although the U.S. does not have an official, overarching resource strategy, security of supply has played a crucial role for years. Several legislations dealt with the issue, including the 1939 Strategic Materials Act, which established a national stockpiling system for the military and was amended several times since, and the

Defense Production Act of 1950 in response to the start of the Korean War to expedite expansion of industrial capacity for many strategic and critical materials.

In the light of President Barack Obama's plans to foster renewable energy supply, the demand for metals like lithium or rare earths is likely to increase. Additionally, rare earths are of great importance to the defense industry. Resource security and supply in the U.S. has – as in the case of Japan – a strong geopolitical component. In April 2010, the Government Accountability Office (GAO) published a report on *Rare Earth Materials in the Defense Supply Chain*. According to its findings, rare earths are crucial for numerous defense systems, such as precision-guided munitions, lasers, communication systems, radar systems, avionics, night vision equipment, and satellites (GAO 2010, p. 27). The rare earths used for such systems are characterized by only very limited substitution possibilities. The GAO report was followed by the *Critical Materials Strategy*, which was published in December 2010 by the Department of Energy (DOE). The DOE's *Critical Materials Strategy* is based in three pillars: (1) *Diversified Global Supply Chains*: taking steps to facilitate extraction, processing and manufacturing in the U.S., as well as encouraging other nations to expedite alternative supplies; (2) *Development of Substitutes*: supporting research leading to material and technology substitutes; (3) *Recycling:* reuse and more efficient use. The strategy calls for an updated assessment of critical mineral resources for production in the United States, regular information on domestic resources, a reduction of permitting delays and other barriers to domestic mineral projects, the promotion of efficient use and recycling of critical minerals as well as identifying viable alternatives, and resource diplomacy to enforce market discipline. An update of the report on critical materials is planned to be published by the end of 2011 (U.S. DOE 2010).

A central component of the strategy is research and development. In fiscal year 2010, the DOE allocated, for instance, \$15 million for research on rare earths and the development of substitutes for magnets. It also invested an additional sum of \$35 million for the development of new batteries that can operate without rare earth metals (U.S. DOE 2010, p. 53). Moreover, the DOE aims at diversifying supply sources, in particularly by increasing the exploitation of domestic reserves. Due to the complicated legal licensing and planning processes, which involve the federal, state and local level, it can take seven to ten years to obtain a mining permit. By way of comparison, it only takes about one to two years in Australia. The DOE aims at reducing permitting delays and other barriers to domestic mineral projects (U.S. DOE 2010, p. 105). Furthermore, the DOE promotes the idea of resource partnerships. Unlike in the German or Japanese case, this does not mean a partnership with resource-exporting countries, however. Rather, the U.S. aims at cooperating more closely with other import-dependent countries to put pressure on China to curtail market distortive practices.

In recent years, U.S. Congress discussed a multitude of legislative proposals which address supply security of metals and minerals. In September 2010, the House of Representatives voted overwhelmingly for the *Rare Earths and Critical Materials Revitalization Act* that planned to facilitate and accelerate the exploration and exploitation of rare earths and other critical materials. A similar bill was introduced in the Senate. However, the bill was not voted on before the midterm elections in November 2010. At the end of each legislative session, all proposed bills and resolutions that have not passed are cleared from the books (Zajec 2010). In early 2011, a multitude of new proposals has been tabled in both the House and the Senate: The *Rare Earths and Critical Materials Revitalization Act of 2011* was reintroduced in the House in February 2011. The *RARE Act of 2011*, proposed in the House Committee on Natural Resources in April 2011, directs the Secretary of the Interior towards putting into effect a global rare earths element assessment. The House *Rare Earths Supply Chain Technology and Resources Transformation Act of 2011* aims at reestablishing a competitive domestic rare earths minerals production industry. In the Senate, the *Critical Minerals and Materials Promotion Act of 2011* was introduced in February 2011. A discussion draft of legislation, the *Critical Minerals Policy Act of 2011*, released in April 2011, intends to revitalize the nation's critical minerals supply chain, directing the U.S. Geological Survey to establish a list of minerals critical to the U.S. economy. It also calls for comprehensive policies to ensure the nation is able to meet its own mineral needs (Thomas.gov 2010).

Last but not least, the U.S. is also interested in resources of the seabed. The country already produces offshore oil and gas. Mineral raw materials are explored, especially around the Pacific islands. According to the U.S. Geological Survey, however, there are no reliable estimations on the actual potential for raw materials within the Exclusive Economic Zone (EEZ) of the U.S. (USGS 2005, p. 60). The U.S. so far has not ratified the UNCLOS. Legally, the U.S. thus has no right to claim an expanded part of the continental shelf or an exploration/mining license in the so-called „area", the seabed of the High Seas (Jarowinsky et al. 2009; USGS 2010).

3.2.2 Canada

Canada is rich in mineral resources, and mining is an important sector of the country's economy. According to Natural Resources Canada (the Canadian ministry for resources), mining accounted for 2.7 percent to the country's GDP, employed 306,974 people, and made up 18.5 percent of Canada's exports in 2009 (Trelawny and Pearce 2009). Canada ranked second in the global production of sulfur and uranium, and was one of the worldwide top five producers of aluminum, cobalt, gem diamonds, refined indium, nickel, platinum-group metals, sodium sulfate, and zinc (Mobbs 2011, p. 5.1). Canada's mineral industry is very export-oriented. In 2009, metals and minerals were exported to nearly 200 countries, of which the top destinations were the U.S. (55%), the EU (16%), China (5%), and Japan (4%). Iron and steel, gold, aluminum, copper, potash, nickel, iron ore, diamonds, uranium, nitrogen, and zinc were the key metals exported. Both exports and imports have grown during the last decade, and for both, China plays a crucial role (Trelawny and Pearce 2009). Canada benefits from increasing mineral and metal prices associated with the global demand.

In 2009, Canada alone accounted for 16 percent of total global expenditure on exploration (rank 1). The Fraser Institute conducts an annual survey on the attractiveness of countries and their mining policies based on a long list of indicators. These include uncertainty concerning the administration, interpretation, and enforcement of existing regulations; environmental regulations; regulatory duplication and inconsistencies; taxation; uncertainty concerning native land claims and protected areas; infrastructure; socioeconomic agreements; political stability; labor issues; geological data base; and security; reliability of legal systems; and trade barriers (McMahon and Cervantes 2011, p. 9). According to this index, Canada's breadth and quality of mineral resources, and its stable political environment makes it the most attractive mining region worldwide. Canadian provinces (Alberta, Saskatchewan, Quebec, and Manitoba) are among the ten leading regions to explore for minerals worldwide (McMahon and Cervantes 2011, p. 11). Canada thus does not face the same problems as the countries discussed previously.

However, this is no reason for complacency for the country. As early as in 1998, Natural Resources Canada published a report titled *From Mineral Resources to Manufactured Products: Toward a Value-Added Mineral and Metal Strategy for Canada*. A strategic document of this format has not been issued since. However, the country applies several strategies to enhance domestic production and to secure supplies. The government aims at improving market transparency and further harmonizing federal and provincial rules in the mining sector. Furthermore, Canada has opted for stockpiling of several metals (such as copper, gold, lead, molybdenum, nickel, silver and zinc) in quantities ranging from 0.5 percent to 4 percent of its total annual production (U.S. DOE 2010, p. 67). Apart from promoting sustainable development and use of mineral and metal resources and ensuring an attractive investment, protecting the environment and public health is the third pillar in the Canadian strategy. Thus, the government faces increasing demands for sustainability of mining projects, and minimizing the environmental and social impact of mining is a serious challenge.

Canada has further launched a new *Geo-mapping for Energy and Minerals* (GEM) program that will run from 2008-2013 and has been stocked with CAN-$100 million by the Canadian Government. The program will pursue mapping of the Arctic and the high north of Canada to find new adequate sites for mining (Natural Resources Canada 2010). GEM is conducted federally by the Geological Survey of Canada (GSC) and the Polar Continental Shelf Project (PCSP), Earth Sciences Sector (ESS), and Natural Resources Canada (NRCan). Exploration used to be limited to terrestrial resources, but some of the big companies, including the Canadian Nautilus Minerals, have expanded their explorations to the oceans and seas. Nautilus Minerals is one of the leading corporations for the exploration of seabed resources. The company has bought or applied for licences for the exploration of massive sulphides in the South Pacific (Jarowinsky et al. 2009, p. 18, 41ff). Canada has been mapping the seabed floor of the Arctic floor for geological data especially since the Russian claims on the area in 2007. In April and May 2009, it has conducted survey flights into the Russian-claimed areas of the Arctic. Canada thus seeks to collect data for the Commission on the Limits of the Continental Shelf to proof its own claims on the Arctic (Braune 2010). In January 2011, Nautilus Minerals became a pioneer for ocean mining when it was granted the world's first deep sea mining lease for an area in the Bismarck Sea in the Pacific Ocean by the government of Papua New Guinea. In the so-called Solwara 1 project, the company plans to extract gold and copper from the rich ore deposits that can be found there (Nautilus Minerals 2011).

3.2.3 China

China is a dominant player on the metal markets both as exporter and importer. In 2009, the country was the world's top producer of aluminum, antimony, barite, bismuth, cement, coal, fluorspar, gold, graphite, iron and steel, lead, phosphate rock, rare earths, salt, talc, tin, tungsten and zinc (Tse 2010). China's position is particularly dominant in the production of rare earths: it accounts for 95 percent of the world production (Tse 2011, p. 1). Moreover, China is an important exporter of many of the elements mentioned above (including indium, rare earths and tungsten). China's minerals sector made up one fifth (22.6%) of the total trade in 2009 (Tse 2010). However, its huge demand places the country somewhat in the same boat as less well endowed countries such as the United States. For many metals, domestic production does not suffice to meet demand (chromium, cobalt, copper, iron ore, manganese, nickel, petroleum, platinum-group metals, and potash); for these, imports were estimated to account for more than 30 percent of domestic consumption in 2009. The country's growth rates of 9.7 percent in the first quarter of 2011 compared to the same quarter in 2010 (Trading Economics 2011) and its large investments in infrastructure have driven demand for metals and made prices soar. China's recent strong economic growth can be largely attributed to the two-year stimulus package of autumn 2008 as well as to the remarkable recovery of world trade in 2010.

China has no official resource strategy; no central strategic document, similar to the resource strategies of the countries reviewed above, can be identified. However, in the so-called "Five-Year Plans", (economic development plans issued by the ruling party), energy and resource issues are said to be important topics. As these cannot be accessed publicly, the exact contents can only be assumed through media and politics. The twelfth Five-Year Plan, covering the years 2011-2015, is aimed at energy and climate issues in particular. Zhou Shengxian, Minister of the Environment, announced in early 2011 that a special Five-Year Plan for heavy metals will be issued (Hsu and Seligsohn 2011). That plan aims at preventing further heavy metal pollution, especially in rural areas of China. Such reforms may have a direct effect on mining corporations and are one step for China towards a more environmentally-friendly domestic industrial sector (Li 2011). Especially the Ministry of Land and Resources of the People's Republic of China aims at research and innovations in the national resource sector. Mineral resources, exploration, mining and trade play a significant role in the strategic communiqués of the ministry. The MLR together with the Ministry of Industry and Information Technology (MIIT) is responsible for developing production plans for the country's strategic commodities, for example setting production quotas (MLR 2010).

Despite the lack of a clearly defined resource strategy, several strategic aspects can be identified concerning China's domestic and foreign raw materials policies. Domestic resource policies: as early as in 1990, the Chinese government declared rare earths a strategic mineral to be protected, resulting in new restrictions on foreign investment in the rare resource sector. Except in joint ventures with Chinese firms, foreign investors were prohibited from mining and restricted from participating in rare-earth smelting projects (Tse 2011, p. 5). While the government had encouraged the export of rare earths in the 1980s, it has gradually reduced the export quota during the past several years to meet increasing domestic demand. In 2006, the government still allowed 47 domestic producers and traders of rare earths and 12 Sino-foreign rare-earth producers to export these elements; in 2009, this number shrunk to 23 domestic rare-earth producers and traders and 11 Sino-foreign rare-earth producers. In 2011, the government authorized only 22 domestic rare-earth producers and traders and 9 Sino-foreign rare-earth producers. Furthermore, the government continuously reduced exports quotas: in 2008, the rare-earth export quota for domestic rare-earth producers and traders was reduced by 21.6 percent and 2.5 percent in 2009. To disincentivize exports, the rebate on exported rare earths was eliminated in 2005 (Tse 2011, p. 6). Even though there are significant reserves of rare earths in other parts of the world, parts of the world, producers so far cannot compete with Chinese production given the country's competitive advantage in low labor costs and low environmental standards. Furthermore, secondary production and processing of rare earths is dominantly based in China. Competing with this sector is also tough as China keeps exports of the end products restrictions-free (Hilpert and Kröger 2011).

Recently, China has started to address environmental problems in the mining sector. For example, the Chinese Ministry for Environmental Protection plans to impose further limits for emissions that are connected to rare earth minerals production (Fang 2011). On May 19, the administration announced to promote the healthy development of the country's rare earth industry including a long list of goals: stricter policies on mining and waste emission standards, regulations to curb illegal mining and smuggling, implementation of mining and production controls, decreasing the consumption rate of rare earth reserves, consolidation of the industry and phase out of inefficient companies with high energy consumption and pollution, and promoting further mining and smeltering innovation (Zhanheng 2011).

Curtailing illegal mining and smuggling is a major component of China's mining policy. Illegal mining, especially in Southern China, is responsible for almost half of the world's supply in rare earths. However, the illegal extraction of rare earths holds for the most part dramatic outcomes for the environment (Elsner 2011, p. 5). Another policy aiming to protect domestic resources is the creation of stocks. Chinese officials announced in February 2010 the authorization of a "strategic reserve" of rare earths. In October 2010 the state-owned enterprise Bautou Steel was allowed to stockpile 300,000 tons of rare earths within 5 years (U.S. DOE 2010).

Such strategic interventions in the resource market not only serve the goal of securing domestic supply. China views metals and minerals not just as economic goods but also as means of power as the often quoted phrase by former Chinese president Deng Xiaoping in 1992 illustrates: „the Middle East has oil, China has rare earth" (Chu 2010). China's price discrimination against foreign customers and its threat to stop deliveries boost raw materials prices and create a market artificially divided between China and the rest of the world. China is plainly seeking to use trade and industry policy to force a shift of production and economic returns in its favour, and is willing to violate international trade rules to achieve that end (Bacchus 2010). Countries that depend on China's rare earths exports, such as the U.S., Japan or South Korea, are now considering mineral stockpiling themselves in order to compensate the shrinking share of Chinese minerals in world markets. In 2010, exports of rare earth minerals already shrunk by 9.3 percent (Areddy 2011). Besides these economic measures, many states try to legally push for more reliable Chinese exports at the WTO.

Foreign resource policies: Not only does China try to restrict production and exports of its mineral resources in order to secure domestic supply; the Peoples' Republic has long since turned to external sources for its resource needs. The government encourages enterprises to invest abroad, especially in the minerals sector. Countries such as Australia, Brazil, Burma, Chile, Indonesia and Mongolia are recommended for foreign investments (Tse 2010). Further, Chinese investments in Africa have risen sharply from 2000 onwards. According to the China-Africa Economic Trade Cooperation report, direct investments of 2009 amounted to $9.3 billion, compared to $490 million in 2003. Areas of investments are predominantly the mining industry, followed by the manufacturing industry (Tradinghouse.net 2010). Investments in developing countries require a certain willingness to take risks. Many of the African resource-rich countries are politically instable. While only few Western companies are willing to take the risk of investing in these countries, Chinese companies are far less hesitant. Backed by the state, they do not need to fear risks of non-payment, for instance. Furthermore, they are under less public scrutiny at home, and thus they do not necessarily have to consider standards as the rule of law, safety requirements for workers and environmental protection measures to the same degree as their Western competitors. Besides investing overseas, China also looks towards marine resources to meet its demand. The Peoples' Republic is one of the eight countries that have bought a license from the International Seabed Authority (ISA) for deep sea exploration. The China Ocean Mineral Resources R&D Association (COMRA) was created in 1991 and has since organized exploration cruises and investments in research and development for resource provision by ocean minerals (COMRA).

4 Conclusion

Reviewing recent studies on the criticality of metals, we find that although the selected studies apply different methodologies, ranging from purely qualitative over semi-quantitative to quantitative frameworks, some elements are found critical in most of them. Rare earths figure most prominently on the list of critical metals. But also elements such as the platinum group metals, indium, niobium, lithium, germanium and terbium appear several times in the listings. Most studies assume an increasing demand for critical metals driven by the dynamic developments in green and information technologies. The search for and development of alternative energy sources creates a new market for resources that were, so far, considered futile by-products of industry ores rather than valuable technology components. Among the literature reviewed, we identified two predominant risks on the resource markets: price risks and supply risks. Price risks derive from long-term and short-term price developments, in particular price fluctuations. As they create uncertainties in the planning process, they pose a severe risk for companies.

In addition, geological and market concentration together with political risks, such as conflicts and strategic interventions, pose risks for the supply security of many metals and minerals. Supply risks may also arise from environmental and social concerns within producing countries.

The overview of country strategies reveals several communalities, at least among the industrialized countries. In general – with the exception of China – supply security is primarily viewed as a task for the industry, not governments, while the latter are responsible for creating a sound, stable and fair governing framework for resource markets. True, there are also variations among the industrialized countries: whereas the German government has shied away from a direct involvement in the resource market, its Japanese counterpart has, for years, taken an active role as a player on the market. While Japan stockpiles certain metals and has a state-owned raw materials corporation (JOGMEG), both ideas are opposed by the German government. But this involvement by no means matches China's intervention in the markets. While China is a country well endowed with natural resources, its high demand and continuous high imports place it somewhat in the same boat as other industrial countries such as the United States.

Despite these differences, resource strategies generally feature three components: (1) reducing dependence on critical metals by improving product efficiency and developing substitutes; (2) enhancing supply by improving market transparency, advancing recycling methods of critical metals, utilizing domestic reserves (including seabed resources), fostering backward integration of industry (by taking over suppliers and investing in the resources sector) and forming resource partnerships with resource-rich countries; and (3) improving market discipline and curtailing strategic interventions by rules setting and enforcement through the World Trade Organisation (WTO) and the G20. Given specific national resource endowments and industrial production structures, countries place different emphases on the various aspects of these three pillars. Germany and Japan, for example, have announced to step up efforts in concluding bilateral partnerships with resource-rich countries, while the United States and Canada will increasingly exploit national reserves. Despite these differences, all countries considered in this study face one common challenge: securing metals and minerals at sustainable prices.

Bibliography

Angerer, G., Erdmann, L., Marscheider-Weidemann, F., Scharp, M., Lüllmann, A., Handke, V., Marwede, M.: Rohstoffe für Zukunftstechnologien: Einfluss des branchenspezifischen Rohstoffbedarfs in rohstoffintensiven Zukunftstechnologien auf die zukünftige. Fraunhofer IRB Verlag, Rohstoffnachfrage (2009)

APS (The American Physical Society), Energy Critical Elements: Securing Materials for Emerging Technologies (2011), `http://www.aps.org/policy/reports/popa-reports/loader.cfm?csModule=security/getfile&PageID=236337` (accessed May 25, 2011)

Areddy, J.T.: China Moves to Strengthen Grip Over Supply of Rare-Earth Metals. In: The Wall Street Journal (February 7, 2011), `http://online.wsj.com/article/SB10001424052748704124504576117511251161274.html` (accessed June 01, 2011)

Bacchus, J.: Hoarding Resources Threatens Free Trade. Wall Street Journal (May 19, 2010)
Bacchus, J.: A Rare-Earths Showdown Looms. The Wall Street Journal (May 20, 2011), http://online.wsj.com/article/SB10001424052748703509104576331010793763864.html (accessed June 01, 2011)
Bardt, H.: Rohstoffpreise und Bedeutung für die deutsche Wirtschaft. Institut der deutschen Wirtschaft Köln (2011), http://www.iwkoeln.de/Studien/IWTrends/tabid/148/articleid/31108/language/en-US/Default.aspx (accessed May 28, 2011)
Behrendt, S., Scharp, M., Feil, M., Dereje, C., Bleischwitz, R., Delzeit, R., Scharp, M.: Seltene Metalle: Maßnahmen und Konzepte zur Lösung des Problems konfliktverschärfender Rohstoffausbeutung am Beispiel Coltan. Umwelt Bundes Amt (2007), http://www.umweltdaten.de/publikationen/fpdf-l/3182.pdf (accessed May 30, 2011)
BGR, Rohstoffsituation Deutschland, Hannover: BGR (2010)
BMWA, Bericht zur aktuellen rohstoffwirtschaftlichen Situation und zu möglichen rohstoffpolitischen Hand-lungsoptionen. BMWA, Berlin (2005)
BMWi, Auswirkungen der weltweiten Konzentrierung in der Bergbauproduktion auf die Rohstoffversorgung in der deutschen Wirtschaft. Bericht Nr. 463 – Kurzfassung. BMWi, Berlin (1999)
BMWi, Rohstoffstrategie der Bundesregierung (2010), http://www.bmwi.de/Dateien/BMWi/PDF/rohstoffstrategie-der-bundesregierung,property=pdf,bereich=bmwi,sprache=de,rwb=true.pdf (accessed May 31, 2011)
Braune, G.: Kanada unterstreicht Ansprüche im Ozean. Handelsblatt (June 08, 2009), http://www.handelsblatt.com/politik/international/kanada-unterstreicht-ansprueche-im-ozean/3193568.html (accessed May 31, 2011)
Bundesverband der Deutschen Industrie, Übersicht über bestehende Handels-und Wettbewerbsverzerrungen auf den Rohstoffmärkten (2011) (unpublished overview)
Bünger, J., Unbehend, O.: Die Rohstoffe der Zukunft. Lohnt der Abbau von Bodenschätzen in der Tiefsee? In: ZDF Abenteuer Wissen (March 11, 2009), http://www.abenteuerwissen.zdf.de/ZDFde/inhalt/13/0,7529485,00.html (accessed May 31, 2011)
Chu, D.L.: Seventeen Metals: "The Middle East Has Oil, China Has Rare Earth". The Trading Report (November 12, 2010), http://www.thetradingreport.com/2010/11/12/seventeen-metals-%E2%80%9Cthe-middle-east-has-oil-china-has-rare-earth%E2%80%9D/ (accessed May 31, 2011)
COMRA (China Ocean Mineral Resources R & D Association), http://www.comra.org/english/eindex.html (accessed May 31, 2011)
Cunningham, L.D.: Tantalum. In: USGS (ed) Metals Prices in the United States Through (1998), http://minerals.usgs.gov/minerals/pubs/commodity/niobium/231798.pdf (accessed May 25, 2011)
Deutscher Bundestag: Unterrichtung durch die Bundesregierung Rohstoffstrategie der Bundesregierung – Sicherung einer nachhaltigen Rohstoffversorgung Deutschlands mit nicht-energetischen mineralischen Rohstoffen. Drucksache 17(3399) (2010)
Dohmen, F., Jung, A.: VEB Rohstoffe. In: Spiegel Online (January 31, 2011), http://www.spiegel.de/spiegel/print/d-76659506.html (accessed May 31, 2011)
Aktuell, D.J.S.: 18 deutsche Unternehmen an Rohstoffpartnerschaft mit Kasachstan interessiert (May 25, 2011) https://www.fis.dowjones.com/article.aspx?

ProductIDFromApplication=212&aid=DJGSTA0020110525e75p00001&r=Rss&s=DJGSTA (accessed May 31, 2011)

Elsner, H.: Kritische Versorgungslage mit Seltenen Erden – Entwicklung "Grüner Technologien" gefährdet? Bundesanstalt für Geowissenschaften und Rohstoffe (2011), http://www.deutsche-rohstoffagentur.de/DE/Gemeinsames/Produkte/Downloads/Commodity_Top_News/Rohstoffwirtschaft/36_kritische-versorgungslage.html (accessed June 06, 2011)

Environment Canada, Government of Canada Announces Decisions on Mount Milligan and Prosperity Gold-copper Mines. News Release, OTTAWA (November 2, 2010), http://www.ec.gc.ca/default.asp?lang=En&n=714D9AAE-1&news=59F03FA9-63AD-4EED-A14F-04BBF32906CF (accessed May 31, 2011)

EurActiv, Commission Unveils Updated Raw Materials Plan (February 3, 2011), http://www.euractiv.com/en/sustainability/commission-unveils-updated-raw-materials-plan-news-501842 (updated February 27, 2011) (accessed May 31, 2011)

EurActiv (2011a) EU to Step Up Raw Materials "Diplomacy" (June 18, 2010), http://www.euractiv.com/en/sustainability/eu-step-up-raw-materials-diplomacy-news-495397 (updated February 27, 2011) (accessed May 31, 2011)

EurActiv (2011b) EU, US, Japan should cooperate on rare earth supply (February 4, 2011), http://www.euractiv.com/en/sustainability/eu-us-japan-cooperate-rare-earth-supply-news-501917 (updated February 27, 2011) (accessed May 31, 2011)

European Commission, Global Europe. Competing in the World (2006), http://trade.ec.europa.eu/doclib/docs/2006/october/tradoc_130376.pdf (accessed June 06, 2011)

European Commission, The Raw Materials Initiative – Meeting our Critical Needs for Growth and Jobs in Europe. SEC(2008) 2741. Brussels (2008), http://ec.europa.eu/enterprise/newsroom/cf/document.cfm?action=display&doc_id=894&userservice_id=1 (accessed May 31, 2011)

European Commission, Commission Staff Working Document accompanying the Communication from the Commission to the European Parliament and the Council. The Raw Materials Initiative – Meeting our Critical Needs for Growth and Jobs in Europe. SEC (2008) 2741, Brussels (2008a), http://www.euromines.org/who_is_downloads/raw_materials_initiative_annexes.pdf (accessed May 31, 2011)

European Commission, EU Requests WTO Panel on Chinese Export Restrictions on Raw Materials. Factsheet. Dispute Settlement Brussels (November 4, 2009), http://trade.ec.europa.eu/doclib/press/index.cfm?id=483&serie=290&langId=en (accessed May 31, 2011)

European Commission, Tackling the Challenges in Commodity Markets and on Raw Materials, Communication from the Commission to the European Parliament, the Council, the European Economic and Social Committee and the Committee of the Regions. COM, 25 final, Brussels (February 2, 2011), http://www.acp-eu-trade.org/library/files/EC_EN_020211_EC_Communication%20on%20commodity%20markets%20and%20raw%20materials.pdf (accessed May 31, 2011) European Commission, Trade. Industrial Goods. Non-Ferrous Metals (2010), http://ec.europa.eu/trade/creating-opportunities/economic-sectors/industrial-goods/non-ferrous-metals/ (last updated December 2010) (accessed May 31, 2011)

European Commission Directorate-General for Trade, Raw Materials Policy, Annual Report (2009)

European Commission Enterprise and Industry, Non-Energy Raw Materials (2011a), http://ec.europa.eu/enterprise/policies/raw-materials/ (accessed June 06, 2011)

European Commission Enterprise and Industry, A European Strategy for Raw Materials (February 18, 2011), http://ec.europa.eu/enterprise/magazine/articles/industrial-policy/article_10958_en.html (accessed May 31, 2011)

European Union (The ad-hoc Working Group on Defining Critical Raw Materials), Critical Raw Materials for the EU. European Commission (2010), http://ec.europa.eu/enterprise/policies/raw-materials/critical/index_en.html (accessed May 28, 2011)

European Union, General Overview of Active WTO Dispute Settlement Cases Involving the EU as Complainant or Defendant and of Active Cases under the Trade Barriers Regulation (2011), http://trade.ec.europa.eu/doclib/docs/2007/may/tradoc_134652.pdf (accessed May 31, 2011)

Fang, Y.: China to impose rare earth resource tax. In: English. news. cn (March 24, 2011), http://news.xinhuanet.com/english2010/china/2011-03/24/c_13796465.htm (accessed May 31, 2011) GAO (Government Accountability Office) Rare Earth Materials in the Defense Supply Chain (2010), http://www.gao.gov/new.items/d10617r.pdf (accessed June 02, 2011)

Goldinvest.de, Seltene Erden – Japan hat eine erste Antwort. In: Goldinvest.de (November 3, 2011), http://www.goldinvest.de/index.php/seltene-erden-japan-hat-eine-erste-antwort-19369 (accessed May 31, 2011)

Heß, R.: Deutsche Suche nach Rohstoffen am Meeresboden. In: Telepolis (January 06, 2010), http://www.heise.de/tp/r4/artikel/31/31824/1.html (accessed May 31, 2011)

Hilpert, H.G., Kröger, A.E.: Chinesisches Monopol bei Seltenen Erden: Risiko für die Hochtechnologie. DIW Wochenbericht 19, 3–10 (2011)

Hilpert, H.G., Mildner, S.A., Lauster, G., Wassenberg, F.: Wettlauf um Metalle. In: Mildner, S.A. (ed.) Konfliktrisiko Rohstoffe? Herausforderungen und Chancen im Umgang mit knappen Ressourcen. SWP-Studien 2011/S 05, pp. 131–170 (2011)

Hiranuma, H.: Securing Supplies of Rare Metals for Environmental Technology. The Tokyo Foundation (2009), http://www.tokyofoundation.org/en/articles/2009/securing-supplies-of-rare-metals-for-environmental-technology (accessed May 31, 2011)

Hsu, A., Seligsohn, D.: What to Look for in China's 12th Five-Year Plan? World Resources Institute (March 02, 2011), http://www.wri.org/stories/2011/03/what-look-chinas-12th-five-year-plan (accessed June 01, 2011)

IW Consult, Rohstoffsituation Bayern: Keine Zukunft ohne Rohstoffe (2009), http://www.deutsche-rohstoffagentur.de/DE/Gemeinsames/Produkte/Downloads/Commodity_Top_News/Rohstoffwirtschaft/36_kritische-versorgungslage.pdf?__blob=publicationFile&v=2

Japan Investor, (The) The Coming Rare Earth Metals Crunch (September 2009), http://www.japaninvestor.com (accessed May 31, 2011)

Jarowinsky, M., et al. (ed.) Studie mit dem Ziel eines Aktionsplans für den Bereich marine mineralische Rohstoffe. Study on demand of the German Ministry of Economics and Technology and the Ministry of Economics of Lower Saxony (July 2009), http://www.jarowinsky-marketing.de/fileadmin/Downloads/MMR_Endbericht_2009.pdf (accessed May 31, 2011)

JOGMEG, Rare Metals Stockpiling Program (2007), http://www.jogmec.go.jp/english/activities/stockpiling_metal/raremetals.html (Accessed June 3)

Kölling, M.: Forschung für die Unabhängigkeit. In: Financial Times Deutschland (January 15, 2011), http://www.ftd.de/finanzen/maerkte/rohstoffe/:rohstoffe-forschung-fuer-die-unabhaengigkeit/50215021.html (accessed May 31, 2011)

Kuo, C.S.: The Mineral Industry of Japan. USGS 2008 Minerals Yearbook, Japan (June 2010), http://minerals.usgs.gov/minerals/pubs/country/2008/myb3-2008-ja.pdf (accessed May 31, 2011)

Li, J.: Project to tackle heavy-metal pollution. China Daily (2011), http://www.chinadaily.com.cn/china/2011-02/19/content_12043264.htm (updated February 19, 2011) (accessed June 01, 2011)

London Metal Stock Exchange, LME Copper Price Graph (2011), http://www.lme.com/copper_graphs.asp (accessed March 9, 2011)

Long, K.R., Van Gosen, B.S., Foley, N.K., Cordier, D.: The Principal rare earth Element Deposits of the United States – A Summary of Domestic Deposits and a Global Perspective. Scientific Investigations Report 2010 – 5220, Reston, Virginia: USGS (2010)

MarketResearch.com, Japan Metals Report Q4 2010. In: MArketResearch.com (2010), http://www.marketresearch.com/product/display.asp?productid=2836790 (accessed May 31, 2011)

McMahon, F., Miguel, C.: Survey of Mining Companies 2010/2011. Fraser Institute (2011)

METI (Ministry of Economy Trade and Industry), Announcement of 'Strategy for Ensuring Stable Supplies of Rare Metals' (July 28, 2009), http://www.meti.go.jp/english/press/data/20090728_01.html (accessed May 31, 2011)

Mining Association of Canada, Contribution of the Mining Industry: A Positive Message to Canadians. Presentation Material for Use by the Canadian Mining Industry (February 2009), http://www.mining.ca/www/media_lib/MAC_Documents/Presentations/2009/02_09_ppMining_contribution__MAC.pdf (accessed May 31, 2011)

Mining Association of Canada, Annual Report 2009 (2010), http://www.mining.ca/www/About_Mining/Publications.php (accessed June 01, 2011)

MLR (Ministry of Land and Resources of the People's Republic of China) (2010), Communiqué on Land and Resources of China 2007 (March 26, 2010), http://www.mlr.gov.cn/mlrenglish/communique/2007/ (accessed June 01, 2011)

Mobbs, P.: The Mineral Industry of Canada. U.S. Geological Survey 2009 Minerals Yearbook Canada. Washington D.C. (2011), http://minerals.usgs.gov/minerals/pubs/country/2009/myb3-2009-ca.pdf

NAS (The National Academy of Science) Minerals, Critical Minerals, and the U.S. Economy. The National Academies Press, Washington, D.C. (2007)

Natural Resources Canada, From Mineral Resources to Manufactured Products: Toward a Value-Added Mineral and Metal Strategy for Canada. Minister of Public Works and Government Services Canada (1998)

Natural Resources Canada, GEM: Geo-mapping for Energy and Minerals (2010), http://gsc.nrcan.gc.ca/gem/index_e.php (last updated October 08, 2010) (accessed May 31, 2011)

Nautilus Minerals, Nautilus Granted Mining Lease. News Release (2011), http://www.nautilusminerals.com/s/Media-NewsReleases.asp?ReportID=437932 (accessed June 07, 2011) NTV (2010) Japan sucht Meeresboden ab (April 26, 2010) http://www.n-tv.de/wissen/Japan-sucht-Meeresboden-ab-article841793.html (January 17, 2011)(accessed May 31, 2011)

Ocean Explorer, Extended Continental Shelf Project, U.S. Department of Commerce, National Oceanic and Atmospheric Administration by the Ocean Explorer Webmaster (August 26, 2010), http://oceanexplorer.noaa.gov/explorations/10ecs/welcome.html (accessed May 31, 2011)

Öko-Institut/ UNEP, Critical Metals for Future Sustainable Technologies and their Recycling Potential. United Nations Environment Programme (2009), http://www.unep.fr/shared/publications/pdf/DTIx1202xPA-Critical%20Metals%20and%20their%20Recycling%20Potential.pdf (accessed May 27, 2011)

Peeling, G.R.: The Canadian Mining Industry: Overview, Issues and the Way Forward. Presentation of the President and CEO of The Mining Association of Canada at EXPOMIN, Santiago, Chile (April 2010) http://www.mining.ca/www/media_lib/MAC_Documents/Presentations/2010/04_12_10_Expomin_2010eng1.pdf (accessed May 31, 2011)

Rehn D: Japan intensiviert Suche nach ‚strategischen' Metallen. In: Germany Trade and Invest, Länder und Märkte (October 01, 2010), http://www.gtai.de/DE/Content/Online-news/2010/23/medien/b2-japan-strategische-metalle.html (accessed May 31, 2011)

Reuters Deutschland (2011), Deutschland und Kasachstan bringen Rohstoff-Abkommen auf den Weg (May 24, 2011), http://de.reuters.com/article/economicsNews/idDEBEE74N0LM20110524 (accessed May 31, 2011)

Stratmann, K.: Brüderle fordert eine deutsche Rohstoff AG. Handelsblatt (November 02, 2010), http://www.handelsblatt.com/politik/deutschland/bruederle-fordert-eine-deutsche-rohstoff-ag/3580192.html (accessed May 31, 2011)

Trading Economics, China GDP Growth Rate (2011), http://www.tradingeconomics.com/china/gdp-growth (accessed May 31, 2011)

Trading-house.net, China baut Handel mit Afrika kräftig aus (December 23, 2010), http://www.trading-house.net/news/china-baut-handel-mit-afrika-kraeftig-aus-21799573.html (accessed May 31, 2011)

Trelawny, P., Pearce, P.: Canadian Mining Industry: 2009 General Review. Canadian Minerals Yearbook (CMY) –2009, via Natural Resources Canada (2009), http://www.nrcan.gc.ca/smm-mms/busi-indu/cmy-amc/2009revu/gen-gen-eng.htm (accessed May 31, 2011)

Tse, P.K.: The Mineral Industry of China. U.S. Geological Survey 2009 Minerals Yearbook China [Advanced Release] (November 2010) (revised November 19, 2010)

Tse, P.K.: China's Rare-Earth Industry. U.S. Geological Survey Open-File Report 2011–1042 (2011)

U.S. DOE (U.S. Department of Energy) (2010) Critical Materials Strategy. Washington D.C. (December 2010)

USGS (U.S. Geological Survey), Marine Mineral Resources of Pacific Islands–A Review of the Exclusive Economic Zones of Islands of U.S. Affiliation, Excluding the State of Hawaii. Circular 1286 (2005), http://pubs.usgs.gov/circ/2005/1286/c1286.pdf (accessed May 31, 2011)

USGS (U.S. Geological Survey), Statistical Summary 2008 (2008), http://minerals.usgs.gov/minerals/pubs/commodity/statistical_summary/myb1-2008-stati.pdf (accessed May 31, 2011)

USGS (U.S. Geological Survey), Copper (2010), http://minerals.usgs.gov/minerals/pubs/commodity/copper/mcs-2010-coppe.pdf (accessed May 31, 2011)

USGS (U.S. Geological Survey), Mineral Commodity Summary 2010, Washington D.C (2010a), http://minerals.usgs.gov/minerals/pubs/mcs/2010/mcs2010.pdf

Vaporean, C.: Copper to Stay High, Avg $9,950 a Tonne in '11-Barclays. Reuters (2010), http://uk.reuters.com/article/2010/12/09/idUKN0924272920101209 (accessed December 9, 2010)

Vasters, J., Buchholz, P., Huy, D., Schmitz, M., Röhling, S., Altfelder, S.: Rohstoffwirtschaftliche Bewertung der Länder Afrikas, Asiens, der Gemeinschaft Unabhängiger Staaten (GUS) mit Georgien und Südamerikas im Hinblick auf die Bedeutung für Deutschland. Deutsche Rohstoffagentur. BGR, Hannover (2010)

Xinhua, China's Rare Earth Producers Push for Clear National Strategy. China Daily (2011), http://www.chinadaily.com.cn/bizchina/2011-01/21/content_11898458.html (updated January 21, 2011) (accessed June 1, 2011)

Zhanheng, C.: Rare Earth Protection Plan. In: China Daily (May 28, 2011), http://www.chinadaily.com.cn/cndy/2011-05/28/content_12596386.html (accessed June 1, 2011)

Zajec, O.: China – Herr über die seltenen Erden (November 15, 2010), http://www.politonline.ch/?content=news&newsid=1641 (accessed May 31, 2011)

Annex I: Measuring Criticality

Report	Raw materials assessed	Definition	Methodology	Results – critical materials
The U.S. Perspective				
The National Academy of Science (2007): Minerals, Critical Minerals, and the U.S. Economy	11 minerals or mineral groups: copper, gallium, indium, lithium, manganese, niobium, platinum group metals, rare earth elements, tantalum, titanium and vanadium.	Defines a mineral as critical if it is both important in use and subject to potential supply restrictions	"Criticality matrix": Two-dimensional matrix (importance of the raw material in use / likelihood of a supply restriction) Short-term perspective	platinum group metals, rare earths, indium, manganese and niobium
U.S. Department of Energy (2010) : Critical Materials Strategy	14 raw materials with importance to green energy technologies	Defines raw materials as critical if they are important to clean energy and subject to supply risk.	Two-dimensional matrix (Importance to clean energy / supply risk) Framework assesses both the short-term and medium-term perspective	Critical in the short run: indium, dysprosium, terbium, europium, neodymium and yttrium Near-Critical in the short run: cerium, lanthanum and tellurium Non-critical in the short run: gallium, lithium, cobalt, praseodymium, and samarium Critical in the medium term: dysprosium, europium, terbium, neodymium, yttrium Near-critical in the medium term: indium, lithium, tellurium Non-critical in the medium term: cerium, cobalt, gallium, lanthanum, praseodymium, samarium
American Physical Society Panel on Public Affairs (2011): Energy Critical Elements—Developing New Technologies to Foster U.S. Energy Independence	No selection criteria specified; included in the assessment are all elements of the periodic system.	"Energy-critical elements" are defined as critical elements with respect to energy-related technologies. ECEs are those elements with the potential for major impact on energy systems.	Qualitative Framework with five criticality indicators: (1) crustal abundance, (2) geopolitical risks, (3) risk of joint production, (4) environmental and social concerns, (5) response times in production and utilization Short-term perspective	Energy-critical (owing to different factors and usage): gallium, germanium, indium, selenium, silver, tellurium, dysprosium, neodymium, praseodymium, samarium, cobalt, rare earths, lithium, lanthanum, helium, platinum, palladium, cerium and other platinum group elements.

The German Perspective				
Behrendt et al. (2007): Seltene Metalle	Rare Metals	Defines rare metals based on three criteria: (1) strong increasing or high price; (2) dynamic and static range of stocks, and (3) concentration of production	Qualitative factor-based approach, embracing six dichotomist factors: (1) value over $500 per kg, (2) price increase of over 100 percent between 2001 and 2004, (3) static range of reserve of over 25 years, (4) the static range of resources of over 50 years, (5) major known deposits are mainly to be found in one or two countries and (6) high concentration of supply and value added chain	High scarcity ratings: antimony and indium. Medium scarcity rating: Cobalt, gold, iridium, palladium, platinum, rhenium, rhodium, ruthenium, zinc, and tin
Angerer et al. (2009): Rohstoffe für Zukunftstechnologien	22 commodities assessed as crucial for 32 future technologies	Defines `vulnerable´ commodities as of particular importance to the national economy and concentrated on a few numbers of countries, whereas their production is commonly located in a political unstable region. Vulnerable commodities are of prospective functional and quantitative importance for the development of future technologies.	Quantitative factor approach for the estimation of prospective demand Time horizon: 2030	Factors for estimated demand increase in 2030, compared to the total production in 2006: Gallium (6.09), neodymium, (3.82), indium (3.29), germanium (2.44), scandium (2.28), platinum (1.56), tantalum (1.01), silver (0.78), tin (0.77), cobalt (0.4), palladium (0.34), titanium (0.29) [list non-exhausting]
IW Consult (2009) Rohstoffsituation Bayern: Keine Zukunft ohne Rohstoffe	37 raw materials	The terms danger, criticality and risk are applied in the context of supply risk based on physical unavailability or price development, leading to a shortage of the concerning commodities in the Bavarian industry.	Risk index with 7 indices: (1) statistical coverage, (2) country risk, (3) three country concentration, (4) three company concentration, (5) future technologies, (6)risk of strategic employment, (7) Substitutability.	Red group (critical): Yttrium, neodymium, cobalt, scandium, wolfram, phosphate, niobium, selenium, germanium, platinum group, lithium, chrome and molybdenum Yellow group (medium-critical): Noble metals (gold and silver), metals (magnesium, manganese, graphite, zinc, titanium, tantalum, copper, and aluminum), the rare metal gallium and industry metals such as fluorite, and barite.
Elsner (2011): Kritische Versorgungslage mit schweren seltenen Erden – Entwicklung "Grüner Technologien" gefährdet?	Heavy rare earths	Critical supply: demand exceeds supply.	Analysis of global supply and demand of heavy rare earths Perspective: 2010 - 2015	Critical supply in 2010: terbium (europium) Critical supply in 2015: europium, dysprosium and terbium

The EU and Global Perspective				
EU Report of the Ad-hoc Working Group on Defining Critical Raw Materials (2010): Critical Raw Materials for the EU	41 non-energy minerals and metals	Defines raw materials as critical, which face high risks with regard to accessibility, i.e. high supply risks or high environmental risks, and be of high economic importance. The access to such materials is likely to be disturbed, whereas expected impacts on the EU economy are significant.	Two-step approach: First step: two-dimensional matrix (economic importance/ supply risk) Second step: materials exceeding a threshold in economic importance are reassessed with regard to environmental country risk. Perspective: 10 years	Critical: antimony, indium, beryllium, magnesium, cobalt, niobium, fluorspar, platinum group metals, gallium, rare earths, germanium, tantalum, graphite, tungsten
Öko-Institut/UNEP (2009): Critical Metals for Future Sustainable Technologies and their Recycling Potential	11 minor metals: indium, germanium, tantalum, tellurium, cobalt, lithium, gallium, rare earths and the platinum metals ruthenium, platinum, and palladium	Defines "green minor metals" as critical, which might become short in supply and increasingly costly due to an increased demand by manufacturers of future sustainable technologies.	Three-factor approach: (1) supply risk, (2) demand growth and, (3) recycling restrictions. Scenarios for three time perspectives: short-, medium- and long-term	Short-term criticality: tellurium, indium and gallium Medium-term criticality/ scenario 1: rare earths, lithium and tantalum Medium-term criticality/ scenario 2: palladium, platinum and ruthenium Long-term criticality: germanium and cobalt

Authors' compilation.

Annex II: Country Strategies

Country	**Strategic documents**	**Key points in resource strategies**
Germany	2007 Elemente einer Rohstoffstrategie der Bundesregierung	• Improvements in material efficiency, recycling and substitution • Linkage of national and European resource strategies
	2010 Rohstoffstrategie der Bundesregierung	Four components: • Enhancing research in resource and product efficiency, in the development of substitutes for critical elements and recycling technologies; • Improving supply by utilizing domestic reserves and forming resource partnerships with resource-rich countries; • Enforcing international market discipline; • Supporting measures that address bad governance and corruption in resource-rich countries.
European Union	2008 The Raw Materials Initiative – Meeting Our Critical Needs for Growth and Jobs in Europe	Strategy based on three pillars: • Fair and sustainable supply from global markets; • Supply from within the EU; • Advancing resource efficiency and recycling technologies; development of subsidies.
	2010 Critical Raw Materials for the EU	• Identification of 14 critical materials, especially for high-tech and green technologies • Linkage of resource politics with regulatory framework for financial markets
	2011 Tackling the Challenges in Commodity Markets and on Raw Materials	Four pillars: • Securing fair and sustainable supply with raw materials from world markets through a new resource diplomacy; • Fostering resource production within the EU through a National Minerals Policy that helps plan and monitor the member states' achievements and actions in the resource sector; • Enhancing research and development for improved recycling and substitution technologies as well as resource and product efficiency; • Resource diplomacy.
Japan	2009 Strategy for Ensuring Stable Supplies of Rare Metals 2010 One-stop system for supply security of the industry	• 31 strategic raw materials identified • Three components: o Focused, Strategic Approach: The government is to evaluate the supply situation and to determine the levels of priority for the different kinds of rare metals; o Four Pillars for Securing Rare Metals: ▪ securing overseas resources, ▪ recycling, ▪ development of alternative materials, ▪ stockpiling; o Development of a Common Infrastructure toward Securing of Rare Metals: human resources development in the resources sector, enhancing the technical capabilities of the resources sector, and integrated efforts (multi-stakeholder process); • Government support for deep seabed resource research.

United States	2010 Rare Earth Materials in the Defense Supply Chain	• Resource scarcity is perceived as a geopolitical risk
	2010 U.S. DOE Critical Materials Strategy	• Identification of 14 critical metals • Three pillars: o Diversified Global Supply Chains: taking steps to facilitate extraction, processing and manufacturing in the U.S., as well as encouraging other nations to expedite alternative supplies; o Development of Substitutes: supporting research leading to material and technology substitutes; (3) Recycling: reuse and more efficient use. Follow-up report on critical materials of the Department of Energy (U.S. DOE) scheduled for 12/2011.
Canada	1998 From Mineral Resources to Manufactured Products: Toward a Value-Added Mineral and Metal Strategy for Canada	Goals: • Promote sustainable development and use of mineral and metal resources; • Protect the environment and public health; • Ensure an attractive investment. Federal measures for resource supply security include: • Exploration support new production sites • Federal support for exports • Transparency in the mining sector • Harmonization of federal and provincial rules • Stockpiling
	2008 Geo-mapping for Energy and Minerals (GEM) program	• Program running from 2008-2013 and aiming at mapping the Arctic and high north seabed for mineral resources
China	No actual strategy document accessible Five-Year Plans may contain recommendations on natural resources in order to boost economic development	Applied measures include: • Export quotas to secure national supplies • New taxation system for rare earth production • Foreign investments in resource-rich countries like Australia, Brazil, Burma, Chile, Indonesia, Mongolia and African states • Investment in research and development for potential mining of marine resources

Authors' compilation.

Annex III: Mining Production 2009

	Resource-dependent countries			Resource-rich countries				
	Germany	EU	Japan	Canada	United States	China	Total of all 6 countries	World Total
Production								
(1) Total minerals production by country								
Iron, steel-alloys (by production in metr. t)	38200	14114206	800	22177708	16631000	284956800	337918714	1153384880
(share of global production in %)	0	1.22	0	1.92	1.44	24.71		
Non-ferrous metals (by production in metr. t)	292050	6527109	7434	4294674	4043920	49032790	64197977	253579750
(share of global production in %)	0.12	2.57	0	1.69	1.59	19.34		
Precious metals (by production in metr. t)	0	1705	8	744	1457	3220	7134	24916
(share of global production in %)	0	6.84	0.03	2.99	5.85	12.92		
Industrial metals (by production in metr. t)	31898287	111285940	6371000	29202199	89151000	154835000	422743426	640943303
(share of global production in %)	4.98	17.36	0.99	4.56	13.91	24.16		
Total (by production in metr. t)	32228537	131928960	6379242	55675325	109827377	488827810	824867251	2047932849
(share of global production in %)	1.57	6.44	0.31	2.72	5.36	23.87		
(2) Most produced minerals (by share of global production)								
(1)	Kaolin (18.70 %)	Perlite (49.24 %)	Tellurium (63.49 %)	Tellurium (25.40 %)	Diatomite (50.31 %)	Rare Earths (97.45 %)		
(2)	Feldspar (18.07 %)	Feldspar (48.70 %)	Perlite (11.38 %)	Potash (21.84 %)	Bentonite (30.17 %)	Antimony (91.35 %)		
(3)	Potash (8.64 %)	Kaolin (30.95 %)	Cadmium (8.71 %)	Titanium (17.72 %)	Boron (27.00 %)	Tungsten (80.07 %)		
(4)	Salt (7.23 %)	Salt (20.49 %)	Gallium (8.33 %)	Diamonds (Gem) (15.89 %)	Molybdenum (22.27 %)	Bismuth (74.47 %)		
(5)	Bentonite (2.4 %)	Gypsum (18.92 %)	Sulfur (5.96 %)	Sulfur (10.22 %)	Kaolin (21.54 %)	Graphite (74.08 %)		

Indicators

Source: World Mining Data 2011. Bundesministerium für Wirtschaft, Familie und Jugend. http://www.bmwfj.gv.at/energieundbergbau/weltbergbaudaten/Seiten/default.aspx. Accessed 10.06.2011

* The following minerals were selected for the analysis:

Iron, steel-alloys: iron, chromium, cobalt, manganese, molybdenum, nickel, tantalum-columbium, titanium, tungsten, vanadium

Non-ferrous metals: aluminium, antimony, arsenic, bauxite, bismuth, cadmium, copper, gallium, germanium, lead, lithium, mercury, rare-earth minerals, tellurium, tin, zinc

Precious metals: gold, platinum-group metals (palladium, platinum, rhodium), silver

Industrial metals: asbestos, baryte, bentonite, boron minerals, diamond (gem and industrial), diatomite, feldspar, fluorspar, gypsum and anhydrite, graphite, guano, kaolin (china-clay), magnesite, perlite, potash, phosphate rock, salt, sulfur, talc (incl. steatite and pyrophyllite), vermiculite, zircon

Hybrid Threats and Supply Chain Safety Management

Marc Oprach[1] and Boris Bovekamp[2]

[1] Federal Armed Forces Transformation Center, Baumschulenstr. 95, 12437 Berlin, Germany
marc.oprach@gmx.net
[2] Federal College for Security Studies, Office of the President, Gethsemanestr. 4, 10437 Berlin, Germany
borisbovekamp@web.de

> "One must keep the dominant characteristics of both belligerents in mind. Out of these characteristics a certain center of gravity develops, the hub of all power and movement, on which everything depends. That is the point against which all our energies should be directed."[1]
>
> Carl von Clausewitz

For almost a week in mid-April of 2010, the ash cloud of the Icelandic volcano Eyjafjallajökull brought the traffic in vast parts of the European airspace to a standstill. While public and media interest mainly concentrated on passenger flights cancelled as a consequence of it, worldwide cargo air traffic was also considerably impeded. The Director General and CEO of the International Air Transport Association (IATA), Giovanni Bisignani, emphasized in a press statement the far-reaching economic consequences: "Aviation drives economies. It supports $3.5 trillion in economic activity annually and 32 million jobs. When it was disrupted for six days in Europe, 100,000 flights were cancelled and $1.7 billion in industry revenue was lost. (…) Flowers from Kenya did not reach their markets. Australian oysters did not reach European kitchens. German factories did not have the parts to assemble their products."[2]

The far-reaching consequences from the cancelled flights highlighted the vulnerability of the complex business and trade flows. Especially highly developed industrial nations must accept that potential state and often also non-state adversaries are spying out these weak points – the hub of all power – and preparing to attack them.

Against this background, the term "hybrid war" may direct attention to a necessary discussion and may be used as the starting point for a purposeful analysis. The concept must focus on state and non-state adversaries who, in a conflict, use the full spectrum of conventional, criminal, terrorist and irregular measures.

M. Essig et al. (Eds.): Supply Chain Safety Management, LNL, pp. 89–100.
springerlink.com

Frank G. Hoffman, the American concept leader, defines the "hybrid threat" as follows:

"Hybrid threats incorporate a full range of modes of warfare, including conventional capabilities, irregular tactics and formations, terrorist acts that include indiscriminate violence and coercion, and criminal disorder."[3]

Using the headline "Hezbollah as a Prototype", he calls the Lebanon war an example of hybrid conflict scenarios.[4] He considers that the dualism of symmetric and asymmetric wars that has evolved from the scientific discourse no longer seems sufficient after the Lebanon war in 2006 because the radical Islamic Hezbollah used complex weapon systems in this conflict which could normally only be expected to be found in the arsenals of nation-state armies. For instance, the firing of a cruise missile, the Iranian version of the Chinese HY-2, at an Israeli fast patrol boat caused the deaths of four Israeli soldiers.[5]

Although the use of a cruise missile by the Hezbollah barely received attention in the European debate on security, the U.S. Secretary of Defense, Mr. Robert M. Gates, refers in an article in the January/February 2009 issue of the journal Foreign Affairs to the strategic dimension of this incident. This military strike by a non-state terrorist group emphatically revealed that, with regard to armament, terrorist groups may draw even with nation-state armies.[6]

Even though the term "hybrid war" is being picked up in almost all U.S. strategy papers, NATO has up to now been struggling to find a common concept.[7] NATO has not yet developed a doctrine for countering this new threat, but defines it as follows: "A hybrid threat is one posed by any current or potential adversary, including state, non-state and terrorists, with the ability, whether demonstrated or likely, to simultaneously employ conventional and non-conventional means adaptively, in pursuit of their objectives."[8]

Not least against the background of the debate that has begun in NATO on the construct of "hybrid threats", it seems necessary to carry on examining the subject in detail. The aim should be to go beyond theoretical analysis and offer practical guidance and, in view of the subject of "Supply Chain Safety Management", to work out the precise practical relevance of the findings. It is obvious that the subject of "hybrid warfare" cannot be approached from a strictly military angle. Instead, here are some scenarios that emphasize the complexity of closely coordinated reactions.

Even ten years after 9/11, there is a lack of imagination with regard both to the "new" forms that threats and attacks can take and to the possibilities that can result from the combination of military and non-military instruments. It must be said that the risk analysis community in particular mostly ponders the intellectual experiences of the conflicts last fought out or observed. It is precisely on account of this that there is the risk of a hybrid adversary who thinks strategically spotting gaps in plans and making systematic use of them.

The special threat effect of attacks on these targets lies mainly in the damage they are expected to cause. Even minor security incidents that are confined to specific areas could immediately initiate global follow-up processes, for instance in air traffic, that have far-reaching time and financial implications and whose effects are almost uncontrollable. The likely targets of these attacks are "critical infrastructures". They are organizations and installations which are important for states and whose failure or degradation would cause sustained supply shortfalls and considerable disruptions to public safety and security or have other dramatic consequences.[9] The following scenarios have been looked at closely in this analysis:

Scenario 1: Attack on port facilities and maritime trade routes
Scenario 2: Attack on airports and air traffic
Scenario 3: Attacks with weapons of mass destruction
Scenario 4: Cruise missile threat
Scenario 5: Cyber attack

The element that links all the scenarios together is that, with reference to the initial quotation of Clausewitz, one must agree with the demand made by Frank G. Hoffman: "We cannot continue to overlook our own vulnerabilities or underestimate the imaginations of our enemies."[10]

Scenario 1: Attack on Port Facilities and Maritime Trade Routes

In hybrid warfare scenarios, potential adversaries analyze the weak points of modern industrial nations and then concentrate on points at particularly great effects can be achieved. In the National Strategy for Critical Infrastructure Protection, the German federal government states that infrastructure is considered "critical" whenever it is of major importance to the functioning of modern societies and any failure or degradation would result in sustained disruptions in the overall system. An important criterion for this assessment is "criticality as a relative measure of the importance of a given infrastructure in terms of the impact of its disruption or functional failure on the security of supply, i.e. providing society with important goods and services."[11]

Shipping and seaborne trade are considered by many analysts to be vulnerable to hybrid attack.[12]

Especially ports contain a number of specific facilities that could be targeted, including terminals, factories, office buildings, power plants, refineries and other critical infrastructure.[13] The economic consequences of a major direct terrorist attack which would block an industrial port like Hamburg, even for just one day, are immense and their far-reaching consequences could barely be estimated.

Another scenario developed in this context aims at the same consequences without causing direct, immediate damage itself. For example, even a dummy sea mine would lead to the blocking of a major port and so help the goal of disrupting the logistic supply and in turn the "lifelines" of a modern industrial nation to be achieved. However, this "low-cost-option" of hybrid groupings will become a real risk factor because of their possibility to acquire older-type mines on the international arms market. Even the temporary blocking of waterways and "choke points" is increasingly becoming a subject of international debate because the waters of the straits are shallow and ideal for mining by either floating mines or mines placed on the sea bottom. During the "tanker war" of the 1980s, the laying of mines was arguably more successful in disrupting shipping traffic than the use of anti-ship missiles.[14]

It is this scenario in particular that reveals the specific core of hybrid threats in contrast to terrorist threats. Hybrid threats incorporate a range of different modes of warfare including conventional capabilities and irregular tactics. Hybrid wars can be also be multinodal - conducted by both states and a variety of non-state actors.[15] Thus, it is also conceivable for a state to use its army, irregular units, militia and loosely allied terrorist and criminal organizations to fight out a conflict conventionally and at the same time by terrorist or irregular means and to utilize all these capabilities against the state, society and economy of his adversary anywhere in the world and in a well coordinated manner. Disruptive attacks against transport and trade routes would be an inherent part of the conflict. It is with this very context in mind that Frank Hoffman refers to the current rearmament activities of Iran and comes to the view that Iran's navy employs "asymmetric and highly irregular tactics that exploit the constricted geographic character of the Gulf. (…) This doctrine applies a hybrid combination of conventional and irregular tactics and weapons to posit a significant anti-access threat to both military and commercial shipping. Hoffman particularly pointed to Iran's large sea mine arsenal, estimated at between 3,000 and 5,000 mines.[16]

But terrorist groupings may achieve an identical effect even without the open or indirect support of a state. If economic loss is the primary objective, hybrid "actors" may seek to carry out different types of attacks, with potentially few human casualties, but significant impacts on critical infrastructure or commerce. They may, for instance, use a "ship with hazardous or dangerous cargo" as a "floating bomb".[17]

The attack on the oil tanker *Limburg* is often cited as an indication of tanker vulnerability. However, the attacks on the French tanker off Yemen in 2004 usually attract most attention in writings on maritime terrorism because it was initiated by al-Qaeda and occurred in the context of 9/11.[18]

The *Limburg* bombing threatened to disrupt the global oil trade and caused considerable consternation among tanker operators. Although the bombing killed only one member of the *Limburg's* crew, it caused insurance rates among Yemeni shippers to rise 300 % and reduced Yemeni port shipping volumes by 50 % in the month after the attack.[19]

Attacks on liquid natural gas (LNG) carriers, liquid petroleum gas (LPG) carriers and chemical tankers could have even greater effects and could cause "catastrophic fires in ports and nearby populated areas" [20] if they occurred in the vicinity of LNG facilities. In this case, too, port facilities could be expected to be closed for several days, with foreseeably far-reaching economic consequences.

Scenario 2: Attack on Airports and Air Traffic

Hybrid scenarios combine "the lethality of state conflict with the fanatical and protracted fervor of irregular warfare". In such conflicts, future adversaries (states, state-sponsored groups, or self-funded actors) will exploit access to modern military capabilities, including (…) man-portable surface-to-air missiles.[21]

On 12 August 2006, 24 Israeli soldiers were killed in the Lebanon War; the worst Israeli loss in a single day. Out of those 24, five soldiers were killed when Hezbollah shot down an Israeli helicopter. The Islamic militant group claimed the helicopter had been attacked with a "Wa'ad" (Arabic for promise) missile. The missiles are reported to be versions of the Russian Strela supplied to Hamas by Iran. The Strela has a range of more than two miles and can reach aircraft or helicopters flying more than a mile above the ground. Deployed close to airfields or landing strips, such missiles could be employed with devastating effect.[22]

Even the threat to civil air traffic from anti-air missiles is therefore a real and by no means wayward scenario due to the availability of these weapon systems.

On January 27 and 28, all the 16 federal states in Germany will conduct the LÜKEx 2010 exercise in cooperation with the federal government. The term LÜKEx stands for „Länderübergreifende Krisenmanagement-Übung/Exercise" (Cross-State Crisis Management Exercise) and refers to a series of exercises concerned with national crisis management in Germany. The subject of LÜKEx 2010 was the perpetration of terrorist attacks involving CBRN means (Chemical, Biological, Radiological and Nuclear Incidents). LÜKEx is not a police exercise concerned with the prevention of terrorist attacks; rather, the exercise scenario is based on the assumption that the police and secret services have failed and terrorist groups have been able to successfully carry out attacks. It started with the simulation of a portable anti-air missile having been fired at an airplane by terrorists from outside the airport.[23]

The consequences of such an attack for air traffic would be immense. Should there also be the necessity to assume that it was not the work of a single attacker and that there must be fear of further attacks, air traffic would be impaired for an unforeseeable period of time. The consequences for passenger air traffic and the economic damage to airlines would have to be taken into consideration, as would the considerable impairments to international cargo traffic.

Scenario 3: Attacks with Weapons of Mass Destruction

One possible factor in a terrorist's reckoning is the employment of nuclear, biological or chemical agents which, in addition to the immediate (lethal) damage they would cause, would have a considerable psychological impact on the people attacked even if their employment remained just an attempt or merely achieved a semblance of success. Besides the use of an anti-air missile, the German crisis management exercise LÜKEx 2010 also simulated the consequences of the employment of CBRN means. All in all, about 2,500 members of the fire, relief and rescue services participated in the exercise, with the aim of gaining practice in the use of the practical disaster relief capabilities in as realistic a setting as possible.

Even NATO's new Strategic Concept makes explicit reference to the danger of terrorism emanating from weapons of mass destruction.

"Terrorism poses a direct threat to the security of the citizens of NATO countries, and to international stability and prosperity more broadly. Extremist groups continue to spread to, and in, areas of strategic importance to the Alliance, and modern technology increases the threat and potential impact of terrorist attacks, in particular if terrorists were to acquire nuclear, chemical, biological or radiological capabilities."[24]

It is noticeable that the debate within NATO is not confined to the threat arising from nation-state nuclear armament efforts and endeavors to increase the ranges of missile systems, but includes hybrid threat scenarios.

Thus, the idea that there is the threat of a nuclear terrorist attack does not seem to be wayward by any means. Stephen Cimbala calls the infiltration of the Pakistani military a threat because it entails the risk of Islamic terrorists acquiring nuclear arms. The threat in this case refers to small bombs which, for example, could reach the American coast inside a container ship in order to be exploded in a big city.[25]

Conceivable consequences of this are mass panicking, people fleeing and, in the extreme case, the complete breakdown of public order. Targets for such terrorist attacks could be all kinds of crowds in public places, in railway stations, at airports and at events. A weapon of mass destruction (WMD) attack on a heavily populated U.S. port or airport could inflict the greatest number of human casualties[26] and have far-reaching consequences for the economic system of a country.

The Sarin attack in Tokyo back in 1995 showed that in this field, railroad traffic must not be neglected either. The danger of warfare gas agents is dreadfully high, and just a few milligrams of Sarin are lethal. In an accident involving a chemicals train loaded with chlorine gas in January 2005 in the U.S. state of South Carolina, 8 people were killed and about 240 suffered from respiratory disorders or burns, even though the accident occurred outside any densely populated area. Targeted attacks at trains loaded with chlorine gas also offer a hybrid aggressor possibilities to achieve extremely great effects with extremely limited actions.

Scenario 4: Cruise Missile Threat

The construct of "hybrid warfare" is inseparably connected to the Lebanon war in 2006 and the cruise missile attack on an Israeli fast patrol boat.[27]

The main attraction of these weapon systems is the possibility they offer to attack a land or sea target directly even if it seems to be protected by fences, access controls or checkpoints.[28]

The employment of these weapon systems by terrorists always appears to be practicable if an area is closed off and so the use, for example, of a bomb deposited in a vehicle seems impossible. If targeted use is made of it, a comparably small explosive device could in these cases have a far more devastating effect.[29]

This explains why not just the armament efforts of other nations were the subject of a cruise missile proliferation analysis by the National Defense Research Institute RAND published in 2008, but also those of violent non-state groupings.[30]

Particular mention must be made of the fact that the construction and employment of primitive remote-controlled unmanned aerial vehicles are already topics of discussion on the internet sites of Al-Qaeda.[31] Citing an emphatic example, John G. Heidenreich, an American security expert, states that it is also possible for a private individual to build a cruise missile with a range of 100 kilometers and a capacity to carry a load of 10 kilograms with a budget of 5,000 dollars – in a garage.[32]

Especially the publicly accessible market for remote-controlled and target-programmable model airplanes could become more and more attractive for terrorist groups.[33]

The threat posed by the use of an ultra light airplane by terrorists as a remote-controlled bomb is also the subject of numerous essays and articles.[34]

Besides these do-it-yourself remote-controlled missiles, attention must also drawn to the devastating consequences of an attack involving the use of a cruise missile that is technologically outdated and can be acquired on favorable conditions on the world market. It is estimated that worldwide there are above all over 75,000 of those easy-to-operate anti-ship guided missiles, which can also be employed against land targets, with their proliferation extending into more than 70 countries.[35]

With the progressive development in technology, the employment of cruise missiles constitutes an independent set of tools for hybrid actors. Therefore, the overall primary objective must be to acknowledge the complementary threat posed by self-made guided missiles and cruise missiles available on the world market and to also discuss the specific aspects of the complex "cruise missile threat".

Scenario 5: Cyber Attack

Besides the violence and mass killing scenarios, the manipulation of the finance and monetary markets and the disruption of economic processes are considered attractive methods for hybrid attackers. Frank G. Hoffman expressly mentions "cyber warfare directed against financial targets" as a possible option for a hybrid actor.[36]

In its new Strategic Concept, NATO also sees non-state terrorist groups as being able to conduct cyber attacks:

"Cyber attacks are becoming more frequent, more organized and more costly in the damage that they inflict on government administrations, businesses, economies and potentially also transportation and supply networks and other critical infrastructure; they can reach a threshold that threatens national and Euro-Atlantic prosperity, security and stability. Foreign militaries and intelligence services, organized criminals, terrorist and/or extremist groups can each be the source of such attacks."[37]

Due to the fact that the sector is largely untransparent for security reasons, targeted activities are difficult to detect as forms of attack in it. Since it is extremely difficult to above all determine the source of such an attack, it must be assumed that nation states avail themselves of the help of terrorist groups in order to disguise their attacks even more. While the question as to the proliferation of anti-air or cruise missiles can be answered quite quickly, the initiation of a cyber attack is comparably difficult to trace back.

Cyber warfare is rated to be an immensely important topic in future and one that needs to be looked more closely and in more detail.

Especially the armed forces were early to spot the explosive nature of this type of threat and have implemented extensive measures. Apart from military command and control and communication systems, military logistics must also be seen as a potential target. As military logistics have become more dependent on computer systems, they have become more vulnerable to cyber attacks.

The risks to military logistics are described in the Joint Concept for Logistics published by the U.S. Department of Defense in August 2010.

"Cyber Risk. This paper proposes a continued great reliance on networked automated information systems. The increasing dependence of DOD on information technologies forebodes catastrophic consequences given disruption or destruction of those technologies."[38]

The awareness that can be seen here regarding possible threats to military logistics allows the conclusion to be drawn that the military logistic systems will be increasingly protected against cyber attacks.

For this very reason and because the economic effect must be counted as a "primary objective" of hybrid adversaries, the actors will primarily concentrate on the vulnerability of civil logistics to cause the greatest possible damage.

Consequences

Warring factions have always sought ways of taking advantage of the weak points of their adversaries. The fact that actors modify and adapt to changes in social, economic and technical environments is a general element and constant of warfare. Thus, the development of "hybrid warfare" merely reflects the fundamental historic fact that the face of war changes and continuously adjusts to changes in conditions. Even in the past, conflicts could never be purely assigned to one of the two categories, symmetric and asymmetric, since usually there was a slant towards both types of conflicts. Yet the employment of "sophisticated" weapons by non-state actors must be seen as a new phenomenon requiring consideration.

Besides the weapon systems used, however, the strategic analysis is the main characteristic of hybrid actors, whose actions are aimed precisely at the weak points of modern industrial nations.

But what are the concrete consequences if in its new Strategic Concept, NATO considers the transport and trade routes to be particularly vulnerable points of our security and our prosperity and if the economic and finance centers come into the firing line of hybrid actors?[39]

On the basis of the five scenarios presented, which are just a selection of the options that hybrid adversaries have, the following consequences can be indentified:

Awareness (Sensitization)

Especially high-linkage interdependent systems without sufficient redundancy are susceptible to attackers who try to cause the greatest possible damage and to generate a complete or at least a large-scale system failure by means of the failure of smaller subsystems.

It must be acknowledged that the targeted employment of complex weapon systems against the indispensable lifelines of modern, strong societies such as the trade and finance centers, port facilities and airports appears to be conceivable.

To counter the hybrid threats, especially the networked threat analysis capabilities, including interdisciplinary risk identification, need to be improved, although not even the best possible analysis provides full immunization against the effect of a strategic surprise.

Active "Countering"

At national level, the problems that still exist in coordination between national security authorities and organizations and the continued underdevelopment in the networked threat analysis capability (including interdisciplinary risk identification and early crisis detection) need to be removed as far and as quickly as possible.

Besides the national optimization of the police and military defense measures, measures for a successful non-proliferation policy above all need to be intensified in the future. Resolute multinational action must be taken to counter the threat of the proliferation of sea mines, cruise missiles and anti-air missiles of all things. Furthermore, above all NATO must be used as a forum for discussion to develop answers to the threats of hybrid warfare.

Preparedness (Seismographics and Emergency Plans)

As in the case of the LÜKEx crisis management exercise, the consequences of an attack must be simulated in a large-scale exercise in order to enable training in the use of practical disaster control capabilities under as realistic conditions as possible. Civilian and large logistics companies must also prepare for these extreme cases.

This is why they are regularly invited to participate in LÜKEx exercises, which have been taking place since 2004, the aim being to familiarize them with the crisis management structures and measures of state *and* private partners, which are a component of national security provision, enable them to practice using and implementing and involve them in their further development. These joint exercises have enhanced the trustful cooperation between the state and the business world and confirmed the belief that crises can only be mastered jointly.

Even though it was not terrorist attacks that led to the cancellation of flights in Europe recently, the ash cloud showed clearly that these extreme situations can only be countered with flexibility and networking. In this case, for example, DHL, the leading logistics company in the world, was able to react quickly. Just a few hours after large areas of airspace had been closed, DHL activated an emergency plan developed by an international team with alternate flight routes and substitute solutions for road transports. This emergency plan enabled large backlogs in commodity transport to be prevented in the interest of the customer, fast and proactive reactions to the continuously changing situation to be taken and the available capacities to be used in the best possible way.[40]

However, it is not possible to guarantee 100 percent protection for infrastructures and its capacities. The present safety first way of thinking must transform into a new "risk culture". This new risk culture needs an open risk communication between the state, the people and businesses as well as cooperation between all the relevant actors in the prevention and handling of events. A new risk culture of this kind, which also includes a greater self-commitment on the part of businesses to actively contribute to the prevention and handling of events, is apt to make our society more robust and resistant in dealing with increasing vulnerabilities.

All in all, the most important conclusion that can be drawn from this must be the perception of these complex threats and the diversity of the necessary answers. American security expert John Heidenreich summed up the dimension of the threat from attacks by hybrid actors succinctly and astutely when he said: "Ignoring the threat will not make it go away."[41]

Footnotes

1 von Clausewitz, C.: On War, Indexed ed., Howard, M., Paret, P. (eds.) (trans.), pp. 595–596. Princeton University Press, Princeton (1989)

2 The eruption of Eyjafjallajökull was a wake-up call for change, by Giovanni Bisignani Director General and CEO, IATA (June 2010)

3 Hoffman, F.G.: Hybrid Warfare and Challenges. JFQ (52), 34–39, 36 (2009)

4 Hoffman, F.G.: Conflict in the 21st Century. The Rise of Hybrid Wars, Potomac Institute for Policy Studies, p. 35 et seq. (December 2007)

5 Gormley, D.M.: Missile Contagion. Survival, 137–154, 138 (August/September 2008)

6 Gates, R.M.: A Balanced Strategy. Foreign Affairs, 28–40, 34 (January/February 2009)

7 Quadrennial Defense Review Report 2010 / Joint Operating Environment 2010 / United States Department of Defense, National Defense Strategy (Washington, D.C, / United States Department of Defense, Quadrennial Defense Review Report (Washington, D.C/United States Joint Forces Command, The Joint Operating Environment, Suffolk, Va (June 2008)

8 Definition by the NATO Military Working Group (Strategic Planning & Concepts) (February 2010)

9 Nationale Strategie zum Schutz Kritischer Infrastrukturen (KRITIS-Strategie), Bundesministerium des Innern (National Strategy for Critical Infrastructure Protection, German Federal Ministry of the Interior) (June 17, 2009)

10 Hoffman, F.G.: Complex Irregular Warfare: The Next Revolution in Military Affairs. Orbis, 395-411, 411

11 Nationale Strategie zum Schutz Kritischer Infrastrukturen (KRITIS-Strategie), Bundesministerium des Innern (National Strategy for Critical Infrastructure Protection, German Federal Ministry of the Interior), June 17, p. 7 (2009)

[12] Bateman, S.: Assessing the Threat of Maritime Terrorism. Issues for the Asia-Pacific Region, Security Challenges 2(3), 77–91, 78 (October 2006)

[13] Caldwell, S.L.: U.S. Government Accountability Office. Statement at the House Committee on Government Reform, Subcommittee on Government Management, Finance, and Accountability hearing on "Securing Our Ports: Information Sharing is Key to Effective Maritime Security" (July 10, 2006)

[14] Bateman, S.: Assessing the Threat of Maritime Terrorism. Issues for the Asia-Pacific Region, Security Challenges 2(3), 77–91, 87 (2006); Raymond, C.Z.: Maritime Terrorism in Southeast Asia: Potential Scenarios. Terrorism Monitor 14(7), 2 (2006)

[15] Hoffman, F.G.: Hybrid Threats: Reconceptualizing the Evolving Character of Modern Conflict, No. 240, Strategic Forum 5 (April 2009)

[16] Hybrid War at Sea: Iran's Great Prophet 5 Exercises (April 26, 2010), http://defensetech.org/2010/04/26/hybrid-war-at-sea-irans-great-prophet-5-exercises-ii/#ixzz1C2Apqej2

[17] Bateman, S.: Assessing the Threat of Maritime Terrorism. Issues for the Asia-Pacific Region, Security Challenges 2(3), 77–91, 85 (2006)

[18] Bateman, S.: Assessing the Threat of Maritime Terrorism. Issues for the Asia-Pacific Region, Security Challenges 2(3), 77–91, 81 (2006)

[19] Parfomak, P.W., Frittelli, J.: Maritime Security, Potential Terrorist Attacks and Protection Priorities. Congressional Research Service, 3 (May 14, 2007)

[20] Parfomak, P.W., Frittelli, J.: Maritime Security, Potential Terrorist Attacks and Protection Priorities. Congressional Research Service, 20 (May 14, 2007)

[21] Hoffman, F.G.: Hybrid Warfare and Challenges. JFQ/issue 52, 34–39, 37 (2009)

[22] Goure, D.: The Next Terrorist Challenge: Anti-Aircraft Missiles, Lexington Institute (October 20, 2010), http://www.defence.pk/forums/world-affairs/1786-hezbollah-shoots-down-israeli-helicopter.html

[23] Cimander, S.: LÜKEx 2010: Terroranschläge mit CBRN-Tatmittel (January 20, 2010), http://www.fwnetz.de/2010/01/20/lukex-2010-terroranschlage-mit-cbrn-tatmittel/; Piper, G.: Lükex 2010: Multiple Terrorangriffe mit ABC-Waffen, (January 16, 2010), http://www.heise.de/tp/r4/artikel/31/31866/1.html

[24] Strategic Concept for the Defence and Security of the Members of the North Atlantic Treaty Organisation

[25] Cimbala, S.J.: Missile Defenses in a "Deuces Wil" Context. Proliferation, Terror and Deterrent Disorder. Comparative Strategy 25(1), 1–18, 2 et seq (2006)

[26] Parfomak, P.W., Frittelli, J.: Maritime Security, Potential Terrorist Attacks and Protection Priorities. Congressional Research Service, 3 (May 14, 2007)

[27] Gormley, D.M.: Missile Contagion. Survival, 137–154, 138 (August/September 2008)

[28] Jackson, B.A., Frelinger, D.R., Lostumba, M.J., Button, R.W.: Evaluating Novel Threats to the Homeland. UAVs and Cruise Missiles, National Defense Research Institute RAND, p. 50 (2008)

[29] Jackson, B.A., Frelinger, D.R., Lostumba, M.J., Button, R.W.: Evaluating Novel Threats to zhe Homeland. UAVs and Cruise Missiles, National Defense Research Institute RAND, p. 20 and p. 31 (2008)

[30] Jackson, B.A., Frelinger, D.R., Lostumba, M.J., Button, R.W.: Evaluating Novel Threats to the Homeland. UAVs and Cruise Missiles, National Defense Research Institute RAND, p. 1 (2008)

[31] Al-Qaeda Online: Understanding Jihadist Internet Infrastructure (2006) quoted from: Jackson, B.A., Frelinger, D.R., Lostumba, M.J., Button, R.W.: Evaluating Novel Threats to the Homeland. UAVs and Cruise Missiles, National Defense Research Institute RAND, p. 12 (2008)

[32] Heidenrich, J.G.: The Cruise Missile Threat and its Proliferation (October 2006), http://www.marshall.org/pdf/materials/478.pdf

[33] Jackson, B.A., Frelinger, D.R., Lostumba, M.J., Button, R.W.: Evaluating Novel Threats to the Homeland. UAVs and Cruise Missiles, National Defense Research Institute RAND, p. 4 (2008)

[34] Majumdar, D.: Poor Man´s Air Force: Terrorists and Light Aircraft, Aviation (September 4, 2008)

[35] Gormley, D.M.: Missile Defence Myopia: Lessons from the Iraq War. Survival, 61–86 (Winter 2003-2004); Gormley, D.M.: New Developments in Unmanned Air Vehicles and Land-Attack Cruise Missiles. In: SIPRI Yearbook 2003, pp. 409–432 (2003)

[36] Hybrid Threats: Reconceptualizing the Evolving Character of Modern Conflict, by Frank G. Hoffman, No. 240, Strategic Forum, p. 5 (April 2009)

[37] Strategic Concept for the Defence and Security of the Members of the North Atlantic Treaty Organisation

[38] United States Department of Defense, Joint Concept for Logistics, August 6, 2010, p. 35.

[39] All countries are increasingly reliant on the vital communication, transport and transit routes on which international trade, energy security and prosperity depend. They require greater international efforts to ensure their resilience against attack or disruption." Strategic Concept for the Defence and Security of the Members of the North Atlantic Treaty Organisation.

[40] Pressemeldung DHL: Ein leistungsfähiges Netzwerk auch in der Krise (May 11, 2010) http://www.dp-dhl.com/de/presse/abonnements/financial_media_newsletter/hintergrund_q1_2010.html

[41] Heidenreich, J.G.: Under the Radar Screen? The Cruise Missile Threat to the U.S. Homeland. Comparative Strategy 23(1), 63–72, 70 (2004)

Political Environment as a Factor of Risk

Carlo Masala

Universität der Bundeswehr München,
Inhaber der Professur für Internationale Politik,
Werner-Heisenberg-Weg 39, 85577 Neubiberg,
Germany
carlo.masala@unibw.de

Abstract. Political environment as a factor of risk are understudied. This paper aims to cut through conceptual clarification, to define both terms and to show in which way they are connected. Some practical examples are given to highlight the theoretical considerations

Keywords: Environment, Risk.

1 Introduction

Risk seems to dominate social life nowadays. Individuals have to live with a high degree of uncertainty regarding their individual development, companies face risks steaming from societal developments and, if business is done abroad, from certain international developments. States are facing risks regarding the behavior of other states. Risk one might conclude is everywhere. Since the end of the cold war, political scientists and especially international relation scholars have discovered "risks" as an analytical category to study international relations. However, it is not clear what risk means, how the concept is defined and what the actual consequences for action are. NATO 2010 New Strategic Concept (2010) identifies a variety of risks (from inner-state conflicts to the spread of weapons of mass destruction) without clarifying why certain developments are considered as risks and not – to use a familiar term from the cold war – as threats.

A similar confusion exists with the term political environment. Although studied since decades political science scholars still do not agree on the exact use of the term political environment. Does it is refer to actual developments or is an environment characterized by an independent structure, which conditions the actions of units within this structure.

The main purpose of this paper is to contribute to a conceptual clarification of two concepts, which in social science and especially in political science are highly contested. Speaking about political environment and risk as well as about political environment as risk requires a clear-cut definition to cut through the confusion existing in the contemporary body of scholarly literature. Therefore, this paper

M. Essig et al. (Eds.): Supply Chain Safety Management, LNL, pp. 101–108.
springerlink.com

comes firstly up with discriminating definitions of both terms and discusses their connections afterwards. In a further step, some empirical examples are given to illustrate how the political environment might be considered as a risk factor. Afterwards some theoretical considerations are presented regarding potential managing strategies for risks emanating from specific political environments.

Although I am not an economist, it seems to me that considerations on political environments as risk factors for certain business activities and for supply chains are relatively new in the discipline and theoretically underspecified. Ghoshoal (1987) distinguished four categories of potential risks but without defining risk itself. Without refining this concept, Manuj and Mentzer (2008) picked his categorization of risks up and focused on possible management strategies. The Global Risk Report by the World Economic Forum (2008) gives political risks a certain, but definitely not a central role in its considerations on supply changes.

To what extent social sciences definitions can be transferred to other discipline is a contested question in itself, which luckily the author of this paper has not to deal with.

2 What Is a Political Environment?

The political environment in which units[1] do operate will have a significant impact on their activities. Usually political environments are understood as actual behavior of governments in domestic politics or relations amongst states in international politics which are influencing the actions and interactions (be they national or transnational) of units. In economy assumptions like "the greater the level of involvement in a foreign markets, the greater the need to monitor the political climate of the countries business is conducted. Changes in government often result in changes in policy and attitudes towards foreign business. Bearing in mind that a foreign company operates in a host country at the discretion of the government concerned, the government can either encourage foreign activities by offering attractive opportunities for investment and trade, or discourage its activities by imposing restrictions such as import quotas, etc. An exporter that is continuously aware of shifts in government attitude will be able to adapt export marketing strategies accordingly" (Exporthelp 2010: Chapter 6) characterize the understanding of what a political environment is.

Such an understanding of the term is a reductionist and further does not help to generalize what a political environment is and how it influences the action of its units within. It is reductionist because it seeks to explain outcomes solely through the behavior of units, leaving aside the effect their environment may have. However, it is not possible to understand politics simply by looking at units. This is because every new observed phenomenon would require the addition of new unit-level variables, which leads to the highly subjective addition and wild proliferation of variables.

[1] I use the term unit (Waltz 1979: 116) to illustrate the fact that my considerations on political environments and risk can be applied to various settings in which firms, tribes, gangs and states do operate.

Furthermore, it neglects the fact that political environments might have their own structure independently from actual processes and interactions taking place within and under such structures. Especially in economic theory, it is quite interestingly to observe the lack of a sophisticated definition of political environment as this discipline operates with the idea of a market which has independently from firms operating in such a market structures, which do regulate and sometimes even punish the firms operating under the influence of such structures (Bade and Parkin 2001).

Borrowing insights from microeconomics Kenneth Waltz in the late 1970s developed a definition of political environments, which contributes to a more scholarly understanding of a) what political environments are and b) how they influence the actions and interactions of the units.

His idea is that a system-level explanation of politics solves some of the problems mentioned afore. By focusing on the structure, or a "set of constraining conditions," (Waltz 1979: 74) of systems one is able to parsimoniously explain why dissimilar units may behave in similar ways. Structures, however, are not direct causes – they act "through socialization of the actors and through competition among them" (Waltz 1979: 74).

If we want to consider how actors will interact, we must look at the system within which they interact. As mentioned before it is not a new concept; economists look at structural constraints: political scientists argue that presidents will behave differently from prime ministers, and so on. Waltz begins by looking at domestic political structures (states) and identifies three important characteristics which create the domestic political environment. The principle by which the system is ordered, the functions that each unit fulfills and, each unit's capacity/ability to act.

In domestic political environments, the system has a hierarchical structure and a clear differentiation of functions. Governments have the monopoly on the use of force and there is a division of labor amongst the executive and legislative branches of a political system.

By analogy, Waltz extends these three principles to the international political environment. The ordering principle is anarchy: if this changed to hierarchy, inter-unit interactions would also change. In anarchy, different units exist in a self-help system: there is therefore no functional differentiation among them. So the two relevant characteristics of the international system are anarchy and relative capability.

States in the international system are like firms in a domestic economy. Every state has the same fundamental interest: to survive. Even if it wants to do other things, it can not do them unless it survives. Waltz argues that in the international political environment states are not the only but the most important actors. He recognizes that other actors exist, but they do not matter to analyze the international political environment since only those units are relevant which are in possession of the greatest capabilities (e.g. military power, economic strength and political stability). In order to justify it assumption Waltz makes use of an economic analogy: If all firms are equally sized, they all matter. If a few large firms, however, dominate the market then economic models need to focus on these. Extending this analogy means we should focus especially on states that are more powerful.

A structure is defined first by the principle by which it is ordered or organized, then by the differentiation and specification of its units, and finally by the distribution of capabilities across units.

Therefore, the ordering principal of the international political environment is anarchy, understood as the absence of central authority. The international system emerges from the "co-action of self-regarding units" (Waltz 1979: 91). In a microtheory, whether economic or political, the motivation of actors is assumed rather than realistically described. Waltz assumes that all states seek on a minimal level to ensure their survival. The real aims of states may be endlessly variable, but in a world without security, survival is the essential prerequisite and thus a useful foundation for the theory.

The second aspect of structure, the differentiation of units, is rendered unnecessary by the condition of anarchy. "Anarchy entails relations of coordination among a system's units, and that implies their sameness. [...] So long as anarchy endures, states remain like units" (Waltz 1979: 93). Which is to say they are "autonomous political units" (Waltz 1979: 95) who face similar tasks.

Waltz gives a discriminating definition of what political environments are. He distinguishes between domestic political environments and international political environments and makes – by analogy – clear what the differences between both environments are. These differences have effects on the second aspect of this paper: to what extend political environments are factors of risk. Before we turn to this, a definition of risk is required.

3 What Is Risk?

It is commonly accepted that political environments nowadays are faced with risks rather than threats. Despite this superficial agreement among scholars there is no agreement in the scholarly community what risks exactly are (McKellar 2010). Some scholars even argued that the term is used inconsistently, for example, as harm (objective harm) as well as expectations regarding the future (expected harm, whether scientifically calculated or not), other authors argue that the conceptual framework is not as blurred and unclear as often claimed (e.g Bonß 1995: 30). Sometimes risk is seen as both, a real risk and a social construction of possible harm (Beck 1992). In modern systems theory risks are understood as being constructed by attributing (expected or observed) negative outcomes to decisions (Ewald 1991: 199). What is missing in literature about risks is the fact that risks, whether they are real or just socially constructed are associated with uncertainty. And it is specifically the factor of uncertainty that distinguishes risks from threats (Daase 2002). A risk is defined for the purpose of this paper as "uncertainty about and severity of the events and consequences (or outcomes) of an activity" (Aven and Renn 2009: 6). Therefore, a political risk is a development, which has the potential to damage the interest of actors, but there is no certainty if and when it will do so. The focus on risk defined as uncertainty reconciles partly positivist and postpositivist research on risk because it places only secondary attention on the question whether risks are objective or constructed. Another distinction is necessary to better understand the concept of political risks. Political risk is of a macro nature when politically inspired changes affect all actors within an environment and they are of micro nature when the environmental changes are intended to affect only some actors.

Since risks are potentially taking place in the future and actors do want to avoid that risks turn into threats for their interest's policies to avoid the leap from risk to threat have to be proactive, meaning that actors have to take measures and design strategies for events that are probably taking place in the future. Nevertheless, this proactive attitude as Daase (2002) has shown does not go without any problems because it entails the risk to produce unintended consequences, which have not been foreseen once strategies were designed. Moreover, actors finally have to deal with the management of unintended consequences produced by their strategies.

After political risks have been defined, the question is still pending to what extend political environments are a factor of risk. This question is addressed in the next section of this paper.

4 Political Environment as a Factor of Risks

The domestic political environment poses insofar a political risk for actors as changes in the structure of the environment itself might affect actors. Secondly, changes of policies might occur which have a direct effect on certain actors within the structure. In democratic domestic environments politics are becoming more and more volatile and demand driven, therefore unforeseeable for actors within the domestic political environment. An example in case is the recent moratorium on nuclear power plants declared by the German government after the earthquake in Japan. Assurances which were given to the nuclear power plant industry after the governing coalition took office in 2009 have been revoked due to a potential risk regarding a similar accident like the one taking place in the nuclear facility in Fukushima. In authoritarian environments risks are a structural component actors operating within have to live all the time with. So on a macro level, a governmental change in democracies poses a risk to all actors within a domestic political environment while the autocratic regime environment is a constant risk for actors. Policy changes that might occur over night in democratic political environments do represent a risk for some actors within the environment. Strategies to mitigate such risks emanating from the policy do not exist since politics is driven mostly by popular demand rather than by knowledge. So the question arises to what extend actors within the environment are ready to bear risks and are able to manage the consequences. However, political risk management is more than just knowing that "bad things can happen" and preparing for damage limitation. Political risk management means enabling the fulfillment of actor objectives in even high-risk political macro and micro environments. In essence, being able to operate in high-risk environments means to be able to show extreme flexibility to accommodate shifting political currents (McKellar 2010). What is true for risks on a micro level applies to risky domestic political environments, too.

On an international level, the political environment poses a risk in itself. Due to the absence of a hierarchical order, the international political environment is characterized by constant risks. If all actors have to act and interact under the condition of a power and security dilemma in which states can not trust each other and the risk of political conflict amongst actors is omnipresent (Mearsheimer 2002). On the micro level of the international political environment risks – after the end of the cold war – have become manifold. A short list serves to characterize this tendency: [2]

Some of the constantly recurring risks that have always been part of the picture are

- International tensions
- Domestic unrest
- Terrorism
- Politically connected criminality
- Bureaucratic morass
- Ethical criticism

The last two decades have seen some significant shifts in risks:

- The rise of political Islam
- Global multi-polarity after the bi-polarity of the Cold War
- Failed and failing states due to weak governance and social fragmentation
- Global asymmetric warfare where minor (terrorist) groups can significantly harm major and more powerful opponents

Finally, future trends are likely to see

- Increased ethical criticism and focus on corporate social responsibility
- Increased confusion in inter-state tensions with a wide range of potential global disputes and power centers
- Increased exposure to civil violence, unpredictable regimes, localized conflicts and civil unrest in new "hot spots"
- Increased exposure to terrorism, accelerated by the rise of Islamist extremism, failed/failing states and new capabilities in asymmetric warfare

Common to most of these risks is the fact that we do not know if and when they materialize. Given the uncertainty about the potential developments of the above-mentioned risks, the situation in the international political environment is much more difficult to manage due to agencies, which can provide some kind of reassurance against the consequences of potential risks.

[2] It is obvious that this list does by no means claim completeness.

How can macro and micro risks stemming from the international political environment by actors be managed? Proactive behavior as shown before always entails the risk of producing unintended consequences. On the other hand, reactive strategies fall short of what is necessary in a risky international environment since they can tackle problems only when they have already occurred which in most of the circumstances (failing states, attack by terrorists with a nuclear device etc.) is far too late.

What is needed is a mixture of proactive strategies which combines elements of cooperation, preparation, compensation, and if necessary intervention. The combination of these strategies leads to a wider variety of choices although social science need to understand better the consequences of unintended consequences and how to manage them in such a way that the main objective of these strategies, namely to prevent risks from developing into threats, is not endangered.

5 Conclusion

This short conceptual piece had a threefold aim. First, it intended to present some discriminating definitions on risks and political environments. Social Science discusses both terms since quite a while but shied away from defining them in a precise manner. By referring to the seminal work of Kenneth Waltz this paper came up with a precise definition and a clear distinction of political environments. In future scholars should distinguish between a domestic (authoritarian vs. democratic) and an international political environment.

In a second step, the paper turned to a definition of risk. It has been shown that there is a debate in social sciences whether risks are objectively existent or socially constructed. The paper proposed to shift attention to the fact that risk, regardless if they are real or constructed, are characterized by uncertainty. Equally important is the distinction between risks on a macro and on a micro level. While the first one refers to risks emanating from the structure of the environment itself, the latter one refers to risks emanating from processes taking place within the structure of an environment.

Thirdly, the paper turned to potential strategies mitigating risks. It made the point that risks can't be managed entirely (neither in domestic nor in international environments) and therefore called upon a) the willingness of actors to bear risks and to show flexibility and adaptability and b) to pursue proactive strategies trying to prevent risks from becoming real threats. However, proactive strategies always entail the risk of producing unintended consequences that are not entirely understood by social scientists.

In conclusion, the paper tried to contribute to more conceptual clarification on issues, which are of real world importance but highly contested in social science literature.

References

Aven, T., Renn, O.: On risk defined as an event where the outcome is uncertain. Journal of Risk Research 12(1), 1–11 (2009)

Bade, R., Parkin, M.: Foundations of Microeconomics. Addision Wesley, Boston (2001)

Beck, U.: Risikogesellschaft. Auf dem Weg in eine andere Moderne. Frankfurt a. M, Suhrkamp (1992)

Bonß, W., Vom Risiko: Unsicherheit und Ungewissheit in der Moderne. Hamburger edn., Hamburg (1995)

Daase, C.: Internationale Risikopolitik. Ein Forschungsprogramm für den sicherheitspolitischen Paradigmenwechsel. In: Daase, C., Feske, S., Peters, I. (eds.) Internationale Risikopolitik. Der Umgang mit neuen Gefahren in den internationalen Beziehungen, Nomos, Baden-Baden (2002)

Ewald, F.: Insurance and risk. In: Burchell, G., Gordon, C., Miller, P. (eds.) The Foucault Effect: Studies in Governmentality, Harvester Wheatsheaf, London (1991)

Exporthelp (2010), `http://www.exporthelp.co.za/modules/1_considering_exporting/env_political.html#ixzz1HWYY966r` (accessed)

Ghoshal, S.: Global strategy: An organizing framework. Strategic Management Journal 8(5), 425–440 (1987)

Manuj, I., Mentzer, J.T.: Global supply chain risk management strategies. International Journal of Physical Distribution and Logistics Management 38(3), 192–233 (2008)

McKellar, R.: A short guide to political risk. Gower Publishing, Farnham (2010)

Mearsheimer, J.J.: The Tragedy of Great Power Politics. Norton, New York (2002)

Waltz, K.N.: Theory of International Politics. Addision Wesley, Reading (1979)

World Economic Forum, Hyper-optimization and supply chain vulnerability: An invisible global risk? In: Global Risks 2008, A Global Risk Network Report (2008)

Internal versus External Supply Chain Risks: A Risk Disclosure Analysis

Christoph Bode[1], René Kemmerling[2], and Stephan M. Wagner[3]

[1] Swiss Federal Institute of Technology Zurich, Senior Researcher and Lecturer, Chair of Logistics Management, Department of Management, Technology, and Economics, Weinbergstrasse 56/58, 8092 Zurich, Switzerland
cbode@ethz.ch

[2] Deutsche Bahn Management Consulting, Supply Chain Management Consultant, DB Mobility Logistics AG, Stephensonstrasse 1, 60326 Frankfurt, Germany
rene.kemmerling@deutschebahn.com

[3] Swiss Federal Institute of Technology Zurich, Professor and Director Executive MBA, Chair of Logistics Management, Department of Management, Technology, and Economics, Weinbergstrasse 56/58, 8092 Zurich, Switzerland
stwagner@ethz.ch

1 Introduction

The globalization of factor and logistics markets, developments in modern information and communications technologies, and increasingly demanding customers are just a few mega trends in the last decade. In order to cope with these challenges many firms first reengineered their internal operational and organizational processes to cut costs, increase product and service quality, and remain agile in fast changing environments. But to stay innovative and competitive many firms recognized that internal improvements are too myopic. Therefore the management of supply chains (SCM) has become very prominent since the 1980s and is now widely regarded as one of the main critical success factors and considered as a key enabler of strategic change and source of strategic advantage for organizations.

As a consequence, compared with the situation a few decades ago, modern firms collaborate differently, especially more closely, with their customers and suppliers. For example in the automotive industry, original equipment manufacturers (OEMs) increased the involvement of their suppliers in the development of new products and processes by setting up strategic alliances and joint ventures with key upstream partners. Furthermore, concepts such as just-in-time (JIT) or just-in-sequence (JIS) require very close collaboration among all players of the supply chain to prevent it from being disrupted (Wagner & Silveira-Camargos, 2012).

But as advantageous and necessary e.g., outsourcing/offshoring of manufacturing activities, low-cost country sourcing, and collaboration with international suppliers in modern supply chains are, the way of working together has exposed networked firms both qualitatively and quantitatively on a higher risk

M. Essig et al. (Eds.): Supply Chain Safety Management, LNL, pp. 109–122.
springerlink.com

level. In this context the management of risks influencing the supply chain and its members, i.e., the discipline of supply chain risk management (SCRM), gained much importance in the last years among practitioners and management scholars. However, due to the little consensus on a definition of SCM (Mentzer et al., 2001; Rossetti & Dooley, 2010), the high degree of heterogeneity among firms and the increasingly changing environment regarding legal requirements, technology, global climate, political situation, etc., ongoing research is constantly necessary to support firms achieving supply chain preparedness.

In the first step on achieving supply chain preparedness, possible causes for supply chain disruptions need to be identified and evaluated with respect to their potential damage and likelihood of occurrence. In addition it is essential for companies to create a risk classification system to group the single risks into risk categories in order to (1) decrease the complexity related to the myriad of possible risks, (2) facilitate the assignment of responsibilities, and (3) create customized risk mitigation measures for each risk class.

This contribution addresses exactly this process of risk classification in the context of supply chain risk management by following two objectives. First, we propose a simple classification system based on the work of Wagner and Bode (2008). This two-level system distinguishes on the top level between (a) *internal-driven* and (b) *external-driven* supply chain risks and on the second level between the five risk categories (1) *demand-side risks*, (2) *supply-side risks*, (3) *infrastructure and operational/production risks*, (4) *regulatory, legal, and bureaucratic risks*, and (5) *catastrophic risks*. Second, we offer first results of a current empirical panel study based on secondary data. These results not only confirm the results of Wagner and Bode (2008) by a different methodological approach but also show that the importance of internal-driven supply chain risks has increased in the last years.

This part of the book is organized as follows. In the second chapter we describe the basic terminology and introduce the supply chain classification system proposed by Wagner and Bode (2008) before adopting it to the two-level supply chain risk classification system in chapter 3. Chapter 4 describes the methodology and framework of our empirical study and selected results are presented in the fifth chapter. The last chapter summarizes our contribution and gives concluding remarks.

2 Supply Chain Disruptions, Risks and Vulnerability

Supply chain disruptions can materialize from inside or outside of a supply chain and can vary greatly in their magnitude, attributes, and effects. Consequently, their nature can be highly divergent. For instance, a delayed shipment of non-critical material is potentially a much less serious impact on the supply chain than is an eight-week labor strike at a single-sourced key supplier. In attempting to differentiate supply chain disruptions from other adverse events in business (e.g., shocks on the financial markets), many scholars have proposed classifications of supply chain disruption in the form of typologies and/or taxonomies[1] (e.g., Cavinato, 2004; Chopra & Sodhi, 2004; Christopher & Peck, 2004; Hallikas, Karvonen, Pulkkinen, Virolainen, & Tuominen, 2004; Manuj & Mentzer, 2008; Norrman & Lindroth, 2004; Spekman & Davis, 2004; Svensson, 2000). The derived categories of supply chain disruptions are usually labeled *supply chain risk sources*, in the sense of being a known source from which supply chain disruptions emerge with a certain probability.

Jüttner (2005), for instance, defined supply chain risk sources as "any variables which cannot be predicted with certainty and from which disruptions can emerge" (p. 122). In this regard, operating a production plant constitutes a risk source, because it is associated with various known risks (e.g., fire). Furthermore, the classifications cover a broad spectrum with respect to the amount of risk sources. For example, Svensson (2000) named two supply chain risk sources (quantitative and qualitative), Jüttner (2005) delineated three (supply, demand, and environmental), and Manuj and Mentzer (2008) proposed eight (supply, operational, demand, security, macro, policy, competitive, and resource).

In addition, a few taxonomies of supply chain risk sources do exist. They are, in contrast to these existing typologies empirically substantiated classifications (Bailey, 1994). Zsidisin and Wagner (2010) examined supply-side risk sources and identified supplier, supply market, and the extended supply chains as sources of risk. Wagner and Bode (2008) proposed a classification in five distinct supply chain risk sources: (1) *demand-side*, (2) *supply-side*, (3) *regulatory, legal, and bureaucratic*, (4) *infrastructure and operational*, and (5) *catastrophic*. Furthermore, while the risk sources demand-side, supply-side, and infrastructure and operational risk are internal-driven with respect to the supply chain perspective, the other two risk sources focus on issues that are rather external-driven to the supply chain. The next sections describe these five risk sources in more detail.

2.1 Internal-Driven Supply Chain Risks

2.1.1 Demand-Side Risk

Supply chain disruptions can emerge from downstream supply chain operations. These include, on the one hand, disruptions in the physical distribution of products to the end-customer which are usually associated with transportation operations, such as a truck drivers' strike (McKinnon, 2006), and the distribution network. On the other hand, demand-side supply chain disruptions can originate from the uncertainty caused by customers' unforeseeable demands (Nagurney, Cruz, Dong, & Zhang, 2005). Here, disruptions may be the results of a mismatch between a company's projections and actual demand, as well as of poor coordination of the

[1] Typologies and taxonomies are *classifications*, i.e., groupings of entities by similarity. A *typology* is theoretically constructed, while a *taxonomy* is derived from empirical data (Bailey, 1994).

supply chain. The consequences of such demand-side disruptions are costly shortages, obsolescence of stocks, poor customer service due to unavailable products or backlogs, or inefficient capacity utilization.

2.1.2 Supply-Side Risk

Firms are exposed to numerous potential supply chain disruptions stemming from the upstream side of their supply chains. Risks reside in purchasing activities, suppliers, supplier relationships, and supply networks. These risks encompass, in particular, supplier business risks, production capacity constraints on the supply market, quality problems, and changes in technology and product design (Zsidisin & Wagner, 2010).

Supplier business risks relate to disruptions that affect the continuity of the supplier and result in the interruption or the termination of the buyer-supplier relationship. This is closely linked with the threat of financial instability of suppliers, and possible consequences of supplier default, insolvency, or bankruptcy (Wagner, Bode, & Koziol, 2009). The financial default of a supplier (e.g., a supplier going out of business) is a common supply chain disruption that can have severe consequences for the buying firm. Another type of disruption can occur when a supplier is vertically integrated by a direct competitor of the customer firm, forcing the termination of the relationship (Chopra & Sodhi, 2004). In buyer-supplier relationships that involve high switching costs for the buying firm, opportunistic behavior from suppliers has also been reported to be a source of supply-side risk (Spekman & Davis, 2004; Stump & Heide, 1996).

2.1.3 Infrastructure Risk

The infrastructure risk source includes potential disruptions that evolve from the infrastructure that a firm maintains for its supply chain operations. This includes socio-technical accidents such as equipment malfunctions, machine breakdowns, disruptions in the supply of electricity or water, IT failures or breakdowns, as well as local human-centered issues (e.g., vandalism, sabotage, labor strikes, industrial accidents) that are addressed within the area of supply chain security (Lee & Wolfe, 2003; Skorna, Bode, & Wagner, 2009; Spekman & Davis, 2004).

2.2 External-Driven Supply Chain Risks

2.2.1 Regulatory, Legal, and Bureaucratic Risk

With the exception of government initiatives for security facilitation such as the Customs-Trade Partnership Against Terrorism (C-TPAT) or Authorized Economic Operators (AEO) certifications (Sarathy, 2006; Zsidisin, Melnyk, & Ragatz, 2005), little attention has been paid to supply chain risks stemming from regulatory and legal conditions. However, in many countries, authorities (administrative, legislative, and regulatory agencies) are a significant factor of uncertainty in the setup and operation of supply chains. Regulatory, legal, and bureaucratic risks refer to the legal enforceability and execution of supply chain-relevant laws, regulations, stipulations, or policies (e.g., trade and transportation laws) as well as the degree and frequency of changes in these rules. Such changes may suddenly lead to violations of (or nonconformance with) laws, rules, regulations, or ethical standards.

2.2.2 Catastrophic Risk

This class encompasses pervasive events which, when they occur, have a severe impact on the area of their occurrence. Such events can be epidemics or natural disasters, socio-political instability, civil unrest, and terrorist attacks (Kleindorfer & Saad, 2005; Martha & Subbakrishna, 2002; Swaminathan, 2003). In many regions of the world, tsunamis, droughts, earthquakes, hurricanes, and floods are a constant threat to the societies and firms located there (Munich Re, 2011). The negative consequences on supply chains are obvious, since production facilities and transportation systems are highly vulnerable to natural disasters. Due to the globalization of markets and a surge in globe-spanning supply chain operations, local catastrophes have increasingly indirect global repercussions.

3 Methodology and Data

In the following we will describe briefly our methodology, data gathering procedure, and content analysis techniques which we used to conduct our panel study.

3.1 Sampling and Time Frame

The sample selection process started with identifying all US companies listed in the Dow Jones STOXX® Americas *600 Index*[2] at the midterm of 2007, 2008 and 2009 respectively. This resulted in a total of 675 companies. Next, two filter steps were applied. First, all firms that were not listed in the index at all three considered midterm dates were excluded which reduced the data set to 422 companies. Second, we excluded all companies belonging to sectors such as media, banking, insurances, real estate, and financial services, because supply

[2] This index contains the 600 largest companies in North America and represents a market capitalization of approximately 11.9 trillion USD as of December 2009. Since its first compilation in July 2003 the index composition is reviewed on a quarterly basis and companies are replaced e.g., due to mergers & acquisitions or failing to permanently meeting the index requirements.

chain management is not a core activity in these industries. As a result, our sample for the empirical analyses contains 219 companies. Table 1 indicates that the sample covers a wide range of industry sectors and company sizes (measured by number of employees).

Table 1 Sample composition

	Count	%
Sector and industry		
Automobiles and Parts	8	1.8
Basic Resources	22	5.0
Chemicals	14	3.2
Food and Beverage	36	8.2
Healthcare	32	7.3
Industrial Goods & Services	92	21.0
Oil and Gas	64	14.6
Personal and Household Goods	38	8.7
Retail	62	14.2
Technology	70	16.0
Number of employees		
Less than 1,000	1	0.4
1,000 – 4,999	20	8.7
5,000 – 9,999	56	24.5
10,000 – 49,999	84	36.7
50,000 – 99,999	33	14.4
100,000 and more	35	15.3

Note. All values are based on data of the year 2009.

The time frame of our study covers the fiscal year 2007 and 2009 of all 219 sampled companies, i.e., in total we analyzed 438 firm–year observations.

3.2 Content Analysis

A content analysis approach was chosen, because the risks are disclosed in a qualitative fashion in item 1A of the 10-K reports; only content analysis is able to handle the quality of such information (Lajili & Zéghal, 2005). Different counting measures can be used, which include ‘word’, ‘sentence’, ‘page’, and ‘the number of lines’ (Rajab & Handley-Schachler, 2009). In this study, ‘paragraph’ was considered as basis as it is the reliable and meaningful coding unit in this type of data source.

A single coder performed the content analysis manually for this study to avoid iteration and repetition. Coding training was provided prior to the commencement of the study by one of the authors who is experienced in applying content analysis techniques. The training consisted of discussing the research objectives and the scope of supply chain risks in item 1A of Form 10-K, defining the coding scheme, and familiarizing the coder with relevant literature regarding risk disclosure, content analysis, and supply chain risk management. To support the coding process, a dedicated software tool was developed.

3.3 Data Source

In order to identify internal- and external-driven supply chain risks and as discussed in the above sections, we focus on "Item 1A: Risk Factors" of the annual 10-K report which each company in our sample has to file to the U.S. Securities and Exchange Commission (SEC) 60 days after fiscal year end closing. These filings are freely available to the public and published on the SEC website via the EDGAR database (http://www.sec.gov/edgar.shtml).

In item 1A of the Form 10-K statement, a company is required to lay out "(...) a discussion of the most significant factors that make the offering speculative or risky" (SEC, 2010, p. 443). Furthermore, "[t]he risk factors may include, among other things, the following: (1) (...) lack of an operating history; (2) (...) lack of profitable operations in recent periods; (3) (...) financial position; (4) (...) business or proposed business" (SEC, 2010, p. 443). As item 1A is the section where the complete list of relevant risks is disclosed, we refrained from looking at other sections within the 10-K report.

3.4 Examples of Disclosed Supply Chain Risk Sources

The proposed risk classification system is depicted in Table 2. In the first column, the top-level risk sources consisting of internal-driven and external-driven supply chain risks are shown. Next, the second column assigns the above described ground level risk sources. Due to the importance of demand-side and supply-side risks, we decided to create sub-categories in order to increase the level of detail in our analyses. The demand-side risk source consists of three categories: (D01) *customer default / credit risk*, (D02) *customer dependence*, and (D03) *other demand-side risks*. This fine-grained classification allows us to disaggregate the rather broad category of demand-side risks. Likewise, the supply-side risks consists of four sub-categories: (S01) *supplier default*, (S02) *supplier dependence*, (S03) *supplier quality problem*, and (S04) *other supply-side risks*.

Table 2 Risk classification schedule

Top-level	Ground-level		Reported risks as quoted in annual SEC filings of sampled companies	Firm
Internal-driven supply chain risks	D01	Customer default / credit risks	*Any difficulties in collecting accounts receivable, including from foreign customers, could harm our operating results and financial condition.*	Nvidia
			In the event that a significant pub chain were to go bankrupt, or experience similar financial difficulties, our business could be adversely impacted.	Molson Coors Brewing
	D02	Customer dependence	*The company may be adversely impacted by the increased significance of some of its customers.*	Campbell Soup
			A limited number of our customers comprise a significant portion of our revenues and any decrease in revenues from these customers could have an adverse effect on our net revenues and operating results.	Juniper Networks
	D03	Other demand-side risks	*Changes in the level of demand for our products could adversely affect our product sales.*	Southern Copper
			The long sales and implementation cycles for our products, as well as our expectation that some customers will sporadically place large orders with short lead times, may cause our revenues and operating results to vary significantly from quarter-to-quarter.	Juniper Networks
	S01	Supplier default	*We rely on business partners in many areas of our business and our business may be harmed if they are unable to honor their obligations to us.*	Electronic Arts
	S02	Supplier dependence	*We are dependent on sole source and limited source suppliers for several key components, which makes us susceptible to shortages or price fluctuations in our supply chain, and we may face increased challenges in supply chain management in the future.*	Juniper Networks
	S03	Supplier quality problem	*We outsource some of our manufacturing. If there are significant changes in the quality control or financial or business condition of these outsourced manufacturers, our business could be negatively impacted.*	Avery Dennison
	S04	Other supply-side risks	*Fluctuations in commodity prices and in the availability of raw materials, especially feed grains, live cattle, live swine and other inputs could negatively impact our earnings.*	Tyson Foods
			We depend on contract growers and independent producers to supply us with livestock.	Tyson Foods
	I0	Infrastructure and operational/ production risks	*The company may be adversely impacted by inadequacies in, or failure of, its information technology system.*	Campbell Soup
			Product liability claims could adversely impact our financial condition and our earnings and impair our reputation.	Medtronic
External-driven supply chain risks	R0	Regulatory, legal, and bureaucratic risks	*The company's results may be impacted negatively by political conditions in the nations where the company does business.*	Campbell Soup
			Our industry is experiencing greater scrutiny and regulation by governmental authorities, which may lead to greater governmental regulation in the future.	Medtronic
	C0	Catastrophic risks	*Global or regional catastrophic events could impact our operations and financial results.*	Coca Cola
			Military action, other armed conflicts, or terrorist attacks.	Halliburton

Finally, for each risk source, Table 2 provides some text examples as they appeared in the reports. These examples clarify the notion of the classification system and the content of each risk source more clearly.

4 Analysis and Results

In the following we present selected results based on the analysis of the risk disclosure in the 2007 and 2009 fiscal year-end filings (Form 10-K) of 219 U.S. companies.

In total, we identified 2.473 distinct risks disclosed in the 2007 annual risk reporting and 3.001 risks in 2009 which reflects an increase of 21.4%. Thus, on average, the firms in our sample reported 11.29 risks in 2007 and 13.70 risks in 2009. This corresponds to an increase of 2.41 risks per firm ($p < 0.001$; two-tailed paired-sample t-test). This development comes along with the negative influence of the financial crisis in 2009. Based on the ground level of our classification system, Figure 1.A shows the frequencies for each of the five risk sources.

Figure 1.A: Frequencies of ground-level risk sources in 2007 and 2009 **Figure 1.B: Relative distribution of ground-level risk sources in 2007 and 2009[a]**

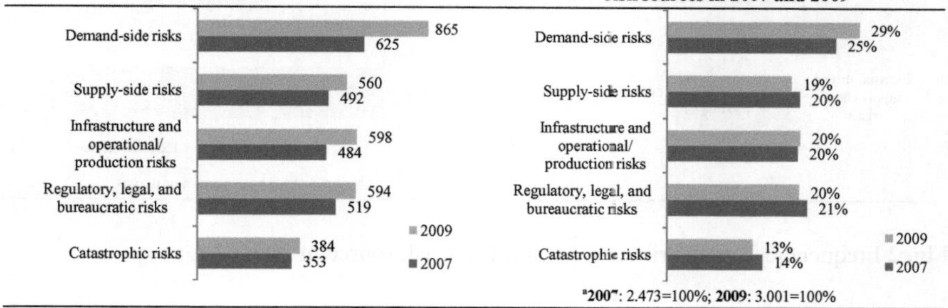

Fig. 1 Frequencies and distribution per ground-level risk source in 2007 and 2009

The amount of disclosed risks in all five categories increased within the two year time-window. However, in relative terms, the changes differ highly among the various risk sources. While the increase for catastrophic risks (+8.8%, $p < 0.05$), regulatory, legal, and bureaucratic risks (+14.5%, $p < 0.001$), and supply-side risks (+13.8%, $p < 0.01$) is rather small and below average (+21.4%), the other two risk sources raised at a higher degree. Demand-side risks (+38.4%, $p < 0.001$) and infrastructure and operational/production risks (+ 23.6%, $p < 0.001$) were reported significantly more often in 2009 than in 2007. This trend might reflect the greater emphasis companies put on the downstream part of their supply chain while at the same time knowing that risks coming from this source can have severe negative impacts for the entire enterprise.

Further, we examined the relative importance of each ground-level risk source. Figure 1.B unveils that the exposure to demand-side risks is not only of highest concern to companies. The weight of this risk source has even increased from 25% in 2007 to 29% in 2009. At the same time, catastrophic risks remained on the lowest awareness level. This confirms the survey-based study of Wagner and Bode (2008) in which companies assigned the least importance to the latter risk source and the highest to demand-side risk sources. The other three sources, i.e., supply-side risks, infrastructure and operational risks and regulatory, legal, and bureaucratic risks share the remaining weights equally and stay constant within the analyzed time window.

Aggregating the empirical data to the top level of our classification system indicates that the number of internal-driven supply chain risks increased by 26.4% ($p < 0.001$) from 2007 to 2009 whereas external-driven supply chain risks increased only by 12.2% ($p < 0.001$) (Figure 2.A). Figure 2.B shows the relative distribution of the two top-level categories and illustrates that internal-driven supply chain risks have slightly increased their weight within the companies' risk portfolio.

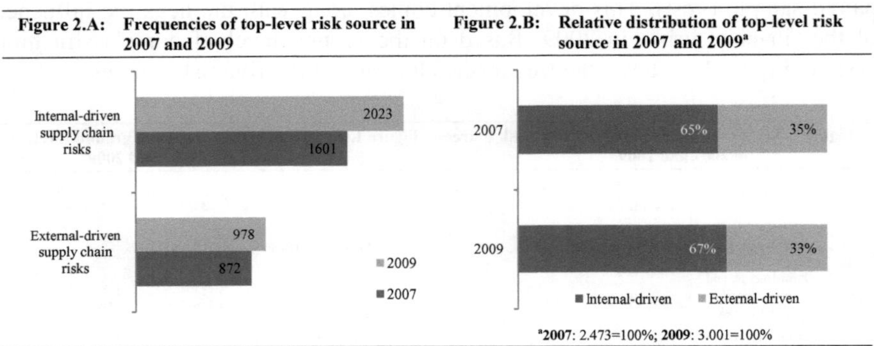

Fig. 2 Frequencies and distribution per top-level risk source in 2007

As described above, we set up our coding schedule in order to unveil more details for the demand and supply-side risk sources. The more detailed view on the data is visualized in Figure 3 and 4 respectively.

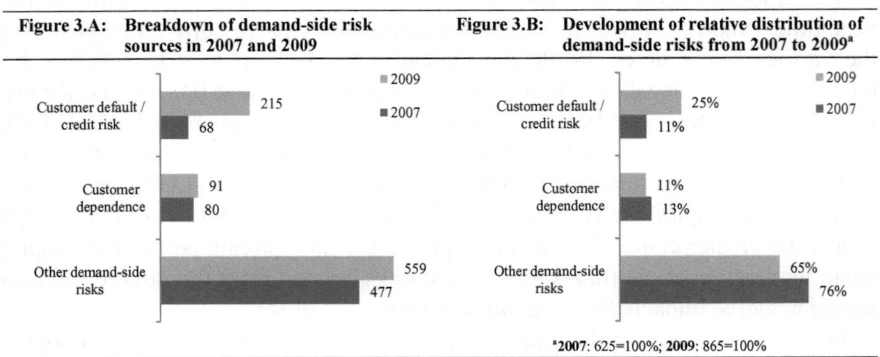

Fig. 3 Demand-side risk sources in 2007 and 2009

Disaggregating the demand-side risks into the categories (1) *customer default and credit risks*, (2) *customer dependence*, and (3) *other demand-side risks* shows that in all three categories the number of reported risks increased from 2007 to 2009 (Figure 3.A). Especially the risks related to the default of customers and customers' inability to pay their obligations experienced an enormous increase by 216% ($p < 0.001$). This trend reflects that companies, even large multinational firms, became more sensitive toward the default of single customers. In fact, 148 of the 219 analyzed companies (i.e., 67.6%) mentioned customer default risk in their 2009 risk reporting whereas in 2007 only 58 companies (26.5%) reported this specific risk. Figure 3.B highlights that, within the demand-side risk sources, more emphasis is put on the risk of customer default in 2009 than in 2007. In 2009 every fourth disclosed demand-side risk is related to the default of customers and the related credit risk whereas in 2007 only 1 out of 9 risks were reported in this category. Additionally the number of risks rooted in the dependence on customers increased slightly in absolute terms but lost share by two percentage points.

Figure 4.A: Breakdown of supply-side risk sources in 2007 and 2009 **Figure 4.B: Development of relative distribution of supply-side risks from 2007 to 2009**[a]

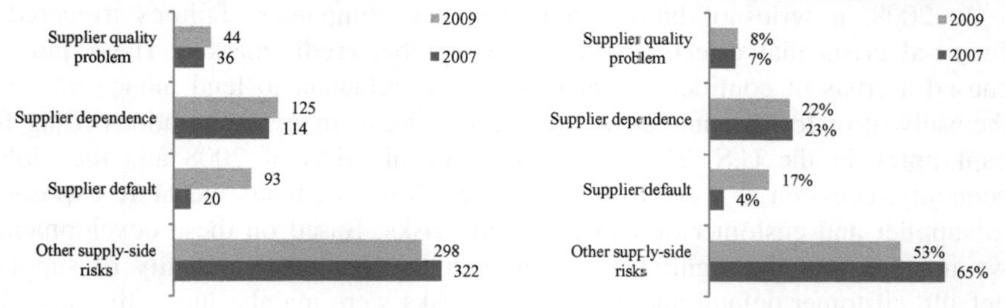

[a]2007: 492=100%; 2009: 560=100%

Fig. 4 Supply-side risk sources in 2007 and 2009

Looking on the upstream part of the supply chain shows that the absolute increase is caused by an increase in the categories (1) *supplier quality problem*, (2) *supplier dependence*, and (3) *supplier default* whereas the (4) *other supply-side risks* decreased slightly at the same time (Figure 4.A). Within this risk source default risks show as well the highest intensification by 365%. Furthermore, only 17 companies reported the risk of supplier default in their 2007 reporting while 80 companies do so in 2009. Although this trend reflects a greater awareness by companies for the negative consequences of supplier defaults, this empirical insight also indicates, that 63% of the companies still not report supplier default in their annual reporting. Analogue to Figure 3.B, Figure 4.B illustrates the mix of reported upstream risks. It is evident that supplier quality problems and the dependence on suppliers remain constant in their significance whereas the supplier default gained weight from the other supply-side risks category.

5 Discussion and Conclusion

The objective of this research was to empirically investigate the supply chain risk disclosures in 10-K reports of U.S firms. Following a content analysis, this study describes and analyzes supply chain risk disclosures of 219 U.S companies over 2 years by summarizing and classifying disclosed supply chain risk related information. Besides legally required, information on the risk situation is mainly demanded by shareholders, potential investors, and other stakeholders such as employees to access and appraise the future performance of the company. Therefore the risk disclosure in annual reports became the main risk communication between firms and outsiders. Referring to our empirical evidence which shows an increased quantity of supply chain risk disclosures from 2007 to 2009 in all five risk sources, i.e., supply-side and demand-side risks, infrastructure and operational/production risks, regulatory, legal, and bureaucratic risks, and finally catastrophic risk, companies increase the amount of information disclosed with regard to the risks faced and their expected impact on future profits in order to more effectively fulfill these demands (Beretta & Bozzolan, 2004). The increasing trends for risk disclosures are consistent with previous researches, e.g., Kajüter and Winkler (2003) and Fischer and Vielmeyer (2004).

In 2008, a series of banks' and insurance companies' failures triggered a financial crisis that effectively halted the global credit market. These failures caused a crisis of confidence that made banks reluctant to lend money amongst themselves, or for that matter, to anyone leading to many corporations filing for bankruptcy in the U.S. Therefore, the financial crisis of 2008 and the global economy recession as a consequence drove a lower demand and more exposures to supplier and customer default and credit risks. Based on these developments, we inferred that the significant increase in the disclosure quantity of supplier default, customer default and demand-side risks were mainly due to the financial crises of 2008.

References

Bailey, K.D.: Typologies and Taxonomies: An Introduction to Classification Techniques. Sage, Thousand Oaks (1994)

Beretta, S., Bozzolan, S.: A framework for the analysis of firm risk communication. International Journal of Accounting 39(3), 265–288 (2004)

Cavinato, J.L.: Supply chain logistics risks: From the back room to the board room. International Journal of Physical Distribution & Logistics Management 34(5), 383–387 (2004)

Chopra, S., Sodhi, M.S.: Managing risk to avoid supply-chain breakdown. Sloan Management Review 46(1), 53–61 (2004)

Christopher, M., Peck, H.: Building the resilient supply chain. International Journal of Logistics Management 15(2), 1–13 (2004)

Fischer, T.M., Vielmeyer, U.: Analyse von Risk Disclosure Scores: Risikoorientierte Unternehmenspublizität der DAX 100-Unternehmen. Zeitschrift für Internationale und Kapitalmarktorientierte Rechnungslegung 4(11), 459–474 (2004)

Hallikas, J., Karvonen, I., Pulkkinen, U., Virolainen, V.-M., Tuominen, M.: Risk management processes in supplier networks. International Journal of Production Economics 90(1), 47–58 (2004)

Jüttner, U.: Supply chain risk management: Understanding the business requirements from a practitioner perspective. International Journal of Logistics Management 16(1), 120–141 (2005)

Kajüter, P., Winkler, C.: Praxis der Risikoberichterstattung deutscher Konzerne. Die Wirtschaftsprüfung 3(5), 217–228 (2003)

Kleindorfer, P.R., Saad, G.H.: Managing disruption risks in supply chains. Production & Operations Management 14(1), 53–68 (2005)

Lajili, K., Zéghal, D.: A content analysis of risk management disclosures in Canadian annual reports. Canadian Journal of Administrative Sciences 22(2), 125–142 (2005)

Lee, H.L., Wolfe, M.: Supply chain security without tears. Supply Chain Management Review 7(1), 12–20 (2003)

Manuj, I., Mentzer, J.T.: Global supply chain risk management. Journal of Business Logistics 29(1), 133–155 (2008)

Martha, J., Subbakrishna, S.: Targeting a just-in-case supply chain for the inevitable next disaster. Supply Chain Management Review 6(5), 18–23 (2002)

McKinnon, A.: Life without trucks: The impact of a temporary disruption of road freight on a national economy. Journal of Business Logistics 27(2), 227–250 (2006)

Mentzer, J.T., DeWitt, W., Keebler, J.S., Min, S., Nix, N.W., Smith, C.D., Zacharia, Z.G.: Defining supply chain management. Journal of Business Logistics 22(2), 1–25 (2001)

Re, M.: Natural Catastrophes 2010: Analyses, Assessments, Positions. Munich Re Publications, Munich (2011)

Nagurney, A., Cruz, J., Dong, J., Zhang, D.: Supply chain networks, electronic commerce, and supply side and demand side risk. European Journal of Operational Research 164(1), 120–142 (2005)

Norrman, A., Lindroth, R.: Categorization of supply chain risk and risk management. In: Brindley, C. (ed.) Supply Chain Risk, pp. 14–27. Ashgate, Hampshire (2004)

Rajab, B., Handley-Schachler, M.: Corporate risk disclosure by UK firms: Trends and determinants. World Review of Entrepreneurship, Management and Sustainable Development 5(3), 224–243 (2009)

Rossetti, C.L., Dooley, K.J.: Job types in the supply chain management profession. Journal of Supply Chain Management 46(3), 40–56 (2010)

Sarathy, R.: Security and the global supply chain. Transportation Journal 45(4), 28–51 (2006)

SEC, 17 CFR § 229.503 (Item 503): Prospectus summary, risk factors, and ratio of earnings to fixed charges, Securities and Exchange Commission, Washington, D.C. (2010)

Skorna, A.C.H., Bode, C., Wagner, S.M.: Technology-enabled risk management along the transport logistics chain. In: Wagner, S.M., Bode, C. (eds.) Managing Risk and Security: The Safeguards of Long-Term Success for Logistics Service Providers. Haupt, Bern (2009)

Spekman, R.E., Davis, E.W.: Risky business: Expanding the discussion on risk and the extended enterprise. International Journal of Physical Distribution & Logistics Management 34(5), 414–433 (2004)

Stump, R.L., Heide, J.B.: Controlling supplier opportunism in industrial relationships. Journal of Marketing Research 33(4), 431–441 (1996)

Svensson, G.: A conceptual framework for the analysis of vulnerability in supply chains. International Journal of Physical Distribution & Logistics Management 30(9), 731–749 (2000)

Swaminathan, J.M.: SARS exposes risks of global supply chains. Journal of Commerce 4(23), 38 (2003)

Wagner, S.M., Bode, C.: An empirical examination of supply chain performance along several dimensions of risk. Journal of Business Logistics 29(1), 307–325 (2008)

Wagner, S.M., Bode, C., Koziol, P.: Supplier default dependencies: Empirical evidence from the automotive industry. European Journal of Operational Research 199(1), 150–161 (2009)

Wagner, S.M., Silveira-Camargos, V.: Managing risks in just-in-sequence supply networks: Exploratory evidence from automakers. IEEE Transactions on Engineering Management 59(1), 52–64 (2012)

Zsidisin, G.A., Melnyk, S.A., Ragatz, G.L.: An institutional theory perspective of business continuity planning for purchasing and supply management. International Journal of Production Research 43(16), 3401–3420 (2005)

Zsidisin, G.A., Wagner, S.M.: Do perceptions become reality? The moderating role of supply risk resiliency on disruption occurrence. Journal of Business Logistics 31(2), 1–20 (2010)

3 Supply Chain Protection

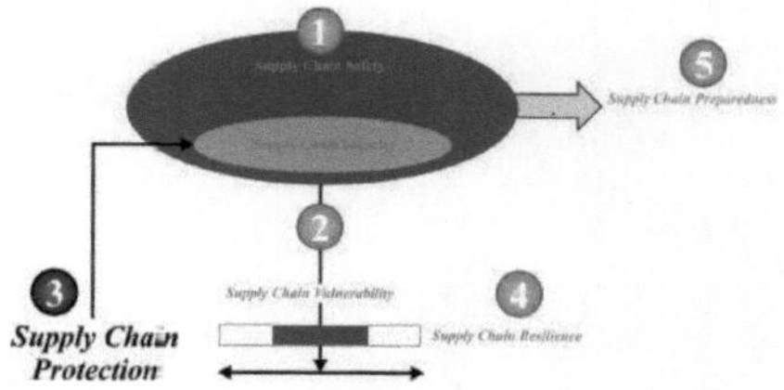

The Secure Process Chain in Aviation Security

Gerhard Wirth

Flughafen München GmbH, Leiter Servicebereich Security, Postfach 23 17 55,
85326 München, Germany
gerhard.wirth@munich-airport.de

1 Introduction

At first it may seem strange when the subject of air security comes up when discussing secure supply chains. Consequently, the question is: What does aviation security have to do with logistics? The answer is quite simple: Regardless of which handling process is involved – "Passenger", "Baggage", "Cargo", or "Maintenance" – at some point it is always interrupted by aviation security, namely whenever the process in question crosses the secured boundary between the public area and the security area of an airport. As a result, aviation security makes a major contribution to the security of these process chains, but at the same time represents a barrier and a challenge, extending from inconvenience to passengers to the impossibility of accurately scheduling processes. The best example at present is the extensive screening of passengers (removal of belts and shoes) to the intensive, time-consuming monitoring of cargo that is making it increasingly difficult to achieve just-in-time delivery.

2 Laws and Regulations Governing Airport Security

<u>Definition of terms</u>

First we must define what we mean by "aviation security".

Varying interpretations of terms such as airport security and aviation security come up on a regular basis. Airport security refers to the responsibility of the airport operator at the airport and the related measures with regard to construction/technology and personnel/organization. Aviation security generally refers to the overall security system in the aviation industry with various involved parties, above all the airport operator, the airlines and the security authorities. Unfortunately, these terms are often used interchangeably even in specialized articles and commentaries.

Also not to be left out of any discussion of "secure process chains", of course, are the tasks of the Federal Police (entry and exit controls) and the customs authorities, which, like all other security processes, also represent barriers in the processes.

Another distinction must be made for air traffic control, whose important role, strictly speaking, belongs to the safety side of aviation, as opposed to security as such.

This article deals primarily with security concerns.

M. Essig et al. (Eds.): Supply Chain Safety Management, LNL, pp. 125–133.
springerlink.com

Policy framework for aviation security

Before the attacks on September 11, 2001, the issue of aviation security was dealt with in Europe mainly at the national level, although of course on the basis of international treaties. After 9/11 the EU rapidly seized the initiative and began to implement uniform aviation security regulations at the EU level. This led to upheavals in various member states, including Germany, the effects of which can still be felt today, and which affect in particular various process chains.

The approach of the EU is a manifestation of the general desire to bring about comprehensive regulations for all aspects of air security and allocate tasks to the authorities of the member states as well as airlines and airport operators. In doing so, the EU does not specify the responsibilities of ministries or authorities. These responsibilities are regulated at the national level, which results in noticeable differences for the passenger within Europe and even within Germany in the assumption of responsibility for controls and their execution – a circumstance that occasionally is met by considerable incomprehension, as indicated below.

In Germany, the concerns of aviation, including aviation security were previously regulated in the Air Traffic Act (LuftVG) and were placed under the jurisdiction of the Federal Ministry of Transport (BMVBS), with the Federal Ministry of the Interior (BMI) acting in advisory capacity with regard to material security requirements. At the international level, the two ministries acted jointly. As a result of the new EU regulations, changes had to be made to the legal groundwork in Germany. This led to the enactment of the Aviation Security Act, which came into force in 2005. The Ministry of Transport retained responsibility in principle for aviation, but the Ministry of the Interior is now designated as the highest authority for aviation security, and in turn utilizes the Federal Police to execute this role. Consequently, the two ministries discharge the responsibilities mainly by mutual agreement. Regrettably, the responsible parties lacked the courage to take decisive action to transfer these responsibilities, so that a "residual jurisdiction" with regard to aviation security has remained with the Federal Aviation Office (LBA), an authority that reports to the Ministry of Transport. This unnecessarily complicates the official coordination and monitoring of security processes.

But it is not only the frictional losses between public authorities that impede the aviation security process in Germany, and thus all handling processes. Under the federalist structure of the EU member state Germany, the country's federal states were already granted far-reaching competencies under the Air Traffic Act, and were specifically granted legal decision-making authority. This means that the implementation of statutory measures – regardless of whether they are EU regulations or national laws – and the monitoring of the implementation by the responsible parties such as airlines and airports – is left to the federal states (with regard to airport supervision) and the Federal Aviation Office (for airline supervision). The degree to which the EU has now imposed EU-wide regulations for aviation security, with the member states more or less under the control of the EU, has increasingly caused the federal government to become sandwiched between pressures exerted from below by federal states, with their extensive authority, and from above by the EU – a situation that will continue to intensify. Conflicts of interest are inevitable and evident, and naturally impact the processes listed above. In countries with centralist systems of government such as France, there is clarity at least in this regard.

Additional policy framework for cross-border traffic

In cross-border traffic, we now distinguish three cases:

- EU / Non-EU
- Schengen / Non-Schengen
- Clean / Unclean

What does this mean?

EU / Non-EU

The "oldest" case is the EU / non-EU situation. Bilateral treaties were signed as far back as the era of the European Economic Community to regulate imports and exports between individual states and allow easier or even free movement of goods. Today the movement of goods is regulated between most EU member states. These states fall into the "EU" category. The situation is different, of course, for states with which there is no agreement on the free movement of goods. They are categorized as "non-EU" states. Residents of EU states experience this distinction at airport duty-free shops. They are permitted to make duty free purchases on flights to and from non-EU states.

The task of the airport operator is now to provide space for the requirements of the customs authorities while ensuring that passenger traffic, particularly for flights arriving from non-EU countries, is routed through the airport so as to make it impossible to avoid passing through customs screening. This is of vital importance, for instance in the case of connecting flights.

Schengen / Non-Schengen:

Similar to cargo traffic, the free cross-border movement of travelers was also later regulated in treaties between EU states. Schengen, a small winemaking town in Luxembourg became world famous in 1985 when five states signed the agreement that bears its name. In the meantime the majority of EU member states have signed the agreement, although it has not yet been implemented by Bulgaria Romania and Cyprus. Another country that has not fully implemented the terms of the agreement is the non-EU state of Liechtenstein. This brings up a further notable aspect, namely the fact that signatories to the Schengen Agreement do not necessarily have to belong to the EU. One such example is Norway.

As a result, so-called non-Schengen passengers have to clear passport control when entering and leaving Schengen countries. Here, too, it is the responsibility of the airport operator – as in the case of customs – to provide the necessary space and traffic flow arrangements.

Clean / Unclean:

Another case is the classification of countries into the categories "clean" and "unclean". When EU-wide regulations were put in place for aviation security after 9/11, the EU expected the defined security standards to be applied in all member countries. The EU ensures that this is the case in cooperation with the member countries through airport inspections conducted by both the EU Commission and the states themselves to assess the implementation of the imposed measures. If serious security deficiencies are found, the airport in question is subject to sanctions and classified as "unclean". It is thus placed in the same category as airports in non-EU countries that fail to meet EU standards.

For airports, this means that departing or connecting passengers classified as secure (i.e. "clean") under EU standards cannot be mixed with non-secure (i.e. "unclean") arriving passengers. Compliance with this requirement must be ensured by the airport operator by setting up appropriate routing.

In this regard, the treatment of flights from Israel and the USA seems particularly illogical. These flights are classified as unclean under the EU definition because the security standards in these countries, although very stringent, are not fully compliant with those of the EU. Nevertheless the EU has entered into bilateral negotiations with other countries to determine how they can meet the EU standards to improve passenger convenience. A priority for the EU in these efforts is its relationship with the USA.

Legal framework in aviation security

As mentioned above, the events of September 11, 2001 led to major changes in the overall conditions and the regulatory environment. Previously the responsibilities were regulated and allocated in the Air Traffic Act, particularly in Sections 19b (Security obligations of airport operators), 20a (Security obligations of airlines) and 29c (Responsibilities of aviation authorities). With Regulation 2320/2002 in 2004 and subsequently with Regulation 300/2008 (which succeeded 2320/2002), the EU Commission created an EU-wide regulatory framework for aviation security. The member states were required to respond and implement the regulation in national law. As a result, the German government passed the Aviation Security Act, which came into force on January 15, 2005. The statutory regulations were removed from the Air Traffic Act and transferred, with additions, to the Aviation Security Act. The "Gospel" of aviation safety now consists of the new Section 5 (Security obligations of aviation safety and security authorities, formerly regulated in Section 29c, LuftVG), Section 8 (Security obligations of airports, formerly Section 19b, LuftVG) and Section 9 (Security obligations of airlines, formerly Section 20a, LuftVG). The regulations are largely unchanged in terms of contents.

There is a fatal flaw here, however: Unlike the member states, where the standards are codified in laws, the responsible EU Commission can amend, supplement or replace its regulations at very short notice. For example, the new EU Aviation Security Regulation 300/2008 came into force in 2008 and replaced the "old" regulation 2320/2002. Various provisions of the legislation were amended, and many new points were added.

Because of the complex legislative processes (coordination between ministries, involvement of the Bundestag and the upper house of the legislature, in which the states have a voice), the member state Germany lags far behind with the implementation of these EU regulations in national law. Conversely, EU law is directly applicable, which results in disruptions and frictional losses within the process chains.

3 Aviation Security – Participating Organizations

As mentioned above, the Aviation Security Act regulates the responsibilities for certain aviation security measures. For airports and airlines the situation is relatively clear. Under sections 8 and 9 of the act, both are responsible for so-called "own security". This involves both structural/technical aspects and personnel/organizational measures. Concretely, it involves all aspects of security in operational areas, access regulations and controls, searches of airport employees, screening of carried items and incoming shipments of goods and also, in the case of airlines, the responsibility for freight security. As a rule, both airports and airlines utilise other service providers to implement the various measures. In the performance of these activities the airports are monitored by the aviation authorities of the responsible states, while airlines are monitored across all states by the Federal Aviation Office.

To a large extent, the situation is made more complex and complicated by the varying allocation of responsibilities on the part of the aviation security authorities. For example, passenger and baggage screening is primarily the responsibility of the individual states. However, most of them have abrogated this responsibility to the federal government, and specifically the Ministry of the Interior, under administrative agreement. The various tasks fall under the jurisdiction of the Federal Police, which in turn have contracted them out to various service providers through tender bidding processes.

But that is not the case in Bavaria, for instance, where the Bavarian state government has retained responsibility for passenger and baggage screening. The responsible aviation authority is the Bavarian Ministry of Economy, Infrastructure, Transport and Technology, while the District Government of Upper Bavaria has jurisdiction over Munich Airport, with the relevant tasks assigned to the Southern Bavarian Aviation Office, which implements these tasks with the aid of a state-owned company.

Thus we see that the aviation security landscape shows considerable diversity, and that jurisdictions and responsibilities are not always immediately clear and recognizable.

In addition to the aviation security tasks outlined here, there are naturally others that arise as a result of different legal requirements. Two examples that should be mentioned here are the Federal Police, which handles entry and exit controls for arriving and departing passengers as well as various security roles, and the customs authorities, which carry out import and export controls on carried goods.

The following chart shows the parties involved in aviation security, the legal basis and the responsibilities for official monitoring.

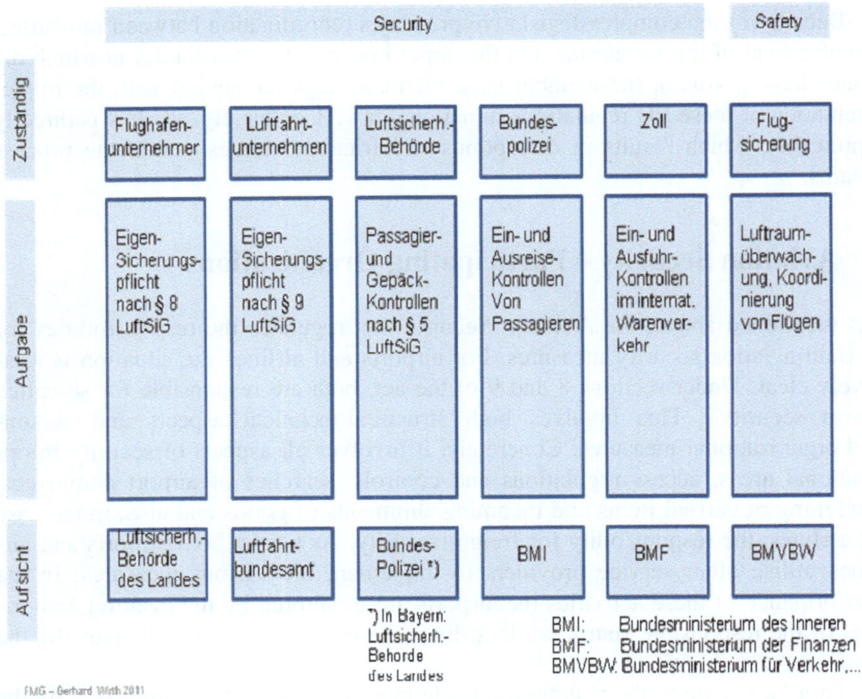

4 Sample Cases of Process Chains

Departing passengers:

In terms of process chains, the departing passenger is the easier case to grasp. The passenger who books a flight must first check in. A relevant factor at this point is whether or not the passenger is only carrying hand baggage or is traveling with check-in baggage. Within the EU, in most cases check-in baggage is now X-rayed in a multi-stage baggage screening system and examined by security staff, and thus made "secure".

The passenger must initially move from the public to the secure area. To permit access to the secure area, the passenger receives a boarding card when checking in. This is no longer exclusively in paper form, but can now be issued to mobile telephones as well. With the boarding card the passenger passes through boarding card screening, for which each airport is responsible in Germany. Depending on the configuration and structure of a terminal, there may be separate screening stations for this purpose (as in Frankfurt), or the airport operator may find a service provider, who jointly performs this screening task with security screening (as in Munich).

Depending on the architecture and functional layout of a terminal, the separation of Schengen and Non-Schengen passengers takes place on entry to the secure area, i.e. in this case the Federal Police are positioned in advance of the passenger and hand baggage screening. This is the set-up in Terminal 1 at Munich Airport, for example. Here the terminal is structured so that an entire waiting area is set aside exclusively for departing Non-Schengen passengers. A passenger that undergoes passport control is deemed to have left the country.

Next the passenger, along with any accompanying hand baggage, passes through the security checks. If the screeners reject the passenger and do not allow him to pass through the screening station, he must then pass through immigration control once again. This meets with incomprehension on the part of passengers, so that complications in the process chain are inevitable.

Due to various incidents and the resulting changes to legal requirements, passenger screening has become highly sophisticated. For example, at one time there was a requirement to weigh or switch on laptops or photographic equipment at passenger screening checkpoints to determine whether the devices had been manipulated. The widely varying ways of performing these checks at German and other European airports led to considerable annoyance among passengers. This was the result of not entirely uniform implementation standards in the federal states and the European member states.

Another example is the different approaches to shoe screening within the EU. At some airports, additional shoe scanners are used, while at others the passengers must remove their shoes and some airports do not carry out special shoe screening at all. This results in confusion and irritation among passengers.

A third example are the amended screening standards for liquids. When the security authorities became aware that certain liquids could be mixed under certain circumstances to produce highly volatile, spontaneously combustible compounds, suitable screening processes had to be developed because the conventional X-ray equipment was unsuitable. The new check placed a considerable additional strain on the screening process, which led to long waiting and processing times at airports in Germany and the EU, which varied depending on how the responsible security authorities were organized.

At this point we should not omit to mention that this example impressively demonstrates that the objectives of policymakers and the capabilities of technology are not always in harmony. After the foiled attacks in London in the summer of 2006, the EU initially banned passengers from carrying liquids (with the exception of a maximum of one liter of liquids in a resealable plastic bag, with each individual container of liquid not exceeding 100 ml; the author mentions this because passengers took an extremely long time to become accustomed to this complicated regulation. Older passengers in particular continually found themselves in extremely upsetting situations at screening checkpoints, in which they faced severe reproaches for carrying liquids in their carry-on luggage, and sometimes found themselves being treated practically like criminals.

This initial regulation was followed by intensive discussions between the EU Commission and the manufacturing sector in which the EU Commission urged the development of better detection technologies. As a result, the screening of liquids carried by passengers, with no quantity limits, was introduced for connecting passengers on April 29, 2011, and is to be extended to all departing passengers as of April 29, 2013.

Experience has shown, however, that the development of such complicated and complex technologies cannot be forced into a timeframe specified by the authorities. Otherwise there is the danger that devices with poor performance capabilities will reach market. The responsible authorities would then have to make substantial investments while accepting the risk that this deficient technology would have only limited applications and, in a worst case scenario, would have to replaced before being fully amortized.

But let us return to the process chain for the departing passenger. A different arrangement of the process chain is possible, namely with passenger and hand luggage screening positioned at the beginning. This model can also be found at Munich Airport, namely in Terminal 2. After all departing passengers have undergone passenger and hand luggage screening, the passenger flow is then divided and routed to two different levels. The departing "Schengen" passengers remain on Level 4 (the same level where passenger screening takes place), and the "Non-Schengen" passengers go to the level above, where exit screening (i.e. passport control) is carried out.

It is thus at the discretion of the airport operator how a terminal is laid out and operated in functional terms, and thus how the process chains are set up. Factors in these decisions are the desire to create transparent processes and the business considerations of the responsible parties.

<u>Arriving passenger</u>

Processes for arriving passengers are much more complex and complicated. Here we must distinguish between passengers arriving from Schengen or Non-Schengen countries, whether or not they are arriving from an EU country, or even from a country classified as "unclean".

Because the "clean vs. unclean" distinction is the most recent case differentiation in the EU, most airports in the EU had varying degrees of difficulty with the implementation. It added yet another dimension to the previously two-dimensional "security landscape", which only had to deal with the "Schengen/Non-Schengen" and "EU/Non-EU" cases, and was thus relatively simple.

The following chart describes the possible combinations of cases that may arise:

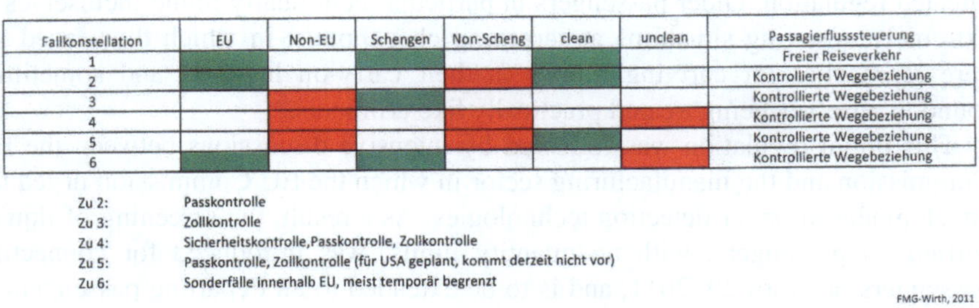

What consequences have resulted from this new EU requirement? In simple terms, as described above, the EU distinguishes between "clean" countries, which meet the security standards of the EU, and "unclean" countries, which do not, and assumes that all airports within the EU are "clean".

For flights from an EU airport to a non-EU country, this is not a problem. In the reverse situation, however, the regulations require EU countries to handle an aircraft coming from a non-EU country, along with its passengers, in such a way that they can be considered "cleared for entry" and "clean" to board a connecting flight within the EU (the principle of unrestricted freedom of travel within the EU). That means that in such a case the passenger and baggage flows must be separated into disembarking, i.e. passengers whose journey will end at this airport, and transit passengers, who must undergo appropriate screening processes in accordance with EU standards before catching a connecting flight. The same applies to baggage.

The available resources were adequate to cope with the previous cases, namely "EU/Non-EU" and "Schengen/Non-Schengen". However, the new "clean/unclean" distinction brought new complications to the processes, and thus placed a serious strain in resources. At Munich Airport this development required construction work to expand Terminal 2, which had started operating only in 2003. Another level was built on top of the terminal pier. It was opened in 2009 and serves as a traffic distribution corridor through which the "unclean" passengers are now routed. The project required investment costs of approx. €60 million. One of the main reasons for the investment was the strong upward trend in US traffic at that time.

Since spring 2011 the EU has recognized the US security procedures, so that the passengers can at least be treated as "normal" EU passengers on entry. However, this also means that the investments described above are in principle superfluous, as the remaining "unclean" passenger traffic can be handled with the existing resources. This is an example which, on the one hand, leads to cost/benefit considerations and, on the other, shows that the authorities deciding on the legal regulations are not necessarily accountable for the resulting economic effects.

The cases examined here show that a passenger airport is a highly complex and heterogeneous structure that reacts with great sensitivity to changes induced by laws and regulations. To be able to manage such an operation, airports use management systems which feed all of the above-mentioned organizational units. This is the only way of ensuring the "just-in-time" positioning of personnel and equipment to handle passengers and luggage, at the right time and in the right location, and thus ensuring a functional process chain.

Protection of Buildings

Norbert Gebbeken

University of the Bundeswehr München, Director, Institute of Engineering Mechanics and Structural Mechanics, Werner-Heisenberg-Weg 39, 85577 Neubiberg, Germany
norbert.gebbeken@unibw.de

Abstract. This chapter deals with the safety of critical built infrastructures against multiple threats due to natural disasters, technical disasters or terrorist attacks. Critical built infrastructures are explained, and a short description of the loading scenarios is given. Today's possibilities to assess and to design critical built infrastructures are presented. This applies for existing infrastructure as also for new that has to be designed.

1 Introduction

The operational capability of a society is based on multiple requirements. One of these requirements is an intact critical built infrastructure which comprises **geotechnical buildings** like dams, dykes, channels, reservoirs, foundations, instable slopes etc, **structures for supply and disposal** like channels, pipe installations, landfills, water treatment, water supply, sewage plant, etc, **buildings for generating and transport of energy** like water reservoir, wind generator, off shore structures, pressure tunnels, masts etc, **governmental-, industrial buildings,** critical manufacturing, banks, community centers, historical buildings, court houses, hospitals, command centers, universities, stadiums, concert halls, computer centers, media buildings, telecommunication, etc, **buildings for transportation infrastructure** like highways, bridges, tunnels, network of waterways, locks, ship lifts, underground stations, multi level highway intersections, rail traffic systems, rail stations etc, or **industrial facilities** like industrial plants, seaports, airports, border control, etc (Figure 1). In other words, every person, wherever he or she is, is permanently confronted with the built infrastructure, not realizing how important it is, because usually it works. But sometimes surprises happen: a fireworks factory explodes and destroys an entire neighborhood, a tanker sinks in the Rhine river blocking the waterway for weeks and endangering the environment, heavy snow in combination with deteriorated roof structures causes the collapse of the roof of an ice rink killing 15 people, heavy rain in 2005 causes flash floods in the alps destroying the entire infrastructure, cutting off villages which were out of water supply, energy, telecommunication etc., heavy storms and ice overload electrical energy transportation masts and let them collapse, earthquakes still cause thousands of casualties, and terrorists are able to transform airplanes into weapons

M. Essig et al. (Eds.): Supply Chain Safety Management, LNL, pp. 135–151.
springerlink.com

hitting buildings. That multiple threats can happen at the same time and their strong interdependency has become reality in the Fukushima disaster. This probably needs to be considered in the future.

In addition our modern industrial processes are interconnected generating multiple interdependencies, and, therefore, vulnerabilities. The "just in time" philosophy is based on the functionality of supply chains. But processes, systems and techniques become instable. Therefore, the society has become vulnerable such that tiny causes can produce big disasters.

Fig. 1 Critical built infrastructures, a system of interdependencies www.ga.gov.au/ausgeonews/ausgeonews200509/cip.jsp

In the following we will define some terms just for clarification.

In the US it is defined (DHS 2010): "Critical infrastructure are the assets, systems, and networks, whether physical or virtual, so vital to the society that their incapacitation or destruction would have a debilitating effect on security, national economic security, public health or safety, or any combination thereof."

Critical built infrastructure comprises all buildings and constructions built by civil engineers, so vital to the society that their incapacitation or destruction would have a debilitating effect on security, national economic security, public health or safety, or any combination thereof.

Safety and security is distinguished such that protective structures provide (passive) safety. Security measures complement the structural safety. Security is provided by guards, cameras, detectors, access control etc. In the following only the term safety is used associated with critical built infrastructures.

Structures that are designed against threats are called Protective Structures. Safety engineers have formed the International Association of Protective Structures for which the author serves as president (www.protectivestructures.org). An overview is given in Thoma & Gebbeken (2010).

2 Extraordinary Loading Scenarios

The safety engineer has to derive the physical loading acting on structures from threat scenarios. They can be categorized as follows

- Natural,
- Technical,
- Manmade.

In the following we concentrate on those scenarios that pose a loading case on structures.

Natural disasters are: earthquake, hurricane, tornado, flooding, tsunami, avalanche, land slide, fire, and others.

Technical disasters are: explosion, impact, flooding, fire, derailing, and others.

Manmade disasters are: explosion attacks, gun fire, missile attack, and others.

Usually we have to design against multiple threats which might cause goal conflicts, e.g. earthquake and missile attack. While natural and technical disasters can be relatively well assessed because the potential is known, manmade disasters cannot be estimated. The terroristic attack is considered as asymmetric whereas the other threats are considered as symmetric. In the following we concentrate on earthquake, explosion effects and impact scenarios.

An earthquake takes place if energy is released due to seismic activity. It takes predominantly place at the interfaces of the tectonic plates (figure 2).

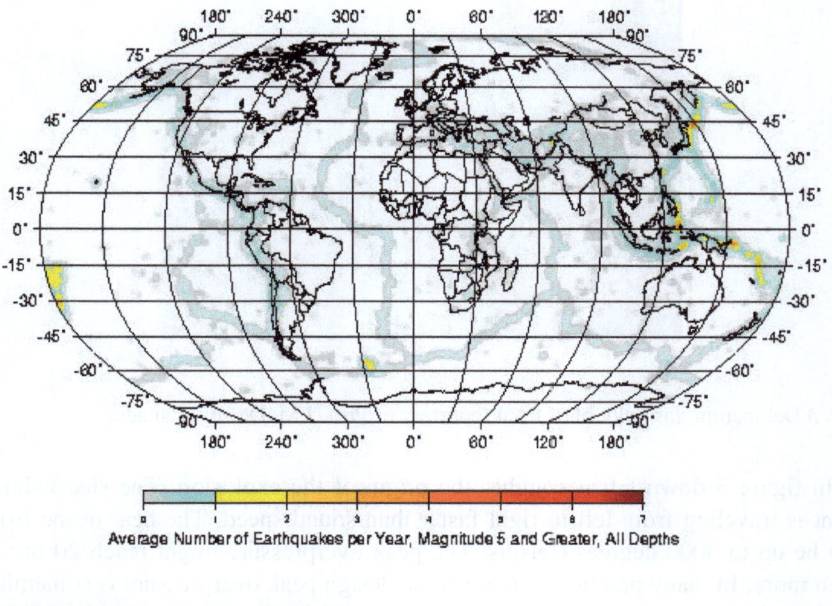

Fig. 2 Earthquake hazard map (USGS)

In media an earthquake is usually categorized in values of intensities or magnitudes (e.g. Richter scale) in order to describe the released energy or the potential for destruction. Due to the basic laws in physics (Euler law: force = mass x acceleration) the engineer needs acceleration data generated by the earthquake in order to calculate earthquake forces acting on the structure (see Eurocode 8, DIN 4149).

Explosions can take place as detonations or deflagrations. Detonations are defined such that the speed of the front of the travelling wave is faster than the sound speed, whereas for deflagrations the front speed is smaller than the sound speed. Industrial explosions are mainly deflagrations, terroristic explosions are mainly detonations. Detonations have to be distinguished in contact detonations, close in detonations, and far field detonations (Gebbeken & Döge 2009). Contact detonations take place if the explosive is in contact with the structure (Gebbeken et al 2004). Contact detonations can destroy a column locally resulting in the loss of the column probably resulting in progressive failure of the building (cascade effect). Far field detonations generate a blast wave not acting locally but acting on an entire structure for which the structure has to be designed. Close in detonation is something in between contact and far field detonations. A secondary effect of explosions are flying fragments (e.g. glass splinters) that act as impactors against people, installations and / or structures (4, 9, 13).

Fig. 3 Detonation, far field, blast front (courtesy of Prof. J.M. Dewey, Canada)

In figure 3 down left is roughly the origin of the explosion. The sharp shock front is traveling from left to right faster than sound speed. The heat in the front can be up to 3000 degrees Celsius. The peak overpressure might reach 20 bar or even more. In many practical situations the design peak overpressure is something around 10 bar. Based on physics or field measurements we obtain a pressure-time-function as given in figure 4.

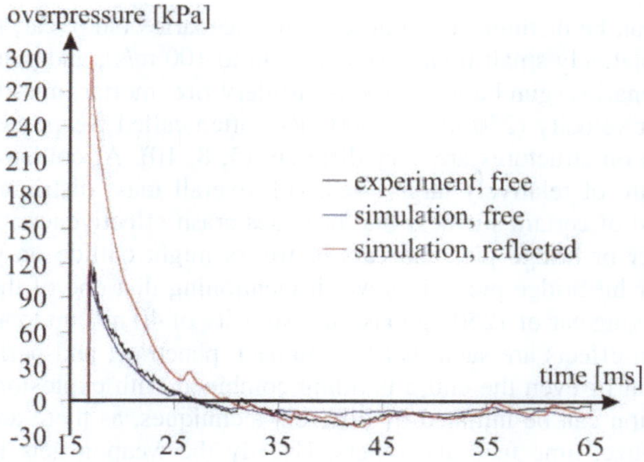

Fig. 4 Detonation, far field, overpressure vs. time

The measuring signal is given in black. The blue curve shows the result of a numerical simulation. The pressure rise time takes only nanoseconds reaching the peak overpressure followed by the overpressure phase which is in the millisecond range followed by the negative pressure phase. The peak overpressure is the difference between the peak pressure p_1 and the ambient air pressure p_0. If the blast wave hits a solid obstacle (building) at the same distance from the measuring point as in the free propagation scenario, a pressure transducer measures the overpressure-time history shown in red. The peak overpressure in this case is approximately 2.8 times that in the free propagation. Consequently the design load is the reflected overpressure as indicated with the red line (Gebbeken & Döge 2009).

A contact detonation at a reinforced concrete beam is shown in figure 5. On the left the ignition of the explosion can be seen. On the right certain damage to the structure is visible. The concrete is almost gone, fragments can be seen, but the reinforcing bars seem to be intact. But there is additional damage like debonding that cannot be seen. The contact detonation acts locally. One of the questions that have to be answered is the question, what is the residual carrying capacity of a damaged structural member?

Fig. 5 Left: Contact detonation, ignition, right: column damaged

Impact can be distinguished into collision scenarios (ship, car, truck, train, airplane) of relatively small impact velocity (up to 100 m/s), and penetration or perforation scenarios (gun bullets, rockets, artillery fire, mortar, missiles, rpg) at very high impact velocity (250 m/s – 2000 m/s), often called weapon effects. The impact effects on structures are very different (5, 8, 10). At collision scenarios the impactors are of relatively large sizes with overall mass distribution, except the engines, and of certain stiffness that provides crash effectiveness. Ships might hit the port pier or bridge pier and cars or trucks might collide on ground level of buildings or hit bridge piers. It is worth mentioning that one of the Tornado load cases is a flying car of 1250 kg mass at a velocity of 40 m/s up to a height of 10m. The weapon effects are such that they impact, penetrate, and perforate the structural element or even the entire building combined with explosions and afterfire. The explosion can be initiated by different techniques, as there are: contact fuze, proximity fuze, time fuze and others. Usually the weapon gets fragmented into thousands of tiny pieces being weapons itself. Due to the ballistic curves these weapons can hit both roofs and exterior walls. If they are applied against technical installations it might be even worse.

One of the suggested design criteria against impact is to add mass to the structure. As we have learned before, mass times acceleration results in force. Now, if the building is in an earthquake region, we have a classical goal conflict. The safety engineer might advice: add mass, and at the same time the earthquake engineer requires less mass. This is why the author and his team have established the integrative approach of multiple threats.

Figure 6 gives an impression of the effect of a truck bomb exploded in front of a building.

Fig. 6 Left: Murrah Federal Building Bombing, right: building partly collapsed (Wikipedia), injuries by floor (Oklahoma State Department of Health)

The Oklahoma State Department of Health explains for the image on the right: "Floor-by-floor breakdown of the injuries/deaths in the Alfred P. Murrah Federal Building from the April 1995 Oklahoma City bombing. Red triangles indicate a fatality, a yellow one indicates a victim was admitted to a hospital, a blue one that the victim was treated and released, a green one that the person was not injured from the blast, and a purple triangle indicates the victim saw a private physician."

Since we have discussed the loading scenarios we might answer the question how to approach the design process.

3 Designing Built Infrastructure against Multiple Threats

Whether an existing infrastructure has to be assessed or a new infrastructure has to be planned, the procedure for a safety engineer is quite the same. Once an owner has decided to build a building or an infrastructure or to assess the existing infrastructure, he has to set up a consulting expert team in order to specify the requirements and to start with the determination of fundamentals. In the following we concentrate on the loading scenarios discussed above.

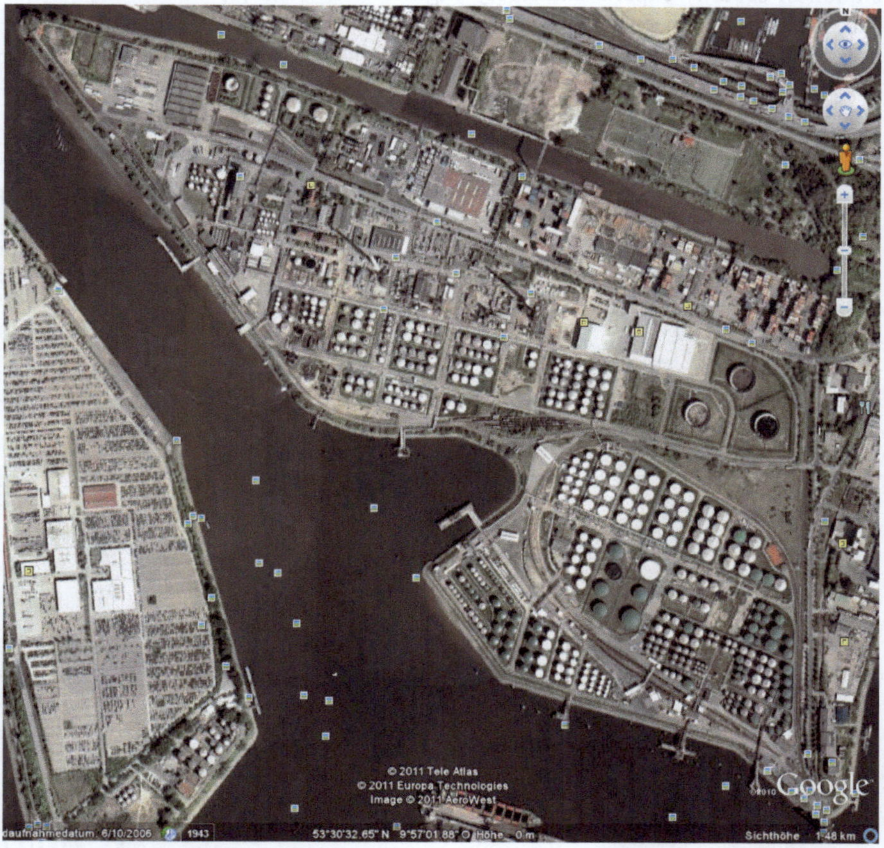

Fig. 7 Site inspection using Google earth

In order to consider natural threats there are standards, codes or guidelines available on a national or international basis (e.g. FEMA 427, STANAG 2280). But the owner is free to assess the acceptable risk and to require more severe load levels. The same holds for technical disasters where the hazard potential is known. When terrorist threats have to be considered, any case has to be uniquely discussed. Here security experts (Federal police, state police, military experts, intelligence) specify the threat. Sometimes also insurance companies play a significant role. Once the threat is defined the safety engineer makes oneself familiar with the overall planning situation (town planning map, development plan, actual situation, site inspection, Google Earth inspection) as shown in figure 7.

Figure 7 provides a very good overview of a specific location. In a next step specific details can be analyzed (Fig. 8) in order to define a finite number of attack points. Theoretically there is an infinite number. This step has to be done in cooperation of security experts and safety engineers. The owner, the security experts and the safety engineer have to agree on the number and the locations of the ignition points. Once they are fixed the explosive threat can be roughly but fast estimated in terms of pressure-time-relations by idealized equations (Gebbeken & Döge 2009).

Fig. 8 Site inspection using Google earth exploring a closer neighborhood

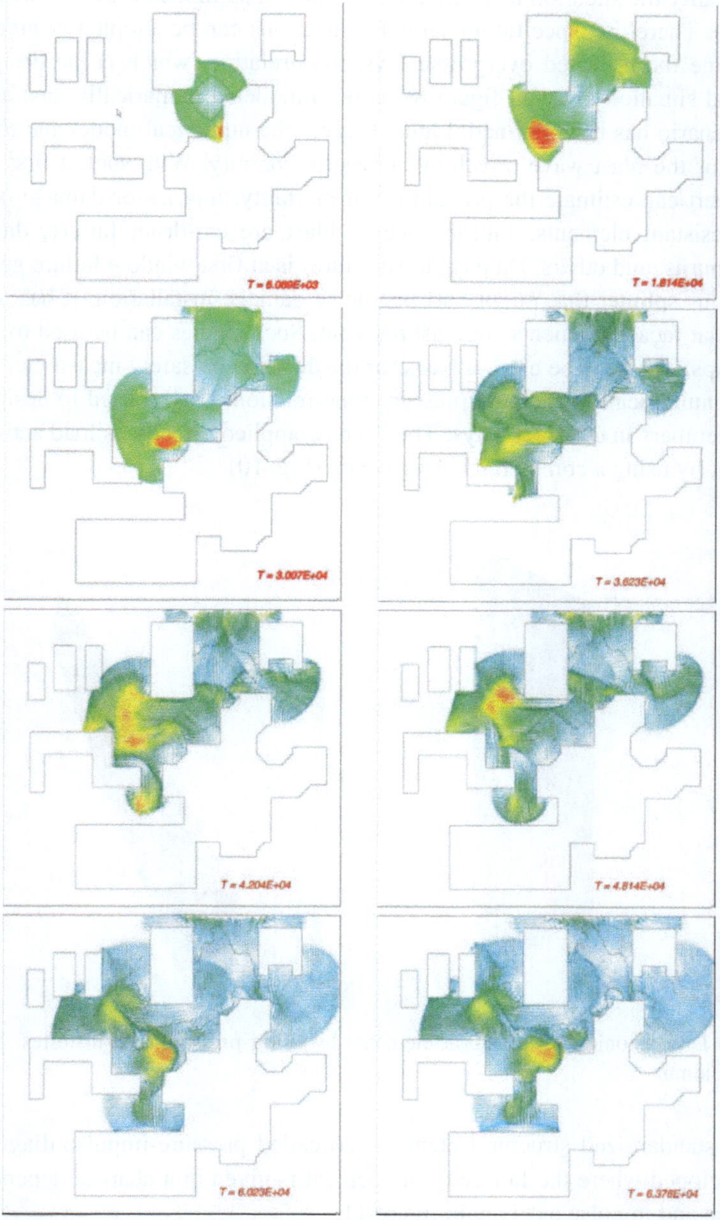

Fig. 9 2-dimensional numerical model of the situation shown in figure 8, blast wave distribution

In reality the situation is much more complex than in idealized model situations (Fig. 8). Therefore, specific numerical simulations can be adopted to numerically determine the reflected overpressure vs. time relation, which is the design load. The real situation given in figure 8 has been modeled numerically, and a detonation scenario has been defined. Figure 9 gives the numerical model and the distribution of the blast wave travelling across the vicinity. With such a first analysis the expert can estimate the probability of mortality, injuries or damage or design blast resistant elements. Injuries due to blast are ear-drum failure, damage to lungs, burns, and others. Damage to structures is at first window failure generating hazardous splinter that might hurt people or damage installation. It has to be ensured that façade elements are blast resistant. Such studies can be used to generate risk maps. They can be used as basis for the decision on safety measures.

The numerically generated pressure-time-relations can be used to design structural members in different ways. They can be applied directly as load acting on an element by using a commercial design code (Fig. 10).

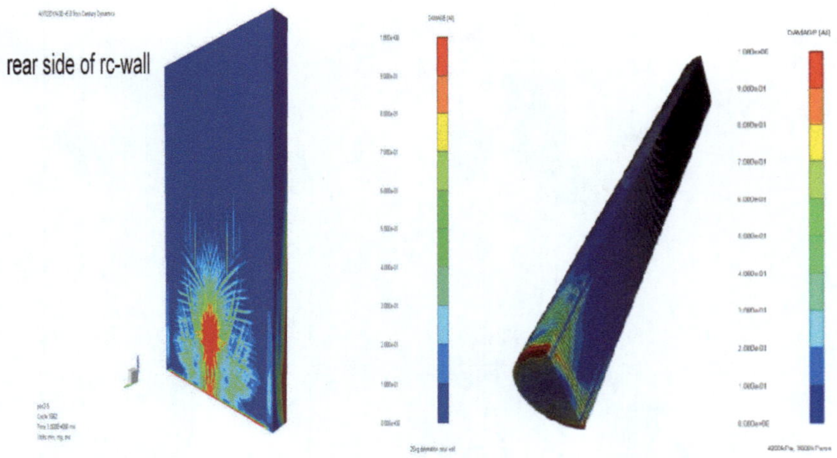

Fig. 10 Dimensioning of structural elements by using pressure-time-histories, left: wall, right: column

For standardized structural elements so called pressure-impulse diagrams can be developed where the failure of an element is given in a chart in dependence of pressure and impulse as given in figure 11.

Such a procedure follows the top down approach, from "global to local". Firstly the overall situation is assessed, and at the end every single structural member and every fastener is designed properly.

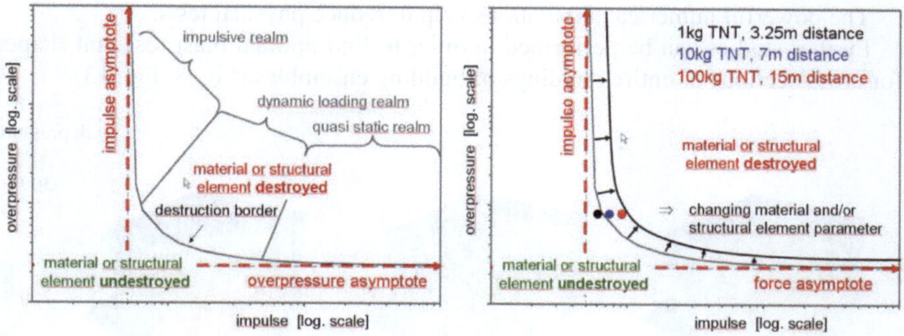

Fig. 11 Overpressure vs. impulse diagram for the dimensioning of structural elements

In the following an example of blast-structure-interaction is given for a facade element. This is a very complex problem that requires very specific knowledge (Fig. 12). For facades see also Teich et al. 2011.

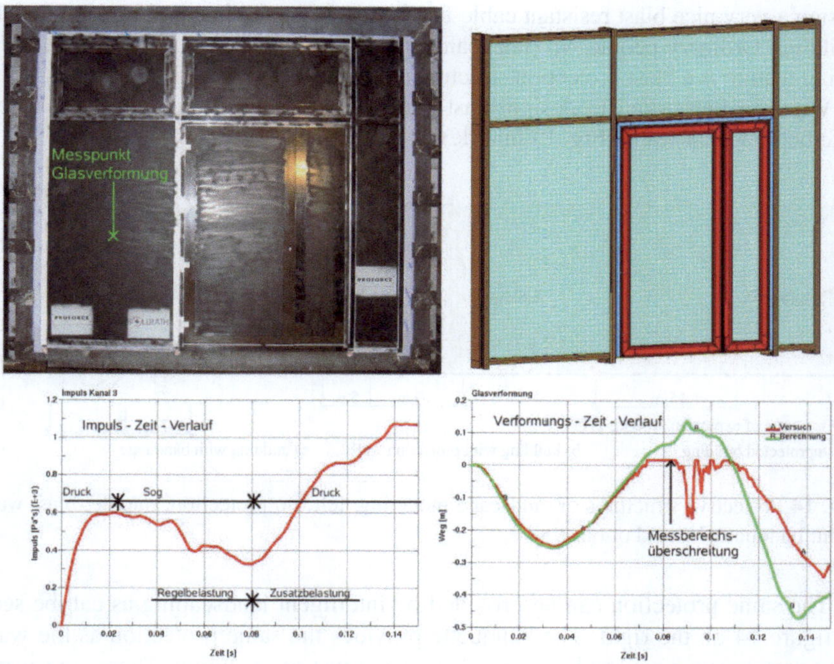

Fig. 12 Blast-Structure-Interaction of a façade, left: Test Bollrath / PRO FORCE 2006, right: numerical representation (MJG Ingenieure GmbH)

The powerful numerical simulations help to reduce physical tests.

Further studies can be performed in order to find optimal blast resistant shapes for cross-sections or entire buildings or building ensembles (Fig. 8, Fig. 13).

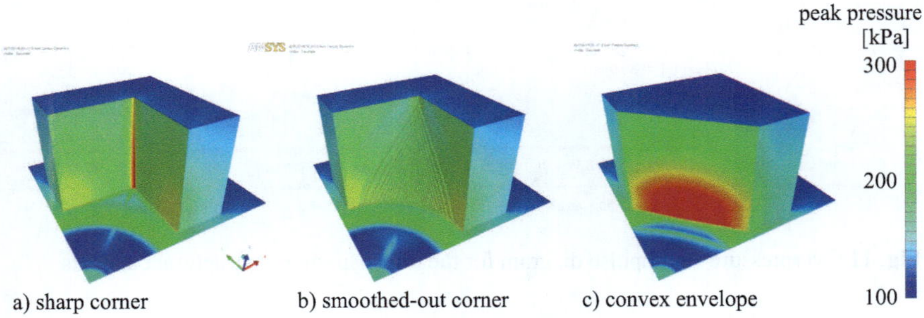

Fig. 13 Numerical studies in order to design blast optimized building envelopes

The three blast pressure contours in figure 13 show clearly that the solution in the middle shows no red color. With such a result the architect and the engineer might design a very nice blast resistant cable-net-façade as an architectural highlight of the building. Ordinary people will not realize that this architectural welcoming and inviting feature is a blast protection structure (Gebbeken & Döge 2010).

We are often asked to design blast walls as perimeter protection. People feel safe behind such walls (Fig. 14 middle).

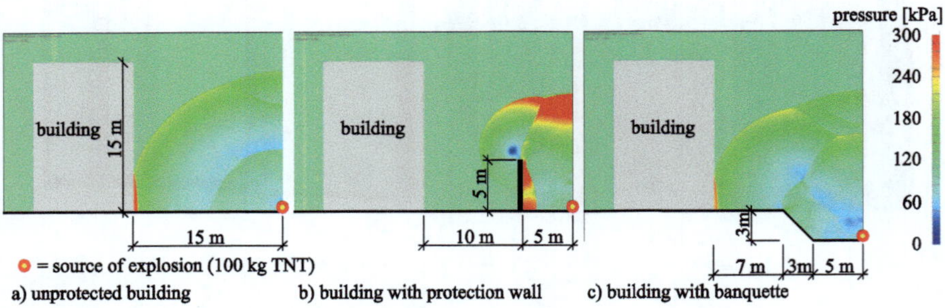

Fig. 14 Protective structures or landscape modeling, left: no protection, middle: blast wall, right: banquette instead of blast wall

The same protection can be provided by intelligent landscaping as can be seen in figure 14 on the right. The banquette provides the same protection as the wall. As a consequence we can learn that there is not just one solution. There are several that have to be discussed in order to find an optimal solution under the given circumstances. It might be that the owner decides to "show force" or to "hide force" and to show an open welcoming scenario. In the latter case we can propose to consider landscape gardening by planting intelligent plants and trees (Fig.15).

Fig. 15 Blast protection by intelligent planting of bushes

We contacted gardeners in order to learn about the resistance of evergreen leaves, branches, and the bio-material fraction with respect to volume. Based on this information we modeled a hedge in the computer in order to numerically study the reduction of blast (Fig. 16).

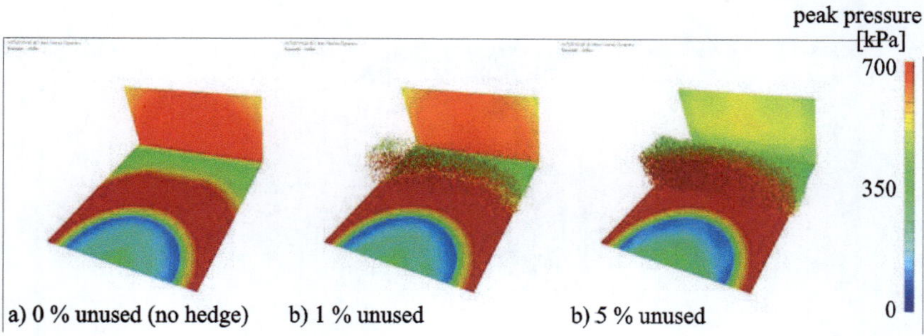

Fig. 16 Blast protection by planting intelligent plants, numerical simulation , left: no protection, middle: 1% biomaterial, right: 5% biomaterial

The numerical result was a reduction in peak pressure and impulse up to 45% which is an excellent result. These numerical studies have to be validated by physical tests in the near future.

Further key elements in the built infrastructure to secure supply chains are highways, bridges, tunnels, underground transportation, railways, waterways, and others in order to provide the physical transportation of goods especially under the "just-in-time" requirement. In the following two elements will be discussed: bridges (Gebbeken & Baumhauer 2006) and underground transportation.

Bridges might be designed such, that they do not have "weak" elements and that they provide redundancy in order to be able to redistribute forces. Properly designed bridges provide robustness. But, the traffic load might increase in terms of density and weight, the presumed life-time of bridges is exceeded and the bridge shows aging characteristics, the bridge is damaged or deteriorated due to various reasons, or an exceptional transportation has to be assessed. These are all cases where the use of a bridge can be restricted or even forbidden affecting the transportation and the functionality of supply chains. In order to assess such extraordinary cases the bridge research team of the University of the Bundeswehr developed a computer assisted tool for the very quick assessment of bridges (Fig. 17).

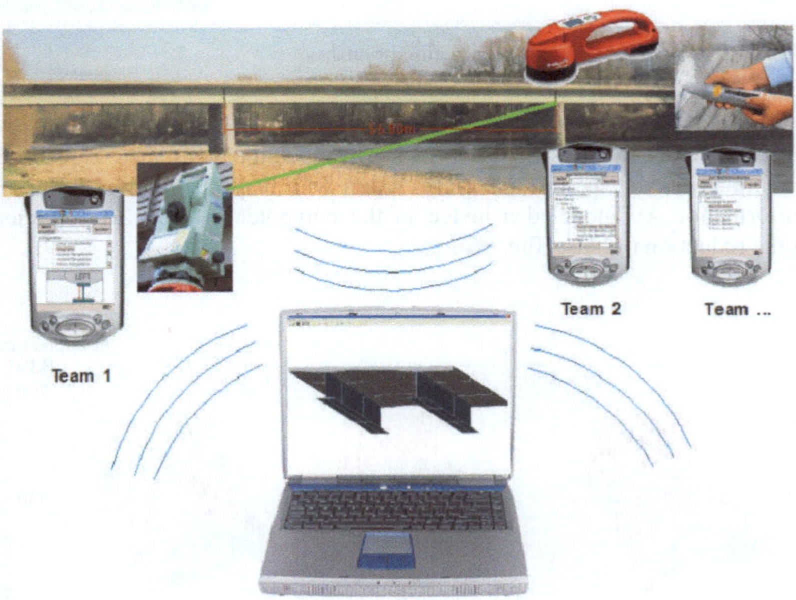

Fig. 17 Bridge assessment system BRASSCO of the University of the Bundeswehr Munich

Figure 17 shows the portable devices that are used for the computer assisted bridge assessment. Field tests and practical applications have shown that an assessment can be completed within several hours. The results can be used by the authorities to decide on further measures.

The underground transportation is a key element of the traffic network especially in larger cities. The infrastructure comprises mainly access buildings, hubs, tunneling system and the trains. Therefore, it is a very complex interacting system in case of an attack. Larcher et al. [2010, 2011] studied underground transportation systems after the London and Madrid bombings. The studies have to focus on different aspects as explained in the following. First the train itself is simulated (Fig. 18).

Fig. 18 Underground train, explosion within the train, damage to the structure

The train has been modelled considering the overall structure, the windows, doors, and the interior. An explosion takes place within the train. Figure 18 shows a sequence of the damage development based on the numerical simulation. Next an entire train is modeled and an explosion attack has been studied for two cases: train outside the tunnel, train inside the tunnel (Fig. 19). As it can be seen from Figure 19 the damage to structure is smaller when the train is in the tunnel. This is due to the blast pressure that is confined in the tunnel providing external pressure. But this does not mean that the threat to the passengers is reduced.

Fig. 19 Underground train, explosion within the train, above outside tunnel, down inside tunnel

Another situation that has to be investigated is the stop of a train in an underground station as given in figure 20. We have to assume an explosion that takes place in the underground station but outside the train. Such a situation can hardly be studied in physical tests. Computer simulation is a powerful alternative.

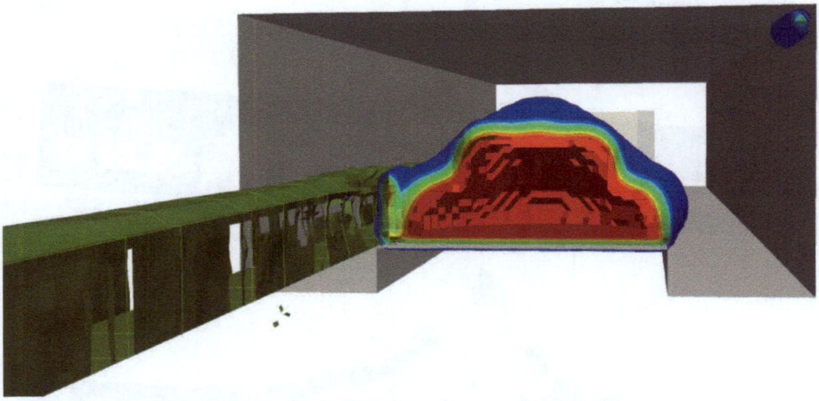

Fig. 20 Underground train station, explosion outside the train

It is a challenge to derive the appropriate numerical models and to carry out the simulation of very sophisticated interaction problems. Figure 20 provides a time step of the numerical simulation. These simulations allow the very detailed study of different phenomena. In figure 20 one can see the development of the blast pressure and the damage to the train structure. The tunnel structure itself was not in the focus, and therefore, assumed to be rigid. The tunnel structure can also be modeled numerically in order to see whether there is any damage to it. For such numerical studies we need computers of very high performance and parallelization features.

As presented in this chapter, the built infrastructures are key elements of physical supply chains. Numerical simulations can help to study threat scenarios or to assess already existing infrastructures. They are also necessary to design new buildings avoiding or reducing physical tests that are usually time consuming and expensive or even impossible to carry out in the 1:1 scale.

References

[1] FEMA 427, RISK MANAGEMENT SERIES, Primer for Design of Commercial Buildings to Mitigate Terrorist Attacks (2003)

[2] Gebbeken, N., Greulich, S., Pietzsch, A., Landmann, F.: The engineering tool XPLOSIM to determine the effects of explosive loadings on reinforced and fibre-reinforced concrete structures. In: Proceedings, MABS 2004, Bad Reichenhall, CD (2004)

[3] Gebbeken, N., Baumhauer, A.: Zur Bewertung und Einstufung beschädigter Brücken (Bridge inspection and assessment). In: Ruge, P., Graf, W. (Hrs.) Neue Bauweisen - Trends in Statik und Dynamik, pp. 133–146. TU Dresden (2006) ISSN 1615-9705
[4] Gebbeken, N., Teich, M., Linse, T.: Numerical Modeling of High Speed Impact and Penetration into Concrete Structures. In: 7th International Conference on Shock & Impact Loads on Structures (SI 2007), Beijing China (2007)
[5] Gebbeken, N., Hartmann, T.: Modellbildung zur Simulation von Stahlfaserbeton unter hochdynamischer Belastung. Beton- und Stahlbetonbau 103, 398–412 (2008) ISSN 0005-9900
[6] Gebbeken, N., Döge, T.: From Explosions to the Design Load. In: Wu, C., Lok, T. (eds.) Shock and Impact Loads on Structures, pp. 25–36. CI-Premier, Singapore (2009) ISBN 978-981-08-3245-2
[7] Gebbeken, N., Döge, T.: Explosion Protection - Architectural Design, Urban Planning, and Landscape Modelling. International Journal of Protective Structures 1(1), 1–22 (2010) ISSN 2041-4196
[8] Hartmann, T., Pietzsch, A., Gebbeken, N.: A Hydrocode Material Model for Concrete. International Journal of Protective Structures 1(4), 443–468 (2010) ISSN 2041-4196
[9] Kennedy, R.P.: A review of procedures for the analysis and design of concrete structures to resist missile impact effects. Nuclear Engineering and Design 37, 183–203 (1976)
[10] Larcher, M.: Development of Discrete Cracks in Concrete Loaded by Shock Waves. International Journal for Impact Engineering 36, 700–710 (2009)
[11] Larcher, M., Casadei, F., Solomos, G.: Influence of venting areas on the air blast pressure inside tubular structures like railway carriages. Journal of Hazardous Materials 183(1-3), 839–846 (2010)
[12] Larcher, M., Casadai, F., Solomos, G., Gebbeken, N.: Simulation terroristischer Anschläge in Massenverkehrsmitteln. Berlin Verlag Ernst & Sohn Bautechnik 88, Heft 4, pp. 225–232 (2011) ISSN 0932-835
[13] Li, Q.M., Reid, S.R., Wen, H.M., Telford, A.R.: Local impact effects of hard missiles on concrete targets. International Journal of Impact Engineering 32, 224–284 (2005)
[14] Teich, M., Gebbeken, N., Larcher, M.: Glasfassaden unter Explosion (Glazing facades under explosions). In: Hofstetter, G., Beer, G. (Hrs.) Baustatik Baupraxis, pp. 397–404 (2011) ISBN 978-3-85125-115-9
[15] Thoma, K., Gebbeken, N. (Hrsg.): BauProtect (Buildings and Utilities Protection). Stuttgart, Fraunhofer Verlag, S.: 1-330 (2010) ISBN 978-3-8396-0151-8

Risk Response Measures for the Management of Theft Risk in Road Freight Transport Chains

Irene Sudy[1], Sebastian Kummer[2], and Ellis Lehner[3]

[1] Hamburg University of Technology (TUHH), Research Associate at the Institute of Business Logistics and General Management, Schwarzenbergstraße 95, 21073 Hamburg, Germany
i.sudy@tu-harburg.de

[2] Wirtschaftsuniversität Wien (WU), Head of the Institute for Transport and Logistics Management, Nordbergstraße 15, 1090 Vienna, Austria
sebastian.kummer@wu.ac.at

[3] Austrian Consulate General - Commercial Section Shanghai, Administration Officer, Shanghai Centre, Suite 514, P.O. Box 155, 1376 Nanjing Xi Lu Shanghai 200040, China
ellis.lehner@advantageaustria.org

1 Introduction

The risk of cargo and vehicle theft is a serious problem in European road transportation affecting all European countries and all partners of the transport chain. According to several studies and reports (TAPA EMEA 2010, Ekwall 2009, Europol 2009, van den Engel and Prummel 2007, ECMT 2002), cargo theft in European road freight transport chains has increased in recent years and cargo theft has become very diligently organized. Cargo theft in road transportation negatively affects the partners in supply chains in various ways. It leads, for example, to the occurrence of disruptions and failures of goods deliveries, which result in additional costs. Besides, the demonstration and provision of risk reducing activities become a prerequisite for co-operation in supply chains and security requirements in contracts and competitive bids are more often defined (TAPA EMEA 2010).

Therefore, the implementation of a comprehensive transport security management becomes increasingly important for providers of logistical services. The development and application of risk response measures facilitate the management of the risk of cargo theft and supports the enhancement of security in transportation.

In this book section, a set of risk response measures for the management of cargo theft in road transportation is developed and categorized according to their abilities to eliminate, reduce, transfer or accept the risk of theft. As a multitude of different security measures and approaches exists for the reduction of theft incidents and its damage, the measures are further categorized according to their capability to reduce the probability of theft occurrence or the extent of a theft incident. In order to get a deeper practical insight and to ensure the practicability of measures proposed, a thorough literature review and personal expert interviews

M. Essig et al. (Eds.): Supply Chain Safety Management, LNL, pp. 153–166.
springerlink.com

with internationally operating logistics service providers as well as insurance companies in Austria and Germany were conducted. This approach allows the alignment of risk response measures that can be found in the risk management literature with the measures applied in practice.

This book section is structured as follows. First, an overview of the current situation of theft of commercial road vehicles and their cargo in Europe is given. Second, the application of steps of the risk management process in general and the risk response measures in particular to cargo theft are explained. Third, risk response measures for managing the risk of cargo theft in transport chains are presented.

2 The Risk of Theft and Organized Crime in Road Transportation – Definitions and Statistics

Especially in an international context transportation chains are to a great extent intermodal with the main leg either by sea, inland waterway navigation, rail or road, whereas pre- and on-carriage is often carried out by road. In this study, the emphasis is placed on road transportation. Besides the actual process of transporting goods on roads, a transport chain comprises additional processes like stuffing, loading, unloading, stripping, transhipment and short-term warehousing (Kummer et al. 2010). The stakeholders involved in transport chains are numerous and range from those who actually perform or organize the transportation to interfacing and influencing actors like insurance companies.

Theft is defined in various ways, depending on the country and their legal frameworks (Aebi et al. 2010). Theft can be defined as "depriving a person/organisation of property without force with the intent to keep it" (Aebi et al. 2010). The association TAPA (Transported Asset Protection Association) defines theft in a more generalised way as "the wrongful taking of property without the owner´s wilful consent" (TAPA EMEA 2010). This study is based on a broad definition of theft as the dishonest deprivation of property with the intention of permanently keeping it (LogSec Consortium 2011). It considers all stealing crimes such as pilferage, robbery, burglary and hijacking (LogSec Consortium 2011, Aebi et al. 2010, Trieschmann et al. 2005, Greenberg 2002, Fennelly 2004).

A major database with crime related data in transportation is administered by the Transported Asset Protection Association for Europe, the Middle East and Africa (TAPA EMEA). Over the last years, the most reported type of road transport related theft attacks is theft from vehicle. Although this indicates that the target of the criminals is the transported cargo and not the vehicle itself, theft of the vehicle has increased over the years (TAPA EMEA 2011). Concerning the type of goods, consumer electronics, food and beverage commodities, tobacco products and computer related products are the theft-prone product groups (IRU 2008; TAPA EMEA 2011). Crime of theft is very carefully organised, especially when high-value goods are the target (ECMT 2002). However, as an interview partner from an international insurance company stated, theft of goods does not necessarily depend on its value. It is dependent on the saleability, usability, disposability and portability of the goods. If goods can easily be sold, they are at high risk of being stolen. In some cases instead of the entire truck load only targeted goods are stolen. However, the risk of theft is not only concerned with the kind of goods carried, but also depends furthermore on the transport route and the level of security (ECMT 2002).

Theft incidents often occur at night or shortly after business hours. According to the TAPA EMEA database, Great Britain and the Netherlands as well as the coast lines of France and Germany are counted among the heavily affected regions in Europe. According to Ekwall (2010), the opportunity of theft depends on the criminal´s ability to use the routines of the target in combination with the lack of security at a certain location. Truck stops at unsecured parking areas represent especially a high risk of cargo and vehicle theft (Europol 2009). Statistics and several interview partners agree that theft attacks in transport processes have become more violent, as threats of force of arms, drugging drivers with narcotics and other modes of criminal action have been reported (TAPA EMEA 2011; Johnson 2010).

3 Application of Risk Management to Theft and Organized Crime in Road Freight Transport Chains

In order to manage theft and organized crime in road transport, the application of the risk management approach is very useful. It is a structured approach and consists of four main steps, namely (1) identification, (2) assessment, (3) response, and (4) monitoring of risks (see Figure 1).

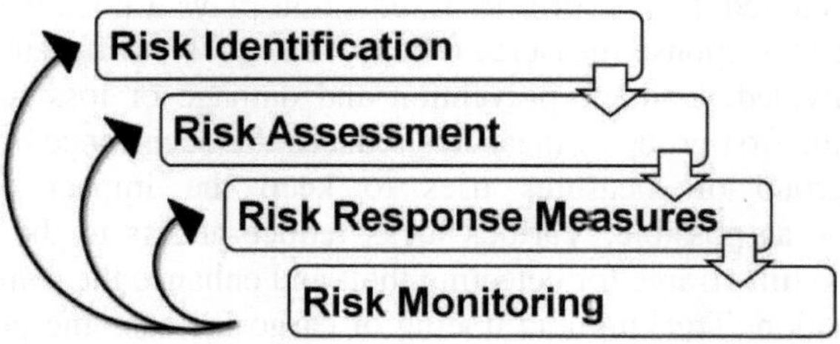

Fig. 1 Steps of risk management

In the first step all kinds of criminal activities associated with theft have to be identified along the road transport chain. In this regard it is very important to pursue a thorough analysis to gain a broad insight into the issue. Analyzing the transport chain both from a process and an institutional perspective is a good way to start. This facilitates the identification of weak points along the transport chain in terms of sub processes as well as companies and organisations involved.

In the second step the identified risks have to be assessed. As not all risk types are equally important for a company, a prioritization or ranking should be made. A common approach is to assess the risks according to their probability of occurrence and their potential severity of loss (Andersen and Terp 2006). Data can be gathered either from external sources, e.g. statistics, or internal sources, e.g. past experiences and forecasts.

In the third step, response measures for the identified and assessed risks are developed, evaluated and selected. The measures can be categorized according to their ability to avoid, reduce, transfer or bear the risk. With regard to the risk mitigation measures, some of them reduce the probability of the occurrence of theft incidents while others reduce the extent of the damage (see Figure 2).

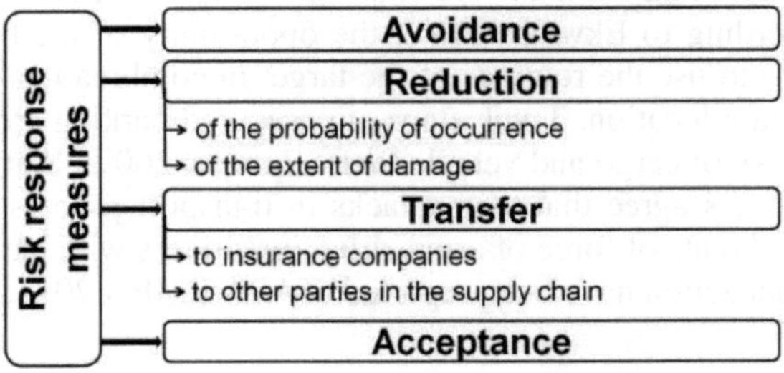

Fig. 2 Risk response measures

The avoidance, or elimination, is the decision not to create a risky situation or to completely eliminate an already existing risk exposure (Andersen and Terp 2006). Thus, by avoiding transportation in certain regions, the risk of theft in these regions can be eliminated. But avoiding, or eliminating, particular risks requires never entering or leaving a certain region and therefore waiving a business opportunity (Andersen and Terp 2006). Risk reduction plays a notably more important role amongst risk response measures (Haller 1981). Two different kinds of measures can be divided: Damage prevention and damage or loss reduction (Haller 1981). While the first group of measures reduces the occurrence of theft incidents, the second group of measures tries to keep the impact of an occurred incident as low as possible. Various locks reduce access to the cargo, whereas seals reduce the time frame for detecting theft and enhance the availability on theft related information. Tracking and tracing of cargo increase the probability of recovering the stolen goods. Risk transfer measures shift the risk of theft to another party in the transport chain who are able to handle the risk exposure better due to certain know-how, core competencies or portfolio balance (Andersen and Terp 2006). This can be done by using contractual clauses, employing subcontractors, outsourcing certain processes to external partners or contracting insurance. However, in most cases, this involves only the transfer of financial risk, whereas indirect effects like lost sales, delayed deliveries or negative publicity remain with the original risk taker (Haller 1981). The extent to which an insurance company covers the risk of theft depends on the terms and conditions. In case of high-risk goods, the risk transfer to the insurer can be quite costly. Another option for risk response is to accept the risk. This means that the risk is consciously taken. Risk acceptance occurs sometimes on an involuntary basis, as the risks which cannot be managed by other risk response measures have to be taken. The provision of sufficient resources to overcome loss or damage without significantly endangering the existence of the company is necessary (Haller 1981).

The objective of the forth step, the risk monitoring, is to evaluate and adjust the risk response measures on a continuous basis and to provide up-to-date information on current risk exposures and risk response measures to the involved partners (Andersen and Terp 2006).

4 Risk Response Measures and Approaches for the Management of Theft and Organized Crime

When looking at theft and organized crime with regard to road transportation, a variety of risk response measures can be found. As described above, they are categorized according to their ability to eliminate, reduce, transfer or accept the risk.

4.1 Elimination/Avoidance of Theft

Eliminating or avoiding risk of theft is a difficult, if not impossible option. Complete elimination of the risk of theft would imply not carrying out any transportation at all – in business practice an unrealistic alternative. Therefore the measures which minimize the occurrence of theft incidents and/or its damage extent prevail.

4.2 Reduction of the Number of Theft Incidents and Their Damage Extent

In theory as well as in practice, a large variety of measures to reduce the number of theft incidents or – once the incident happened – to reduce the extent of damage can be found. The measures can be categorized according to the asset that should be protected:

- Securing vehicle, trailer and container
- Securing cargo
- Securing facilities
- Securing data
- Organizational security measures

Security devices and anti-theft measures for protecting the vehicle, trailer or container – and therefore also the cargo transported – are numerous and range from locking devices and alarms, tracking and tracing to convoying. The second group of security devices secures the shipment itself and not necessarily the vehicle, trailer or container. Another group of security measures protects the facilities where the cargo is stored or where the vehicles are parked. Data protection of cargo related information plays a critical role in preventing insiders from passing on the information to individuals with criminal intent. Organizational security measures comprise co-operation and partnerships with other companies as well as memberships in security initiatives.

4.2.1 Securing Vehicle, Trailer or Container

An overview of measures for the reduction of theft risk on vehicle, trailer or container level is given in Figure 3.

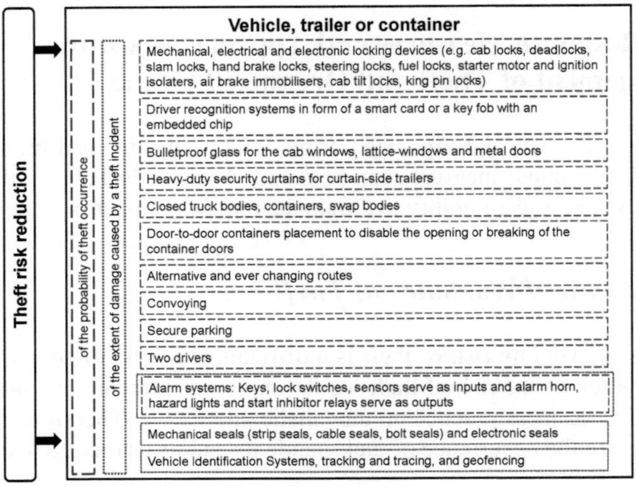

Fig. 3 Measures for the reduction of theft risk on vehicle, trailer or container level

A wide variety of mechanical, electrical and electronic locking devices for securing the vehicle, trailer or container can be applied. The purpose of locking devices is to deny entry to the truck cab, engine, transport unit or load compartment of a vehicle and to prevent a thief from driving away (Erjavec 2005).

Generally, all newly manufactured cabs are equipped with a cab lock that is locked or unlocked by the use of a key or remote control. Even though locking devices decrease the risk of theft (Sutorius 2009), the vehicle will remain exposed to the risk of theft if the driver leaves the door unlocked. To avoid this, the cab lock can be connected to an integrated alarm system by using *deadlocks* or *slam locks* (ECMT 2002). When using slam locks, the cabin door locks automatically when slamming the door. In this case the driver requires a key to unlock the vehicle again and to regain entry. Slam locks are often used for trucks of parcel carriers and other multi-drop delivery vehicles and are usually installed at the time of manufacture (ECMT 2002). Additional security for the cab can be provided by mechanical devices such as *hand brake locks* and *steering locks.* For transports which require high levels of security, access to the cab can additionally be hampered by applying bulletproof glass for the cab windows and lattice-windows or metal doors, thus increasing the safety of the driver (ECMT 2002).

Another group of security devices aims to prevent the truck from being driven away. Examples are *fuel locks, starter motor and ignition isolators* as well as *air brake immobilisers*. Driver Recognition Systems in form of a smart card or a key fob with an embedded chip prevent the movement of the vehicle in absence of the chip (ECMT 2002).

In order to prevent access to the engine compartment, *cab tilt locks* are used, which lock the tilt mechanism of a lorry cab. *King pin locks* prevent the coupling of the towing unit.

As far as the load compartment of a truck is concerned, curtain-side trailers are especially vulnerable to theft as the trailer curtains can be easily cut open with a sharp knife (ECMT 2002). Even though closed truck bodies decrease the risk of theft, curtain-side trailers are very common due to the ease of access to the cargo load from all sides of the trailer. To decrease the vulnerability of curtain-side trailers, heavy-duty security curtains of different degrees of attack resistance can be obtained and fitted to the alarm system of the vehicle. If small wires embedded in the curtain are cut, the alarm will be triggered (ECMT 2002). Containers have the advantage of being placed door-to-door on a truck, which hinders the opening of the container doors.

Another risk mitigation measure is the alteration of truck routings in the transportation network (Chopra and Meindl 2007). Convoying is another measure to reduce the risk of trucks getting attacked on the roads. Escorted convoys are usually used in high-risk areas but downsides like higher co-ordination efforts and traffic impacts have to be considered (Byrne 2010).

The use of secure parking areas for rests and other waiting periods is considered as one of the most practical measures to be taken to reduce the risk of theft (Vigilant 2010). As the number of secure parking locations and its capacities is still limited and the geographical distance between secure parking facilities can be very long, the planning of long-distance transportation involving rest periods and other stops only at secure parking areas is almost impossible. Secure parking areas have a higher number of security features in place than non-secured parking areas, e.g. CCTV, guards, lights and fences. Different organizations like TAPA, IRU, the German Insurance Association and the European Union support the development of secure parking areas. They assess parking areas and rank them according to the number of implemented security measures. Even though the parking fees of the secure parking areas are higher, some road hauliers prefer rests and stops at secure parking areas especially for high-risk shipments. In case no secure parking areas are available, two drivers can be deployed to make sure that the vehicle is never left unattended during stops and rest times.

Alarm systems are used to prevent or delay the attempt of vehicle and cargo theft by deterring thieves by audible siren and/or immobilising the vehicle and prevent the vehicle from being driven away (ECMT 2002). Keys, lock switches, monitoring, and movement sensors as well as alarm sensors (e.g. panic buttons) serve as inputs. Alarm horn, hazard lights and start inhibitor relays are the outputs. Alarm can be triggered in various ways, e.g. by door opening, switching on the ignition, or moving inside the vehicle (Denton 2004).

In road transportation, containers or swap-bodies are often used as transport unit. The container doors are usually designed to have both locking devices and seals. Padlocks provide the lowest level of security. Locks may also be linked to an audible sounding alarm system to alarm the driver in case of a theft attempt.

Contrary to locking devices, the main function of mechanical and electronic seals is not locking and protecting against unauthorized access. Mechanical seals such as strip seals, cable seals or bolt seals can be easily opened with a bolt cutter. Electronic seals may provide identification, documentation of unauthorized access, integrity of the shipment and assignment of liability in case goods have been stolen (Ulrich and Kückelhaus 2010).

Vehicle Identification Systems, electronic tracking and tracing of vehicles as well as geofencing are rather after-theft measures, as the intended purpose of these measures is not the prevention of theft attacks but rather quick response to theft. Detection and recovery of the stolen vehicle and its cargo load are enabled by providing real-time notification of law enforcement authorities and the owner of the stolen vehicle and of the cargo on the location of the truck.

4.2.2 Securing Cargo

Not only the transport vehicle and its trailer or transport unit, but also the shipment itself can be protected against theft with various measures (see Figure 4).

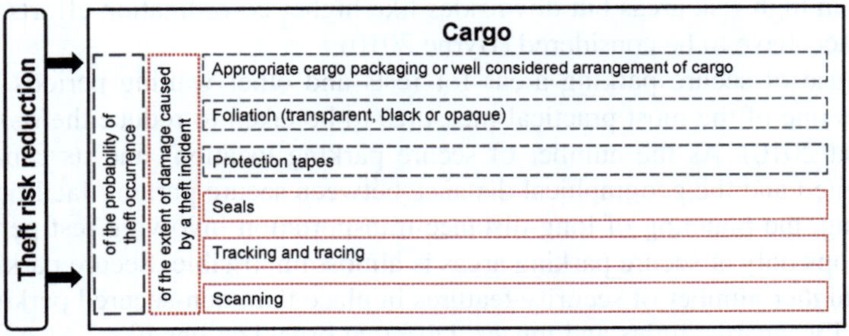

Fig. 4 Measures for the reduction of theft risk on cargo level

Security can be increased by appropriate cargo packaging or well considered arrangement of cargo loads. The story of an interviewed transport insurer serves as an example: Protective gloves of a company disappeared frequently while being in the transport chain. It was not obvious, where and how they disappeared. To solve the problem, the company started to pack the gloves not in pairs anymore, i.e. they shipped the gloves for the right hand and the left hand in separate packages. Short time afterwards, the theft incidents declined abruptly, because: What do you do of no feasible use of only right-hand or left-hand gloves.

The application of pallet foliation (transparent, black or opaque) and protection tapes supports the reduction of cargo theft. The usage of transparent foliation reveals the goods being transported, which prevents cargo insusceptible to theft from being stolen. Black or opaque foliation does not reveal any information about the content of the cargo load. In the case of theft-prone cargo, some experts recommend the usage of black or opaque foliation in combination with protection ties or tapes to secure the cargo. But this measure can encourage thieves to steal the cargo since usually only high value goods are protected in this way.

Similar to the application on vehicle or transport unit level, mechanical and electronic seals are not very useful for mitigating the risk of theft. However, they help in detecting a theft as seals can provide documentation of unauthorized access (Ulrich and Kückelhaus 2010).

Tracking and tracing systems on cargo level are also mostly applied for detection purposes. Various technologies detect misloadings and theft shortly after occurrence. The devices are attached to the goods and send in pre-defined time intervals information about the location of the goods. The system can also detect whether the goods are "off time" (e.g. when the truck is stuck in traffic jams) or if the cargo is "off track" (e.g. in case of misloading or theft).

With regard to theft, scanning of loads can be used to inspect the content of a truck load (Donner and Kruk 2009) and to check if the entire or part of the load got stolen during the journey.

4.2.3 Securing Facilities

Facilities like distribution centres are at high risk of theft (Europol 2009). Days prior to the theft incident, criminals often keep areas around the facility under surveillance. Hence it is important to secure premises to minimize risk of theft (TruckPol 2011). Figure 5 provides an overview of measures for the reduction of cargo theft, which can be applied on facility level, according to their ability to reduce the probability of theft occurrence or the extent of the damage caused by a theft incident.

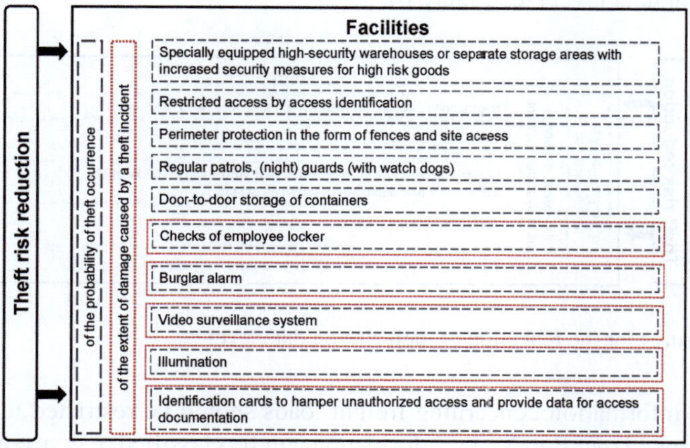

Fig. 5 Measures for the reduction of theft risk on facility level

With a layered security approach, the premises can be classified according to the necessary security level. For high-risk goods, specially equipped high-security warehouses and/or separate storage areas with increased security measures should be generally used. More than one security system should be in place to make sure that in case of the protection system failure, another system is still in operation and/or can report and/or remove the breakdown. Access to the premises where the cargo is located should be restricted by access control mechanism (Europol 2009). Identification cards hamper unauthorized access to the facilities and provide data for access documentation. Useful measures to avoid or detect intruders are fences, site access and its control with appropriate illumination and video surveillance systems, which constitutes perimeter protection. Regular patrols, (night) guards (with watch dogs) and burglar alarm can provide further security (Europol 2009).

As far as the storing of containers is concerned, placing containers door-to-door can disable the opening or breaking of the container doors.

To prevent employees, visitors or subcontracted staff from getting involved in theft, easy-to-see video surveillance cameras placed in high security areas, in the parcel area or at loading bays can be installed. They are highly effective in deterring people from taking the opportunity to steal. Checks of employee lockers are another precautionary measure that serves as a signal. Practical experience shows that revealing that the premises and the freight are well protected and secured works as a deterrent to potential thieves and helps raise security awareness of employees.

4.2.4 Securing Data

The issue of information leakage is a critical factor concerning cargo theft (Özberk 2010; van den Engel and Prummel 2007). Nowadays, thieves already know in advance which truck they want to attack as important information about cargo and route has leaked from one or more involved parties of the transport chain (van den Engel and Prummel 2007). Therefore, it is necessary to apply theft reduction measures on data level (see Figure 6).

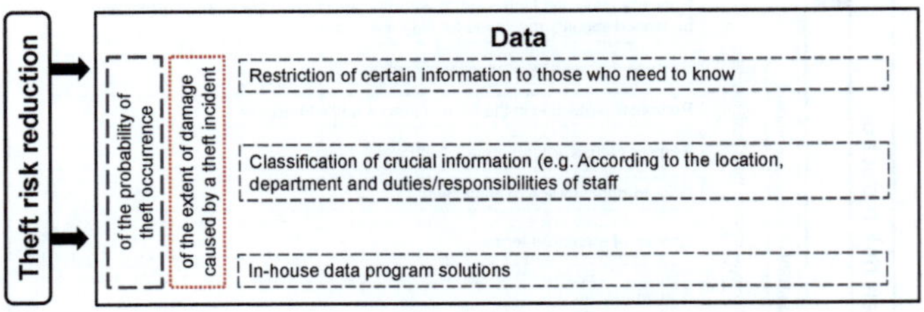

Fig. 6 Measures for the reduction of theft risk on data level

Certain information concerning freight loads should be restricted to those who need to know (TruckPol 2011). Information can be classified, e.g. according to the location, department as well as duties and responsibilities of the staff. Different application levels can be provided and information not being relevant to certain staff (e.g. terminal workers or truck drivers) can be retained (Özberk 2010; van den Engel and Prummel 2007). The interviewed internal auditor of a logistics company stated that – in comparison to open systems – in-house data program solutions also help to keep information restricted.

4.2.5 Organizational Security Measures

Organizational security measures reduce the risk of theft on an overall level. Figure 7 shows the variety of organizational measures that can be applied.

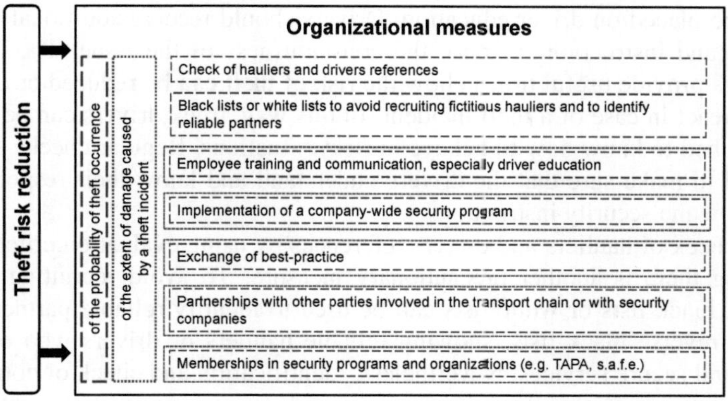

Fig. 7 Measures for the reduction of theft risk on an overall level

Partnerships with other parties involved in the transport chain or with security companies can be seen as a threat reducing measure as the joint design and planning of transport processes support the efforts to reduce cargo theft.

In addition, memberships in organizations aiming at increasing the security of road transportation such as the Transported Asset Protection Association (TAPA) provide several benefits. Conferences and meetings serve as a platform for the exchange of best practice, sharing experiences, mutual learning and information enrichment (Johnson 2010). Memberships can be used as a networking opportunity to gain a deeper knowledge on the tactics of criminals and receive information on new technologies, since basically all members are confronted with similar issues. Various organizations provide databases on theft and crime in road transportation, and offer security trainings for the employees. This availability facilitates the planning of travel routes and stops.

Some national and international organizations offer certificates that offer their members the possibility to communicate their security efforts to customers and other external partners in the transport chain, thus serving as competitive advantage. It raises not only the awareness within the entire transport chain, but also among employees such as drivers, who may be directly affected by theft incidents. Moreover, in tendering procedures, an adequate level of transport security is increasingly requested, particularly for the transport of theft-prone goods.

Some companies implement their own security programs. These programs vary according to the needs of the companies and e.g. focus on specific types of cargo that are particularly exposed to theft.

Other crucial anti-theft measures are employee training and communication. Trainings help to implement security in the daily routine of the employees (TruckPol 2011). The trainings should start immediately after an employee has been hired and be held on a regular basis. Trainings support the communication among all employees to recall the issue of cargo theft. As far as communication is concerned, multiple channels of communication should be used, e.g. articles in company magazines, folders and flyers or via intranet.

As the truck drivers are directly and particularly exposed to theft, an emphasis should be placed on driver education. Drivers should receive additional trainings, manuals and instructions to raise the consciousness of the issue. The driver instructions provide useful tips on how the risk of theft can be reduced or on how to act and react in case of a theft incident. In this way, truck drivers can be prepared and coached to know how to behave in risky situations. Regular checks should be included to make sure that the drivers understand and know their responsibilities and follow the security instructions.

The check of hauliers and drivers references by using evaluation criteria before recruiting them is another precautionary measure. To avoid recruiting fictitious hauliers, black lists or white lists can be used to identify reliable partners. White lists are reverse black lists, showing reliable hauliers or drivers who offer high quality and secure services. Nevertheless, identification and check of correct shipping documents at any time when trucks and drivers arrive at the gates of the company´s premises should be assured (Europol 2009).

4.3 Transfer of Theft Risk

The risk of theft can also be transferred to other parties. Risk of theft is commonly transferred to insurance companies. The calculation of the premium depends on the type of goods and/or transport routes. In case of theft-prone goods and high risk transport routes, some insurance companies insure only with certain restrictions or require the application of appropriate security measures, e.g. a second driver in order to avoid unoccupied truck stops, or no recruitment of hauliers via internet in order to prevent bogus companies. The insurance companies usually advise their customers on the application of security methods since security is also in the interest of the insurer.

Apart from transferring the risk of theft to an insurance company, it is also possible to transfer the risk to other partners in the transport chain. This is for example possible through contractual liability transfer via Incoterms. Incoterms clarify at which point in the transport chain the responsibility is transferred from one partner to the other.

4.4 Acceptance of Theft Risk

Risk acceptance implies that companies take the risk of theft consciously with the willingness to bear the consequences (Haller 1981). Despite numerous security measurement possibilities, it is economically unreasonable to prevent or transfer cargo theft entirely. Which kinds of risks are born to what extend depends largely on the willingness and the possibilities of the company (Kummer et al. 2010).

5 Summary and Conclusions

Because of rising theft incidents over the past years resulting in annual economic losses of billions of Euros, cargo theft is becoming increasingly important for road transportation.

The application of risk response measures is an important step in the risk management process which helps increase security and resilience of road transport chains. Risk response measures manage the risk of theft in road transportation in various ways. Basically, some have a preventative effect and minimize the probability of a theft attack (e.g. locks) and some have a reactive effect in case the theft incident already occurred and help to minimize the extent of the damage (e.g. seals, tracking and tracing, geofencing). The various possibilities are categorized according to their ability to eliminate, reduce or transfer the risk. The security measures can be applied at different levels, namely vehicle and trailer, cargo, facilities, data and transport chains in general.

Expert interviews revealed that companies proactively manage the risk of cargo theft in road transportation by using a multi-layered approach that comprises a set of measures to reduce the probability or the extent of theft on different levels within their company. As an additional measure, insurances to transfer the risk of theft are commonly applied to reduce the negative impact of theft incidents. The risk of theft can hardly be totally eliminated. However, it can be reduced by numerous anti-theft security measures as described in this paper.

The costs of the risk management measures are still an issue, as their benefits are not always obvious and the costs cannot always be transferred to the customer as there is often no willingness on behalf of the customer to accept higher prices.

The costs of theft do not merely consist of the value of the stolen products. The costs include much more like investigation costs, administrative costs, product replacement, high insurance premium, contractual penalty payments, lost sales, lost reputation, and lost customers. These indirect costs arising from theft have to be taken into account (Caroll 2010).

In conclusion, transport companies apply a combined approach of measures of various types of risk response measures. As far as the risk mitigation measures are concerned, interviewed companies emphasized the importance of employee training and education of drivers as well as raising of risk awareness.

References

Aebi, M.F., et al.: European Sourcebook of Crime and Criminal Justice Statistics, 4th edn. WODC, Den Haag (2010), http://www.europeansourcebook.org/ob285_full.pdf (accessed April 22, 2011)

Andersen, K., Terp, A.: Risk Management. In: Andersen, T.J. (ed.) Perspectives on Strategic Risk Management, pp. 27–46. Copenhagen Business School Press, Copenhagen (2006)

Byrne, S.: Truck convoys – an effective risk management tool? Eurowatch Presentation at the TAPA Conference in Budapest (2010)

Carroll, J.: Cargo Crime – the insurers' view. In: Vigilant. The monthly cargo crime update for members of TAPA EMEA (2010)

Chopra, S., Meindl, P.: Supply Chain Management. Strategy, Planning, and Operation, 3rd edn. Pearson Prentice Hall, Upper Saddle River (2007)

Denton, T.: Automobile electrical and electronic systems, 3rd edn. Elsevier Butterworth-Heinemann, Oxford (2004)

Donner, M., Kruk, C.: Supply Chain Security Guide. The International Bank for Reconstruction and Development/The World Bank, Washington (2009), http://siteresources.worldbank.org/INTTRANSPORT/Resources/336291-1239112757744/5997693-1252703593834/6433604-1256564181444/guide_full_version.pdf (accessed July 19, 2011)

ECMT, Crime in Road Freight Transport. OECD Publications, France (2002)

Ekwall, D.: Analysing the official statistics for antagonistic threats against transports in EU: a supply chain risk perspective. Presentation at the TAPA Conference in Budapest (2010)

Ekwall, D.: The displacement effect in cargo theft. International Journal of Physical Distribution and Logistics Management 39(1), 47–62 (2009)

Erjavec, J.: Automotive Technology: a Systems Approach, 4th edn. Thomson Delmar Learning, New York (2005)

European Conference of Ministers of Transport (ECMT), Crime in Road Freight Transport. OECD Publications, France (2002)

Europol, Cargo Theft Report. Applying the Brakes to Road Cargo Crime in Europe, The Hague (2009)

Fennelly, L.J.: Handbook of loss prevention and crime prevention, 4th edn. Elsevier Butterworth-Heinemann, Oxford (2004)

Greenberg, J.: Who stole the money, and when? Individual and situational determinants of employee theft. Organizational Behavior and Human Decision Processes 89(1), 985–1003 (2002)

Haller, M.: Risiko-Management und Versicherung. No. 13 des Versicherungswirtschaftlichen Studienwerks. Gabler, Wiesbaden (1981) (in German)

IRU, Attacks on Drivers of International Heavy Goods Vehicles. Secretariat General. International Road Transport Union (IRU), Switzerland (2008)

Johnson, P.: A vital contribution to protecting trusted brands. The monthly cargo crime update for members of TAPA EMEA (2010)

Kummer, S., Schramm, H.J., Sudy, I.: Internationales Transport- und Logistikmanagement. 2nd eds. UTB, Vienna (2010) (in German)

LogSec Consortium, The LogSec Roadmap (2011), http://www.logsec.org/images/upload/file/docs_logsec-roadmap-finalpublic.pdf (accessed November 21, 2011)

Özberk, B.C.: Mapping the Flow of Theft Endangered Goods in EU. Master thesis, University of Boras (2010)

Sutorius, P.W.: Europol report raises the profile of cargo crime in the political arena. In: Vigilant. The Monthly Cargo Crime Update for Members of TAPA EMEA (2009)

TAPA EMEA (2010), http://www.tapaemea.com (accessed on October 28, 2010)

TAPA EMEA, IIS spreadsheet as (February 25, 2011)

Trieschmann, J.S., Hoyt, R.E., Sommer, D.W.: Risk Management and Insurance, 12th edn. Thomson, South-Western (2005)

TruckPol, Steer Clear of Truck Theft (2011), http://www.truckpol.com/downloads/steerclear.pdf (accessed March 30, 2011)

Ulrich, K., Kückelhaus, M.: Mobile Solutions for Secure Supply Chains. DHL Solutions & Innovations Presentation at the TAPA Conference (2010) (in Bonn)

Van den Engel, A.W., Prummel, E.: Organised Theft of Commercial Vehicles and their Loads in the European Union. European Parliament. Directorate General Internal Policies of the Union. Policy Department Structural and Cohesion Policies. Transport and Tourism, Brussels (2007)

Vigilant: The future of secure parking in Europe. The monthly cargo crime update for members of TAPA EMEA (2010)

Security of Supply Chains from a Service Provider's Perspective

Karl Engelhard[1] and Christian Böhm[2]

[1] Hellmann Worldwide Logistics GmbH & Co. KG, Executive Director/Member of the Main Board, Ludwig-Erhard-Str. 7, 28197 Bremen, Germany
karl.engelhard@de.hellmann.net

[2] Hellmann Worldwide Logistics GmbH & Co. KG, Head of Competence Center Military Logistics, Ludwig-Erhard-Str. 7, 28197 Bremen, Germany
christian.boehm@de.hellmann.net

This chapter looks at the requirements logistics service providers are facing today, with respect to supply chain security. Taking Hellmann Worldwide Logistics as an example, it will demonstrate how logistics service providers react to these requirements. First, todays risks and relevant requirements will be described in respect to the consequences for logistics service providers and supply chains. Following this, we will look at examples of some concrete preventive measures toward security in the supply chain, taken by Hellmann Worldwide Logistics.

1 Security in the Supply Chain from the Perspective of the Logistics Service Provider

1.1 Requirements Logistics Service Providers and Supply Chains Are Facing Today

Supply chains have developed into global networks that span the globe. There is hardly an industry, which could deny the importance of worldwide sourcing for the manufacturing of its goods and their inter-regional and/or international distribution thereafter. Thus, logistics companies are expected to set-up and operate appropriate global supply chains. This requires worldwide people networks and information (systems) networks on the one hand, and worldwide freight transportation networks comprising all modes of transport on the other hand.

There are numerous examples available, which illustrate the increasing importance of global supply chains operated by logistics service providers.

One such example is the development of worldwide container transport, which increased almost five-fold between 1990 and 2009.[1] More than 500,000 people are employed in freight forwarding and warehousing industry in Germany.[2] They generated over 76 billion EUR turnover in 2008.[3]

---

[1] Clarkson (2009), Container Intelligence Monthly.

[2] Bundesagentur für Arbeit '*German Federal Employment Agency*' (2009).

[3] Deutscher Speditions- und Logistikverband ‚*German Forwarding and Logistics Association*' (2011), website.

M. Essig et al. (Eds.): Supply Chain Safety Management, LNL, pp. 167–178.
springerlink.com

Those involved in the supply chain include manufacturers, exporters, freight forwarders, warehouse-keepers, customs agents and road haulers as well as importers and finally the end-user.

The number of people involved in a supply chain can be anywhere from a handful to many hundred. The vast number of logistics companies worldwide gives some idea of how many businesses and people are involved in operating global supply chains. This number increases considerably if state employees are added to the calculation. It is therefore obvious that the resulting complexity gives rise to great planning, and above all, operational challenges for logistics service providers such as Hellmann Worldwide Logistics.

Hellmann Worldwide Logistics employs approximately 9,000 people in over 200 subsidiaries worldwide, and is responsible for the transport of around 14 million shipments per year. Including all of the companies in our partner networks, a total of 16,500 people in 157 countries carry out logistics services for customers of Hellmann Worldwide Logistics across the globe.

Of course logistics companies are involved in constant competition with each other. Above all, customers require that their logistics service providers transport goods fast, economically, and of course secure.

How long a shipment takes depends on the mode of transport chosen, but moreover on the capacity and usability of the respective transport infrastructure such as airports, seaports as well as roads. Weather conditions can also severely restrict transport by road, air, rail and sea, thus greatly influencing the duration of a supply chain.

Illustration 1 Theft during transport is a serious security risk

Providing security of a supply chain, and thus secure the goods being transported, is an essential prerequisite to logistics service companies. Preventing hazards and risks within the supply chain has therefore been fundamental to all considerations facing logistics service providers since the beginning of freight forwarding industry.

1.2 Dealing with Hazards and Risks

How to deal with hazards and risks has always been of major importance to logistics companies. Thus the implementation of preventive measures is not new to them. Only a secure supply chain is a sustainable supply chain.

Nevertheless, the dangers to the supply chain and the approach to risks have changed in recent years. There appear to be more sources of increased danger. On the one hand, the fact that world trade is ever more closely knit, bringing with it an increased worldwide shipping volume, is one possible reason for the increase in damage. On the other hand, there is the increase in extreme weather conditions, a threat to seamless supply chains.

However, the risk of supply chains falling prey to targeted and intended attacks has especially risen. A rise in criminal and terrorist motivated attacks can be observed. According to a recent European Parliament study, total damages of 8.2 billion EUR are incurred along the entire supply chain within the EU per year. This can be broken down as 6.72 EUR per shipment.[4] It is also apparent that 60% of all theft in transit takes place during stops and 15% of these thefts are the result of hijacking. Both facts demonstrate the extent of the threat from criminal activities.

The threat of terrorism has also increased, which has led to a change in attitudes to the security of supply chains, especially since the terrorist attacks of September, 11 2001 on the World Trade Center in New York and, more recently, the parcel bombs, which are said to have originated in Yemen.

Numerous legal requirements and certification standards have been added to the existing security measures since 9/11. Many international organizations and national states have passed new safety regulations or tightened existing norms in the field of transport and logistics. The commercial sector has also reacted to these new threats with its own initiatives and security requirements.[5]

An excerpt of important initiatives and legal regulations from the standpoint of logistics service providers can be found as follows:

- ISO 28000
- C-TPAT
- AEO (Authorized Economic Operator)
- EU and UN terror lists
- FDA Bioterrorism Act
- "Framework of Standards to Secure and Facilitate Global Trade" (SAFE)
- Technology Asset Protection Association (TAPA)
- International Ship and Port Facility Security Code (ISPS Code)
- Importer Security Filing (ISF, also 10+2-Rule)
- Business Alliance for Secure Commerce (BASC)
- Container Security Initiative (CSI)

The examples of regulations and initiatives listed here should be assessed differently with respect to their level of maturity, how binding they are legally, and scope. Some of these are only valid for certain modes of transport (e.g. sea freight) or only for import or export in certain regions or states (e.g. USA). It is of the utmost importance for the logistics service industry that when establishing regulations, the legislator co-operates internationally in order to find legal rulings, which are non-overlapping and, hence, are applicable in practise and truly serve to increase security.

---

[4] Comp. Study by European Parliament (2007).

[5] Sicherheitsstrategie in der Wirtschaft *(Commercial Security Strategy)* (2009).

To summarize; the customer expectations regarding cost-efficiency and high performance, on the one hand and security requirements – whether initiated internally or by the legislator, on the other hand, both have risen.

From the perspective of logistics service providers these developments are not ease to bring in line with each other, as demonstrated in a survey by the consulting company SCI Verkehr.[6]

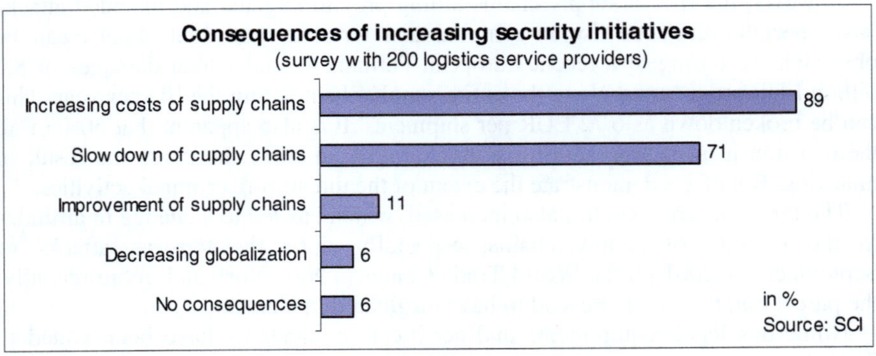

Illustration 2 Consequences of heightened security

Logistics companies are expected to satisfy requirements from both sides. This presents a huge challenge, which can be met by various means.

Hellmann Worldwide Logistics began many years ago to establish preventive measures within the company, in order to guarantee security in the supply chain and counteract planned attacks. The establishing of risk analyses, processes, security and protection measures as well as continuous employee training has been in process for decades.

Logistics companies such as Hellmann Worldwide Logistics have a vested interest in secure supply chains and protecting customers' goods from unauthorized access. Since 2002, many logistics companies have systematically established themselves at various locations throughout Germany, in particular due to a strict German ruling on gross corporate negligence.

Below is a list of some of the measures taken by Hellmann Worldwide Logistics. The focus of this chapter is the hazard to goods posed by planned theft, robbery, ambush en route, as well as terrorist attacks on vehicles and transport infrastructure, and misuse for illegal purposes.

2 Preventive Measures for Supply Chain Security

Hellmann Worldwide Logistics began many years ago to implement various measures in order to comply with legal requirements and the company's own security standards. The primary objective of these measures is to be proactive and preventive. At the same time, these measures present possibilities to react fast and

---

[6] 6 LOG.Kompass *(Annual German Transport Report)* (2011).

purposefully, if an incident occurs, and by analysing the incident to draw conclusions, which help prevent similar incidents in the future.

These measures can be broken down into three categories.

1. Organizational Measures

These measures primarily concern the approach to the subject of security, through deployment of personnel and material resources, and determining appropriate goals. Thus the organization can develop the required sensitivity for security and apply measures practically. Acquiring status certificates and authorizations for adherence to compliance regulations is also important for this area.

2. Physical Security Measures

These measures cover the physical protection of the fixed infrastructure (e.g. buildings, warehouses) and the mobile infrastructure (transport vessels, e.g. containers) and should counteract specific and planned attacks. Security technology plays an increasingly significant role in this area, alongside more traditional methods such as general plant protection.

3. Digital Security Measures

These are measures, which serve to protect the information and information systems used in the running of the company and in controlling supply chains.

2.1 Organizational Measures

In order to practically apply security measures in the supply chain, it is important to firstly create the necessary structures within the organization.

Hellmann Worldwide Logistics has already laid the foundations for these measures through creating specific operational areas and committees as well as compiling handbooks and codes of practice for security. This made it possible to awaken the necessary awareness of questions of security and to embed them in the company. Effective security management can only then be defined and implemented when the people within the company deal with risks and hazards on the one hand, and with external requirements on the other.

Hellmann Worldwide Logistics has already implemented a range of measures over the years. These are, e.g.:

- Implementing risk management teams
- Creating a risk management handbook
- Checking the EU and US anti-terror lists via operative order systems
- Status of "Regulated Officer" in aviation security
- Customs & Compliance Manager
- Status according to C-TPAT in the USA
- Audits, Risk analyses by location
- Testing of sub-contractors / supplier management
- Authorizing of personnel checks and access checks
- Process description for high-security goods

These measures are regularly subject to critical examination and adapted accordingly.

2.1.1 Authorized Economic Operator (AEO)

As has already been mentioned, numerous participants, organizations and people interact in operating a supply chain. This, coupled with the increasing rate of globalization, prompted the World Customs Organization (WCO) to create a worldwide framework for modern, effective risk management in customs administration, in the form of the so-called "Framework of Standards to Secure and Facilitate Global Trade" (SAFE).[7]

Implementing the status of AEO is a fundamental part of this security initiative. Its objective is to ensure a consistent international supply chain from the manufacturer of a product to the end-user. Negotiations are currently underway with partner countries (especially the USA, China, and Switzerland), which should lead to worldwide recognition of this status.

Authorized Economic Operators have a special status. They are considered to be especially reliable and trustworthy and can thus avail of certain privileges in customs clearance.

The European Union implemented these political security considerations as European law in 2005, with the appropriate change to the Customs Codex (VO EG No. 648/2005), and substantiated them with the publication of the provisions for implementation (VO EG No. 1875/2006) in December 2006. As of January 2008, companies based in the European Union and involved in customs processes may apply for this status, with which they may avail of special treatment during customs checks and /r simplifications according to customs regulations.

Hellmann Worldwide Logistics was awarded the status of Authorized Economic Operator. This accreditation demands a comprehensive spectrum of security measures. These individual measures will not be treated in any further detail here.

2.1.2 Traceability

The concept of traceability is based on the EU Regulation No. 178/2002 and is a further step toward security in the supply chain, for instance, in the food industry.

Commercial policy and economic considerations, regarding damage limitation in certain incidents during manufacturing and distribution processes for foods and animal feed, were deciding factors in introducing traceability measures.

The objective of the traceability concept is to facilitate the identification and selection of damaged goods within the supply chain, and so to enable their extraction from it. Thus contaminated foodstuffs or goods with production faults can be intercepted on their way to the end-user and recalled.

Hellmann Worldwide Logistics has adapted its warehouse processes and warehouse information systems in order to satisfy these requirements.

For this reason, as soon as a Hellmann warehouse receives goods, their every movement within the warehouse is documented through scanning the goods and storage/picking & packing area.

Furthermore, manufacturers today, electronically transfer comprehensive records of every product to the logistics service provider, parallel to the actual

[7] German Customs Website, (`http://www.zoll.de/b0_zoll_und_steuern/a0_zoelle/l0_zugelassener_wirtschaftsbeteiligter/index.html`)

physical delivery of the goods, for further processing by the logistics service provider's IT systems. These records provide important details regarding best-before-date, batch number and the shipment's Serial Shipping Container Code (SSCC).

Illustration 3 Every movement within the warehouse is documented

Access to this information with a view to seamless documentation of all movements within the warehouse (and in a broader sense, the supply chain) facilitates the steering of the flow of goods at the push of a button.

2.2 *Physical Security Measures*

2.2.1 Access and Theft Prevention in Cross-Dock Warehouses

Hellmann Worldwide Logistics has carried out risk analyses for all of its locations. Regular audits ensure that changing circumstances at a location are also documented and that respective measures can be adapted.

All subsidiaries are equipped with external security systems and access-control systems. Furthermore, the Plant Protection Department monitors the premises and can react immediately where anomalies arise.

Hellmann Worldwide Logistics has equipped all forwarding warehouses in Germany with intelligent camera systems since 2002. A single cross-dock can contain up to 400 cameras, which display each grid from up to three sides.

Thus the entire supply chain within the warehouse –from delivery to dispatch – is visible and documented.

The aims of this measure are twofold: to protect against theft and unauthorized access to packages, and to improve process-security through quickly finding erroneous shipments and detecting insufficiently packed goods. The theft rate has been tending toward zero since the system was installed.

Illustration 4 Up to 400 cameras are in use in one cross-dock warehouse

2.2.2 Transportation Security through Securesystem

While it is possible to permanently install security systems within the bounds of one's own premises – as described above – access restriction and camera systems facilitate the upholding of desired security standards, the circumstances on the road are quite different.

As soon as a product leaves the business premises, e.g. in a container, proactive security becomes much more difficult to ensure.

One possibility is a combination of positioning technology, sensor technology and seal monitoring in order to close this security gap. Hellmann Worldwide Logistics made a decisive contribution as a partner of EADS Atrium, in developing the SecureSystem, and has been using this system for sea freight since 2009.

This is a surveillance system for containers, which is available worldwide and around the clock, in real-time. The technical elements are a mobile surveillance unit, which is placed inside the container and equipped with sensors and radio transmitter. This mobile surveillance unit continuously exchanges data with the control center via GSM or alternatively by satellite network.

Furthermore, the mobile surveillance unit is equipped with Bluetooth, allowing it to exchange data with mobile terminals (handhelds) close to the container and, if required, read data by way of special electronic certificate and or open or close the container.

The functionalities of the system allow a container to be located geographically (tracking & tracing), interior inspection using various sensors, as well as authorization and checking of opening and closing activities.

Geographic locating allows the container to be exactly pinpointed and monitored in real-time. This facilitates adherence to prescribed routes or registering deviations from this route. The position of a container in expansive harbour facilities can also be determined exactly.

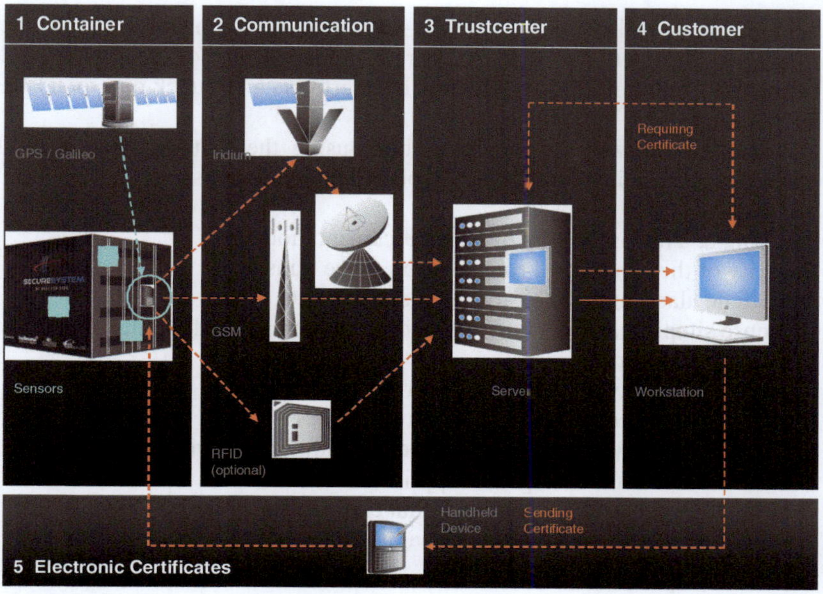

Illustration 5 SecureSystem increases security in the supply chain

The relevant sensors allow the interiors of the containers to be monitored. The most frequent use is for informing of a change in a predetermined temperature as well as undesired and unauthorized opening, e.g. by drilling the container walls.

The control and documentation of all authorized and unauthorized sealing and opening activities allow for a completely new security concept. Electronic keys for authorized access are issued centrally from the control center.

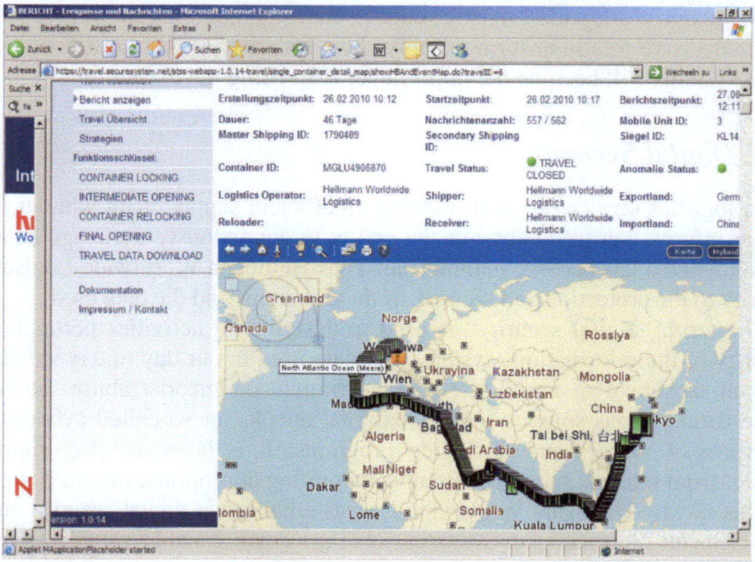

Illustration 6 SecureSystem enables global tracking & tracing

SecureSystem makes use of satellite-supported communication based on the Iridium Satellite network and is thereby independent of ground-controlled infrastructure. Thus, eliminating the risk of interference or attack on ground-based infrastructure.

Hellmann Worldwide Logistics has been using the system since 2009. The advantages for those involved in the supply chain are obvious:

- Tracking & Tracing information and documentation
- Integrity of the container throughout the entire transport
- Immediate alarm at digression from planned route
- Immense simplification of customs documentation with additional information.
- Documentation for insurers
- Monitoring of individual quality factors (e.g. temperature, moisture)
- Use of electronic certificates for controlled access
- Immediate alarm when container is opened without authorization
- Use of data from further logistics applications
- Graphic illustration of all data on digital cards and overviews
- Creation of e-certificates for individual transports / containers
- Complete transport documentation

The advantages in increasing security in the supply chain are obvious. Apart from the advantages listed above, it is possible that systems such as SecureSystem can practically implement the Container Security Initiative (CSI), instigated by the US authorities, and all associated comprehensive scanning obligations for import containers, if the integrity of the container is guaranteed from its loading point.

This could mean that the necessity to rescan in the USA can be dispensed with, which means great savings in both time and money, while at the same time increasing security.

Hellmann Worldwide Logistics expects that such systems as SecureSystem will play a decisive role in ensuring supply chain security in the future. It is likely that these systems will become the norm rather than being the exception.

2.3 Digital Security

For a logistics service provider, running an efficient supply chain without intensive use of information technology or the ability to continuously exchange data among those involved has become unthinkable. Thus Hellmann Worldwide Logistics gives priority to the protection and security of these systems and the data used.

The term, "digital security" is used to describe all activities pertaining to the security of the informations systems and data used in our day to day work, as well as their treatment in terms of possible criminal or terrorist abuse. Such threats come from various sources such as external attacks, or so-called cyber-attacks in the form of viruses and damaging programs. It is however also important to protect from internal sabotage attempts by putting appropriate measures in place.

This is why Hellmann Worldwide Logistics has made digital security one of the core elements of the company's successful IT governance and IT compliance concept.

IT compliance focusses on those compliance requirements made of the company's IT systems. These mainly encompass information security, availability, storage of data and data protection. Logistics service providers are subject to numerous legal regulations. Non-adherance to these can mean high fines and liability. Further rules are enforced by EU regulations, international conventions and trade practice.

The three main areas of IT compliance at Hellmann Worldwide Logistics are:

1. Confidentiality: Are our systems and data safe from external access?
2. Integrity: Is all data completely protected?
3. Availability: Are our systems and data always available when and where we need them?

Hellmann Worldwide Logistics has implemented a management system for security of information (ISMS), based on the internationally recognized ISO/IEC 27001 standard, to ensure the appropriate handling of risks.

To facilitate this, Hellmann Worldwide Logistics has allocated concrete roles to its employees. These roles are:

- Sponsors from the committee 'Information Security Management'
- Quality Manager & Auditor – Auditing (certified Auditor from TÜV-*German technical control board-* Rheinland)
- Chief Information Security Officer -Technical Security (Vocational training by the University of Regensburg)
- Project Manager & Auditor -ISMS Organization (certified Auditor from TÜV Rheinland)

The importance of digital security is increasing. Hellmann Worldwide Logistics has successfully counteracted 5 worldwide virus outbreaks since 2005.

However, digital security is not for free. IT-security spending is estimated to make up between 7 and 10% of the total IT budget.

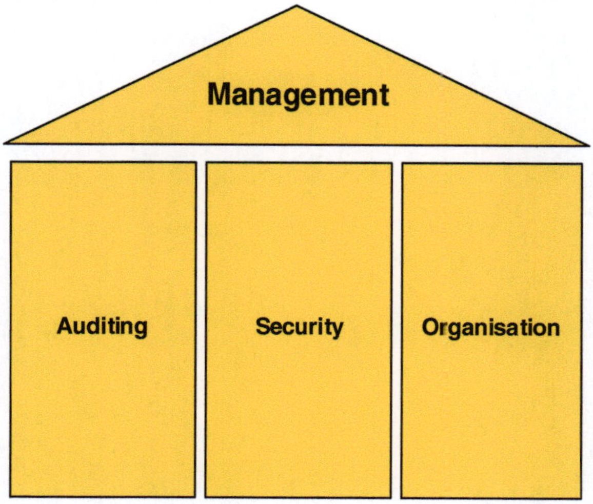

Illustration 7 Elements of Hellmann's information security management system

3 Summary

Today's logistics service providers orchestrate complex supply chains in order to manage the global flow of goods. On the one hand, they are required to satisfy customer expectations of thrift, speed and security, and have to meet legal requirements on the other hand.

Creating security in the supply chain has always been a focal point among logistics companies. In recent decades, heightened risk of terrorist attacks and danger for data and IT systems have been added to dangers of theft and hazardous weather conditions.

Logistics service providers can – or where legally obliged, must – implement preventive measures to guarantee security in the supply chain. Much can be undertaken in the areas of organization or management, physical security and digital security.

Hellmann Worldwide Logistics has implemented many such preventive measures. Numerous measures have been undertaken to guarantee the physical security of fixed and mobile infrastructure as well as the introduction of various organizational measures such as establishing structures for managing risks. Where digital safety is concerned, strong measures have been undertaken in order to protect data and systems from misuse.

The importance of security in supply chains will continue to grow. The increase in the flow of goods, brought about by progressive global division of labor and regional concentrations of raw materials, coupled with the increase in threats from criminal or terrorist sources will be the two most influential factors in this.

Logistics companies such as Hellmann Worldwide Logistics will be paying close attention to such developments and choose the correct measures to provide the highest standards of security, while simultaneously guaranteeing economic viability. Because only a secure supply chain is a sustainable supply chain.

Cyber Security: Challenges and Application Areas

Gabi Dreo Rodosek[1] and Mario Golling[2]

[1] Universität der Bundeswehr München, Chair for Communications Systems and Internet Services, Werner-Heisenberg-Weg 39, 85577 Neubiberg, Germany
Gabi.Dreo@unibw.de
[2] Universität der Bundeswehr München, Research Associate at the Chair for Communications Systems and Internet Services, Werner-Heisenberg-Weg 39, 85577 Neubiberg, Germany
Mario.Golling@unibw.de

1 ICT as the Underlying Technology

Our life is moving to the digital world. The changing landscapes of life are forcing us to change the way we deal with things. Living in an always on and always connected world makes our life easier and more convenient. Smart homes automatically adapt to the environmental conditions with respect to heating and cooling; smart cars adapt to traffic conditions and situations (e.g., accident or congestion); smart health systems monitor our health condition and smart entertainment adapts to our mood. A lot of "smart things" make our lives more convenient. Social networks, online banking, e-health, e-marketplaces are other examples of the networked society.

The foundation of such a networked society is the information and communication technology (ICT). Bandwidths of 100 Gbit/s and higher, billions of devices and "smart things" exchanging data in an ubiquitous way as well as concepts such as cloud computing and virtualization promise easily accessible computing and storage power that is located somewhere in the network and accessible from anywhere.

Regardless of these advantages, such trends include potential risks with respect to security and privacy as well. Similar to protecting our physical identities in the real world, we need to protect our identities in the digital word as well. The protection of our resources (mobile devices, data, and infrastructures) against cyber criminals is becoming a central issue. Inadequately protected resources are often an underestimated risk factor especially since technology is moving at ever faster rates and regulation is struggling to keep up.

Since ubiquitous connectivity means also widespread vulnerability, the task to protect relevant "assets" in the digital world is becoming even more demanding. Due to the convergence towards "All-over-IP" and "IP-over-All" networks, it is necessary to deal with threats from the "classical" Internet also in new application areas. Threats such as worms, viruses and Trojan horses have to be addressed in completely new areas such as smart grids, vehicular ad-hoc networks or supply chains. It is known that organizations running such critical infrastructures constantly face cyber attacks.

The widespread usage of ICT as the underlying technology for various application areas, as depicted in Figure 1, is considered to be a paradigm shift. Some

M. Essig et al. (Eds.): Supply Chain Safety Management, LNL, pp. 179–197.
springerlink.com

application areas are revisited in the following discussion. Certainly, such an overview cannot be exhaustive but it should give an impression of the new challenges we have to face.

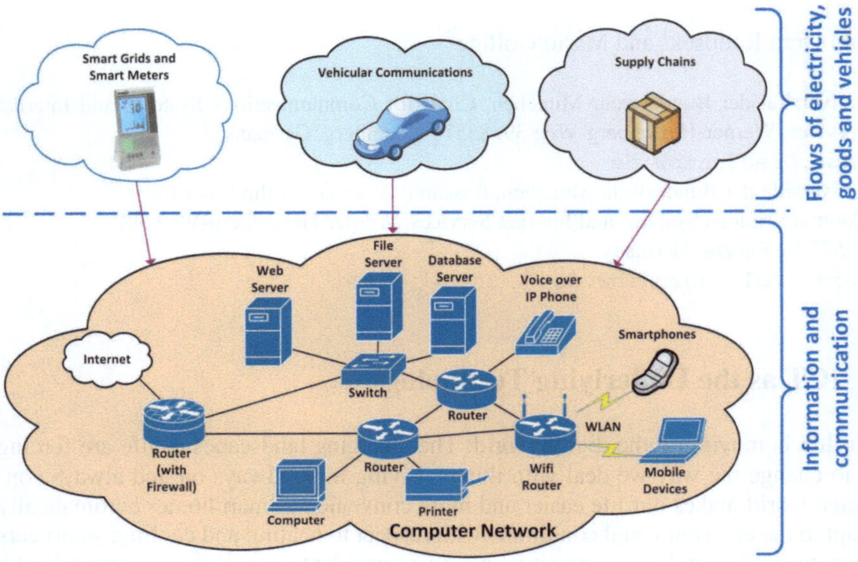

Fig. 1 ICT support for various application areas

1.1 Smart Grid and Smart Meter

Smart grids are the next-generation electrical power system that uses ICT in the delivery, consumption, and management of electrical energy. The flexibility and manageability of smart grids enable the intelligent on demand control of power flows from various suppliers (e.g., offshore wind farms) to every consumer, allowing higher utilization even during high demand periods. Furthermore, an optimization of power grids by applying predictive instead of reactive control is achievable due to measurement-based technology (i.e., wide area measurement system, WAMS). A precondition, however, is the usage of flexible and secure communication networks as the basis of an efficient network management of the power grid. In other words, sophisticated communication networks and network management approaches are the enablers for smart grids.

Security and privacy issues have also been recognized as the central aspects to be addressed and solved towards the development of smart grids. Figure 2 visualizes the problem area and identifies some potential threats in the smart grid environment.

Smart meters are devices that meter the consumption of electricity (they can be easily used for measuring gas and water consumption as well) and forward the information backwards to the energy providers. The introduction of smart meters offers a number of potential benefits to households and energy providers. Getting detailed and accurate data about the energy consumption helps to manage the energy usage locally at homes but also in the whole smart grid. However, the potential of hacking a smart meter is currently quite high, and could be considered as a new hacker's playground. Several approaches are possible such as manipulating the meter wirelessly by using software radio to emulate a variety of communication devices or sniffing on wireless communications. Another possibility is to download malware on the smart meter and thus spread it through the network. Furthermore, accessing data about the consumption of electricity on smart meters could result in generating user profiles. Smart grids use automated meters and two-way communication as well as advanced sensors to improve electricity distribution and efficiency. If a hacker controls thousands of smart meters, he could simultaneously shut them down. Massive blackouts are possible, manipulation of data from sensors with the WAMS could cause a dramatic increase or decrease of power, leading to wrong rerouting of electricity flows to specific consumers or increase the import of electricity, although not needed. Smart grids could be vulnerable to hackers. Since smart grids are a critical infrastructure, it is necessary to think about cyber security before putting an unsafe smart grid online. Several projects and papers are addressing the security aspects in smart grids (e.g., [1], [2], [3]).

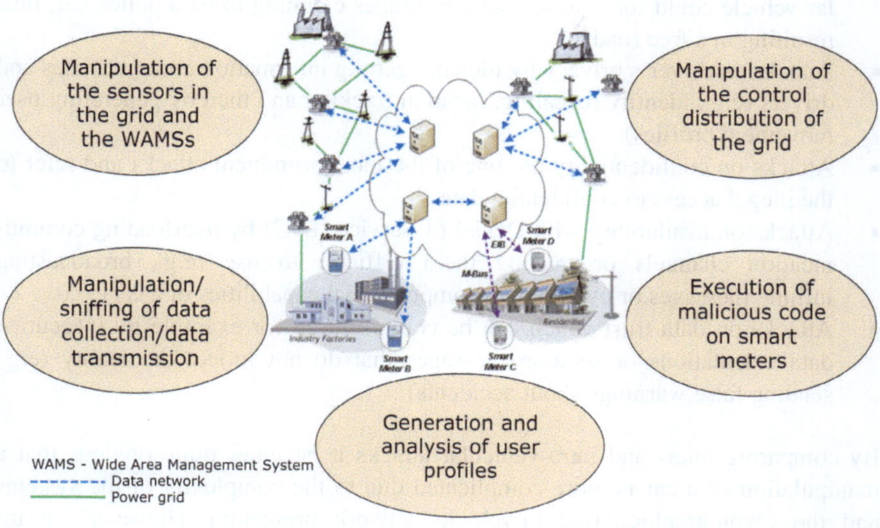

Fig. 2 Possible threats in a smart grid

1.2 Vehicular Communication

Vehicle-to-vehicle (V2V) and vehicle-to-infrastructure (V2I) communications are other scenarios of a wide deployment of wireless communication technologies that are changing our lifestyle. Increasing the safety of traffic as well as enabling a number of applications to send and receive traffic conditions and congestions are some specifics. The so-called VANETs (Vehicle Ad-Hoc Networks) connect cars to other cars, to the infrastructure and even to pedestrians. Such an exchange of data enables the drivers (i) to get a better understanding of the traffic situation and (ii) to be informed about abnormal situations as soon as possible. This data is used also by car assistance systems to take preventive actions such as automatically slowing down. Besides, the drivers have access to personalized location- and situation-based or situation-aware services and applications over the network.

Despite the numerous advantages of VANETs, several challenges with respect to security and privacy requirements need to be addressed as well. Due to the specific nature, vehicular communications are targets of a wide range of exploits that can be divided into (i) inter- and (ii) intra-vehicular communication. A crucial requirement is that life-critical information cannot be inserted or modified by a malicious person although a malicious node would become a part of the network. An analysis of attacks on inter-vehicle communication systems is given in (e.g. [4]), and summarized as follows:

- Attacks on identification and authentication where an attacker pretends to be another entity (e.g., stealing other entity's credentials) or tries to show the false possession of attributes to get some benefits. As a consequence, a regular vehicle could for instance send messages claiming to be a police car, thus resulting in a free road.
- Attacks on driver's privacy by illegally getting information about vehicles and drivers (e.g., identity revealing, location tracking and thereby generating user movement profiles).
- Attacks on confidentiality are one of the most prominent attacks and refer to the illegal access to confidential data.
- Attacks on availability with Denial of Service (DoS) by overloading communication channels or making them difficult to use (e.g., broadcasting infinite messages or overloading computational capabilities of a car).
- Attacks on data trust which can be compromised for example by inaccurate data calculations or sending messages that do not reflect the reality (e.g., sending false warnings about accidents).

By comparing inter- and intra-vehicular attacks it becomes quite obvious that a manipulation of a car is more complicated due to the complexity of the systems and the cryptographic-based in-vehicle network protection. However, an in-vehicle security module needs to protect the interface between in-car networks and the wireless communication systems by controlling external access to in-car networks, onboard control units and sensors. "Classical" security mechanisms from the Internet such as firewalls and intrusion detection systems are used for these purposes. Several publications and projects address the security and privacy aspects in VANETs (e.g., [29], [30]).

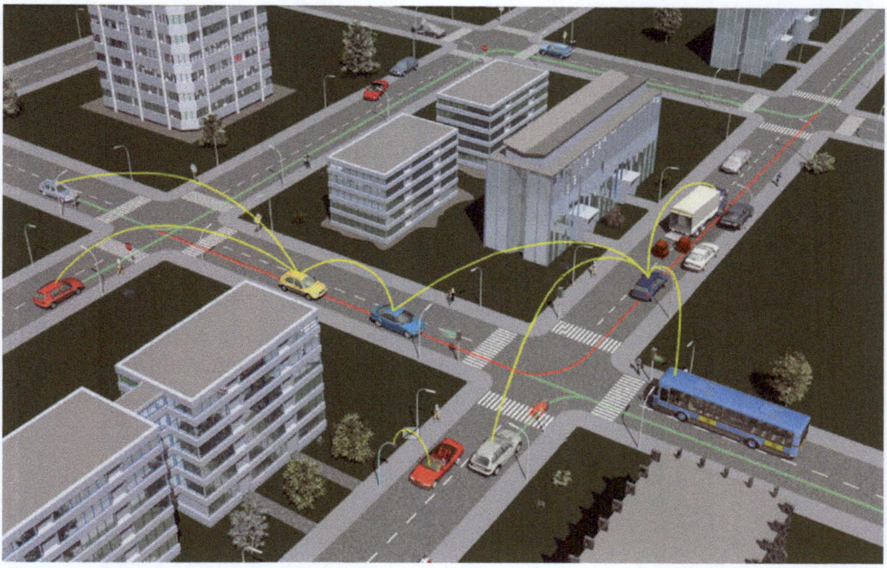

Fig. 3 VANET communications [31]

2 Threats of Cyber Security

Regarding IT security and protection of information there are three basic principles that must be ensured [7]: confidentiality, integrity, and availability (see also Figure 4).

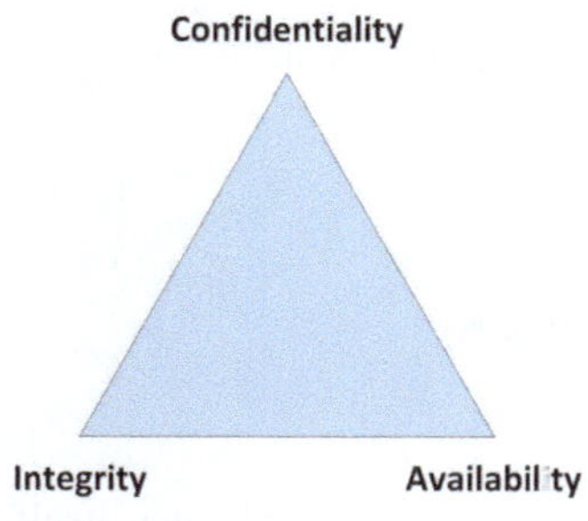

Fig. 4 CIA triad

Confidentiality is the ability to hide information from people who are unauthorized to read or modify it and thus needs protection from unauthorized disclosure. Cryptography and encryption methods are an example of an attempt to ensure confidentiality of data transferred from one computer to another.

Integrity is the protection of information from unauthorized, unanticipated or unintentional modification. This protection against undetectable modifications for example makes sure that messages are not actively modified during transition.

Availability is the access to information whenever needed. Therefore, the information technology resources must be available on a timely basis to meet mission requirements or to avoid substantial losses. High-availability systems aim to remain available at any time and prevent service interruptions due to power failures, hardware failures or system upgrades.

These IT security goals are threaten by numerous attacks. Some examples of these attacks are shown in Figure 5 and described through the rest of this section. Other classifications of attacks can be found in the literature as well. Depending on the objective for example, attacks can also be categorized with respect to the intent that can be either criminal, military, industrial espionage or others.

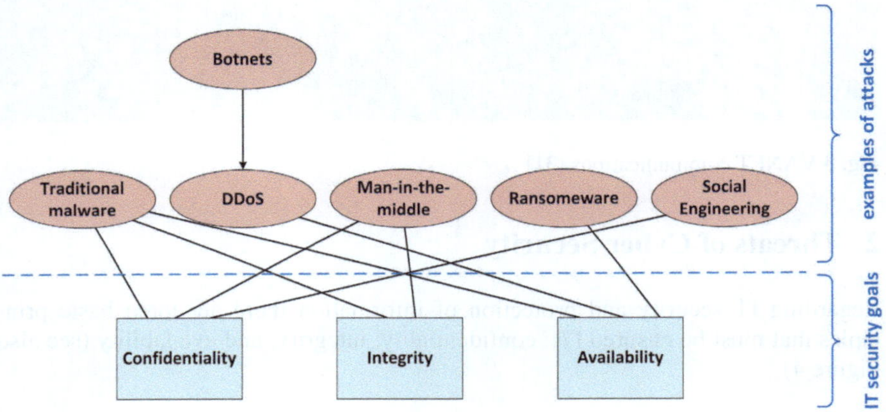

Fig. 5 Examples of attacks on IT security goals

2.1 Malware

2.1.1 Traditional Approaches

Viruses

In analogy to biology, where a virus replicates itself inside the living cells of organisms, the term computer virus describes a sequence of instructions that has been written to change the functionality of a computer without permission or observation of the user [8]. Just like the biological paradigm a computer virus always requires an infected host program / host environment to run, e.g., an office program (here often macro viruses) or a game that is infected (strictly speaking, a computer virus outside a host program is not viable). Often, during the execution of the virus a (possibly modified) copy of the virus is written to a new storage area (e.g., a file) that previously did not contain this command sequence (reproduction) and therefore, for example, a new file is infected.

Worms

A computer worm is a program that self-propagates across a network exploiting security or policy flaws in widely used services [9]. Unlike viruses that need a host program to spread themselves worms are fully self-executable programs. Worms are considered as malware and are evolving towards complex, distributed and intelligent programs such as Stuxnet, discussed in more detail later on.

Trojans

Programs that are disguised as useful applications but, without the knowledge of the user, perform a completely different negative function are called Trojans. Usually - to inspire confidence – a faked functionality causes the download and installation of the program. Unlike a computer virus, the Trojan horse lacks the property to reproduce itself independently. According to a study by Trend Micro the predominant majorities of all newly discovered traditional malwares were Trojans and thus have outplaced viruses and worms significantly [10]. Very often, additional malicious software is downloaded and installed by a Trojan horse.

A specific type of malware that is installed on computers and collects information about user's computing is spyware (e.g., key loggers). An overview of viruses, worms and Trojans can be seen in Table 1.

Table 1 Overview of viruses, worms and Trojans

	Viruses	Worms	Trojans
Brief description	A virus spreads by copying itself into uninfected files and adapting them so that the virus is executed when the host program is started	Executable program with the ability to reproduce	Program is disguised as a useful application, but contains hidden functions, often used specifically
Reproduction/ spreading	Infection of other files / boot sectors; users must perform an action, so that the virus is activated	Fully automatic self-reproduction (by sending itself as an attachment; use of USB auto-run etc.)	no
self-executable program	no (Viruses need a host-program)	yes	yes

2.1.2 Recent Developments

Stuxnet: A Proof-of-Concept of a Digital Weapon?

The massive migration from the classic model of isolated systems to a system-of-systems model where the infrastructures are intensifying their interconnections through communication networks has been seen also in the area of SCADA systems. Supervisory Control and Data Acquisition (SCADA) systems are widely used in industrial installations to control and maintain field sensors and actuators. Attacking industrial systems by computer worms is something new. Stuxnet, which experts claim for attacking Iranian nuclear industrial complexes, thus represents a new dimension in the development of worms. First, due to the specific target (i.e., Siemens SCADA systems), and secondly due to the complexity by using Windows zero day exploits (e.g., [11], [27]). Both aspects may lead to the conclusion that Stuxnet represent a new area of digital weapons.

Availability of exploit toolkits

Until few years ago attacks on computers were in the domain of relative few specialists. But since 2006 more and more of so-called "assault kits" are available on the market that allow virtually anyone to distribute malware on the Internet [12]. Although couple of malware kits were already available in the 1990s, primary the commercial distribution of the WebAttack Toolkit and its ability to exploit vulnerabilities in the browser caused the number of newly discovered malware variants to double annually (starting from 2006) [12].

The exploit toolkit Neosploit is one of the better-known attack kits that online criminals use to inject malware in the computers of Web users [13]. Another current toolkit, which has a powerful set of exploits and is spreading like wildfire is black hole [12]. At present, it is the most prevalent exploit toolkit in the wild [14].

Installation of malware through Drive-by-Downloads

Today, computers are very often infected by so-called drive-by downloads. These are attacks where the user accidentally downloads malicious software on his computer without actively initiating the download. Such an attack can occur just by viewing a web page or reading an e-mail. Frequently, vulnerabilities in browser configurations like those in script languages such as JavaScript are exploited to execute malicious code on the client machine. Although various mechanisms are already in place to prevent this effect, such as the so-called sandbox principle in Java, in which the code is executed in an isolated (storage) environment without or with very limited access to objects outside the area, new vulnerabilities are emerging recently. Frequently, such drive-by-downloads are disguised as useful applications (e.g., Adobe Flash).

2.2 Distributed Denial of Service (DDoS)

In a so-called distributed denial of service attack (DDoS) IT systems are made unavailable, for example by the use of an unmanageable mass of parallel data requirements. A common approach is the saturation of the target machine with external communications requests so that it cannot respond to legitimate traffic completely or responds so slowly that it is effectively of no practical use. Usually, DDoS consists of the combined efforts of many people to prevent web sites or web services to function efficiently (temporarily or indefinitely). In order to create a high traffic load situation many distributed client systems are involved to block the data connections or the IT systems involved. DDoS attacks have been a well-known phenomenon for quite some time and recently became the center of public attention through events such as the Estonia Affair in the second quarter of 2007, followed by the conflict in the former Soviet Republic of Georgia in the fall of 2008 and the DDoS attacks on MasterCard, Visa and others in 2010. Depending on the time and the criticality of IT systems respectively applications effected, the resulting damage can range from loss of production, revenues or reputation to supply bottlenecks for individuals or companies.

2.3 Botnets

The term "Botnet" refers to a composition of infected computers that are remotely controlled by an attacker (so-called bot herder) to perform for example distributed attacks (e.g., DDoS) on provider infrastructure or Internet services (e.g., [18]). This combination of several thousand computers gives the attacker the possibility to have access to processing power and bandwidth that exceed those of many conventional Internet access providers by orders of magnitude. The word “Bot”, derived from “Robot”, consequently describes a program that runs without human interaction usually completely invisible to the user.

For central, coordinated remote control, mostly Internet Relay Chat (IRC), a protocol originally developed for pure text-based chat communications, is used (as depicted in Figure 6). After infection, the bot program is executed that opens an IRC connection to a server afterwards. The IRC server is used as a relay station so that the Bot programs can communicate with each other as well as receive commands from the attacker. In order to control the membership to the Botnet and to minimize external influences the IRC channel is usually protected with a password.

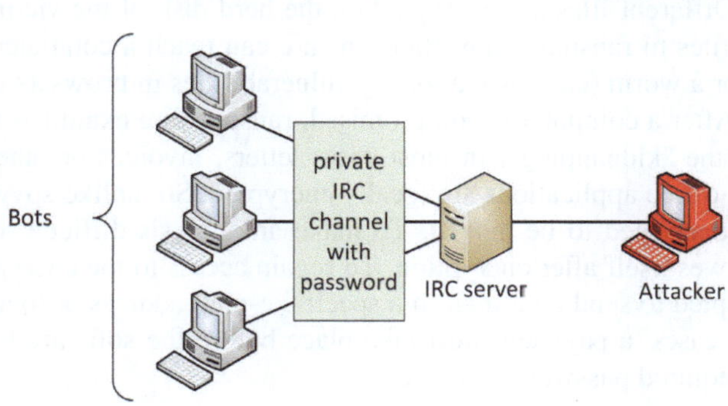

Fig. 6 How Botnets work

Due to the fact that the traffic is handled completely by the central IRC server (which represents the bottleneck) recent Botnets make use of decentralized peer-to-peer (P2P) communication more frequently [17]. P2P, which has gained public awareness with Napster, Gnutella or eMule, has the advantage that any computer (theoretically) can be transmitter and receiver at the same time that makes detection and deactivation more difficult. Even after the failure of one peer, the functionality of the Botnet continues as the Botnet reorganizes itself automatically. In recent years the rise of Botnets with criminal intent is reported. The Russian Business Network, which experts claim for up to 60 percent of all Internet crimes, offers for example computing power for virtually all purposes [19].

2.4 Man-In-The-Middle (MITM)

The term man-in-the-middle attack generally classifies attacks where the attacker is logically or physically located between the communication partners and thus has full control over the data traffic in between. Depending on the counterpart, the attacker faces the corresponding opponent without almost any possibility of detection.

Nowadays, the encryption of traffic is increasing. Thus, security extensions like https are widely used to provide secure end-to-end-communication in many critical applications. Unfortunately, without mutual authentication, i.e., without exchange of digital certificates, many of them are vulnerable to man-in-the-middle attacks. Recently, some attack methods have become public that enable also man-in-the-middle attacks of security extensions [24].

2.5 Ransomware

The word ransomware and the related phenomenon of "crypto viruses" appeared in the mid-nineties [15]. It highlights a particular class of malicious software that requires a payment in exchange for a stolen functionality. The basic procedure is as follows. Different files are encrypted on the hard disk of the victim in order to decrypt the files to ransom again. Ransomware can reach a computer like a computer virus or a worm (e.g., by exploiting vulnerabilities in browsers or the lack of a firewall). After a computer is compromised, ransomware examines the appropriate data for the "kidnapping". In most cases letters, invoices or other documents created with office applications are locally encrypted. So, unlike spyware no large amounts of data need to be moved. To make an analysis difficult, ransomeware usually removes itself after encryption. To regain access to the encrypted data, the user is prompted to send an e-mail to a specific e-mail address or to access a website. In both cases, a payment must take place before the software to decrypt the data or the required password is offered.

2.6 Social Engineering

The times of large-scale virus attacks are mostly passed. Some of the biggest threats to the security of corporate networks nowadays are targeted attacks [22]. Here, in contrast to earlier times, the design of the attack is specifically tailored to individuals or organizations (e.g., a trustworthy e-mail containing personal data, apparently sent by a person that is in close contact with the victim). Thus, on the one hand the probability that the victim actually opens the e-mail is increased, and on the other hand, existing protective measures are easier to be got around. Therefore the attacker starts with identifying potential victims. If the target group is determined, sensitive personal data is collected and categorized. Hereby, Web 2.0 with its social network sites like Facebook, LinkedIn, Twitter or Xing, is a huge treasure trove. Due to the personal data, an individualized e-mail containing malicious payload is generated and sent to the victims [23]. If the victim opens the payload, than the computer can be used and controlled (as visualized in Figure 7).

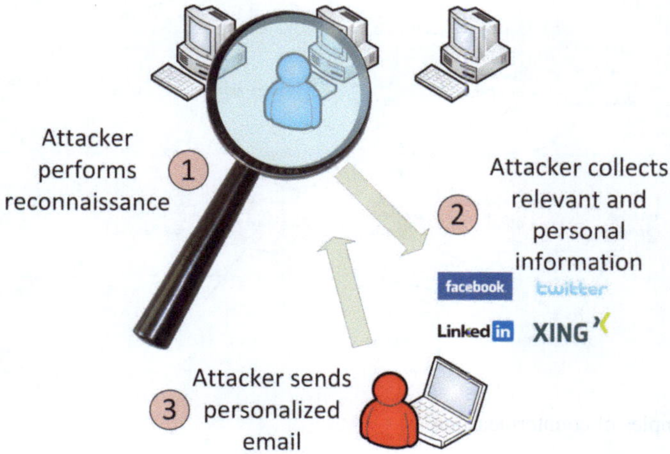

Fig. 7 Targeted attack [21]

Since 2005, an increase of targeted attacks on federal agencies and industrial espionage can be observed. Public attention was especially gained in 2007, where numerous computers in federal ministries and the German Chancellery were infected with spyware from China as a result of a targeted attack.

Recently, some methods have emerged that allow an even more sophisticated profiling, enabling an attacker to start more advanced targeted attacks or to improve the efficiency of spam campaigns [16]. The central idea behind these methods is always the same. The profiles of the different social networks are evaluated by special procedures and automatically linked between each other to enrich the information. It has been demonstrated that – based on a list of about 10.4 million e-mails - the automatic user profiling of more than 1.2 million user profiles including the linking between different social networks is possible [16].

3 Countermeasures

Cyber security is not only a technical issue. Numerous sources confirm that particular organizational and personnel countermeasures are of increasing importance, and less technical protection. Security awareness of the employees is certainly an essential aspect (e.g., not bringing USB sticks to the corporate network).

Based on the attacks described in the previous section, this section covers some examples of possible countermeasures. For this purpose, the countermeasures have been divided in technical and human/organizational measures, as depicted in Figure 8.

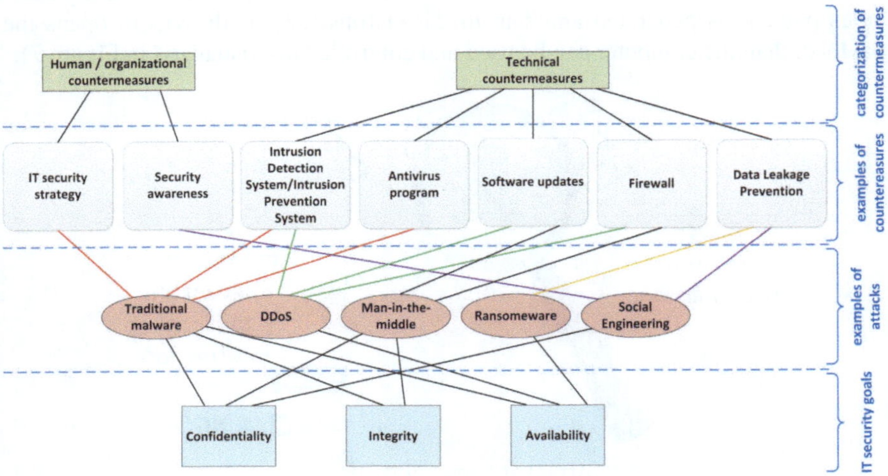

Fig. 8 Examples of countermeasures

3.1 Human Respectively Organizational Countermeasures

Orthogonal to the classification of various attacks, a second division into internal and external perpetrators seems to be important to the overall understanding, as a non-negligible number of perpetrators are insiders [25]. Therefore, the issue of insider threats will be discussed at this point before starting with the analysis of countermeasures.

The significance of insider threats

Regarding the protection against industrial espionage and information flows out of the company, many businesses focus only on protection against attacks from outside [26]. However, very often the perpetrators are within the company as visualized also in Figure 9 [25]. In times of rising fears to loose one's job, permanent growing workload and often a lack of appreciation of performance, many employees are increasingly willing to enrich themselves at the expense of the company they are working for. Loyalty to the employer is no longer always natural. A loss of wages is thus compensated by a small additional income more often.

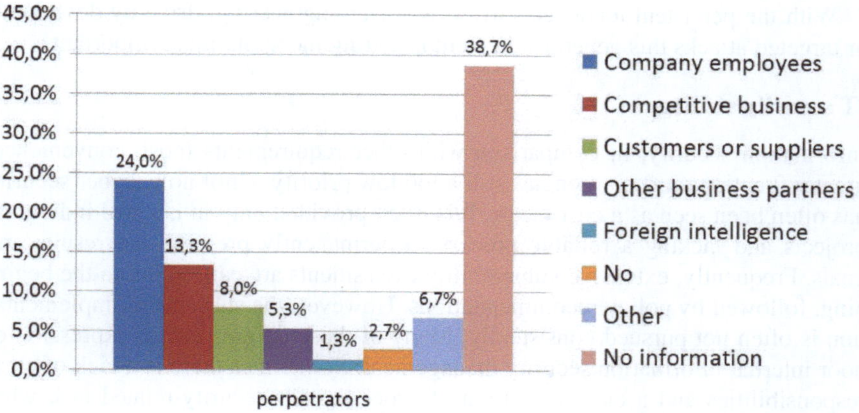

Fig. 9 Perpetrators of espionage in the German economy [25]

Interviewed in a study conducted in German companies about types of employees who were specifically responsible for the espionage, first and foremost, the clerks (which usually have many access rights including the access to sensitive documents and information) with 31.4% were identified, followed by skilled workers with 22.9% and 17.1% within the management [26]. These three areas together caused about two thirds of the entire data leakage of the company.

The management on the second level is also surprisingly vulnerable to espionage. In contrast to the senior management which was involved only very rarely in the criminal actions (with 2.2%), the second level management is so much more tempted to enrich themselves at the expense of their employer by selling information.

Protection against industrial espionage has "top priority" as well. Delegating it to subordinate bodies would be a wrong signal. All employees should be included in the information protection as they can all be victims of an attack.

Security awareness

According to the European Network and Information Security Agency (ENISA), awareness of the risks and available safeguards is the first line of defense for the security of information systems and networks [32].

The defense of human interference with the aim to get unjustified access to data or things (social engineering) is not easy to accomplish since the attacker exploits positive human qualities such as the desire to help in emergency situations. In such cases a solution can only be non-technical such as making identity and legitimacy guaranteed without any doubt before any further actions are performed.

Social engineering for example can only be defended if the employees are informed about possible tricks of attackers and have learned to deal with potential threats. In the context of security awareness, it is, for example, important (i) to be careful when talking about business issues outside the company, (ii) not to publish too many private details in social networks or (iii) not to surf to unknown web sites or downloading and installing unknown software or software from an unknown source.

With the persistent tendencies towards social engineering, drive-by downloads or targeted attacks this general statement is getting more and more important.

IT security strategy

Information security, in comparison with other requirements (cost, convenience, greater functionality...), often has still a too low priority. Until now, cyber security has often been seen as a cost factor. It is often provided only in isolated individual projects and lacking a reliable process to permanently preserve the results and goals. Frequently, extensive vulnerability assessments are carried out in the beginning, followed by policy recommendations. However, the subsequent implementation is often not pursued consistently. Many of these deficits are an expression of poor internal information security management. Sometimes there is a lack of clear responsibilities and a clear definition of processes for security-related task whereas sometimes agreed measures are not checked on a regular basis. Nevertheless, the insight is growing that an inadequate cyber security strategy (along with security processes and mechanisms) is certainly costing more.

3.2 Technical Countermeasures

The protection of IT goals is a combination of technical and non-technical countermeasures. For example, the risk of drive-by downloads cannot be reduced significantly by avoiding the "backyard" of the Internet. Sometimes, even public websites from magazines or television advertisements contain active elements that are executed by third parties. Some of the websites are set up with many elements that are brought together from many computers (done automatically by the browser). One of these computers can be hacked and loaded with malicious software.

Based on the attacks described in section 2, selected technical countermeasures are discussed in the following.

Antivirus programs

An antivirus program detects, blocks, and possibly eliminates known viruses, worms and Trojans. Unfortunately, due to the continuous development and the unpredictability of the “evil intelligence”, there is practically no virus scanner on the market that protects against all imaginable kinds of viruses, worms and Trojans. Signature-based approaches which are state-of-the-art have their limitations due to their reactive nature (i.e., signatures need to be known in advance). Novel, anomaly-based approaches need to be developed. However, antivirus programs offer a very good protection nowadays especially against known viruses and thus are a must for every computer.

Software updates

Through continuous software updates, vulnerabilities can be eliminated or functions expanded. Similar as with antivirus programs, regular software updates are a must for ICT systems (computers, switches, routers, ...). An update is especially necessary when vulnerabilities are known to have an impact on the safe operation of the network, when failures occur repeatedly or a functional enhancement of safety/technical requirements is necessary. However, in most cases patches are released in a cumulative way on some specific days (patch days), allowing to exploit the unpatched vulnerabilities.

Firewalls

A firewall is used to restrict network access based on sender or destination address or services being used. The firewall monitors the traffic going through and decides based on defined rules to determine whether certain network packets are blocked or not. By this, it tries to prevent unauthorized access.

Depending on the location of the firewall, the distinction between a personal firewall (also called a desktop firewall) and an external firewall (also known as network or hardware firewall) is made. In contrast to the personal firewall the software of an external firewall runs on a separate device that connects networks or network segments to each other. However, firewall rules need to be updated as soon as new services appear. Some peer-to-peer services such as Skype even avoid firewall filtering. New services, the necessity to provide firewalls also on smart phones are examples of the new challenges in firewall research.

Intruder: Intrusion Detection System/Intrusion Prevention System

Firewalls do not detect attacks. The detection of attacks against a computer system or computer network is the responsibility of Intrusion Detection Systems (IDS) [28]. When the pure generation of events is enhanced with additional features in order to protect against a specific attack, then the system is called an Intrusion Prevention System (IPS). There are three types of IDS/IPS:

A *host-based system* needs to be installed on each monitored system. However, the term "host" must not be misunderstood. In this context a host system is meant to be the one, where an IDS/IPS is installed, and not as a synonym for a computer.

A *network-based system* attempts to capture and analyze all packets on the network and reports suspicious activity if detected. These systems also attempt to identify patterns of attacks within the network traffic [28].

Hybrid systems combine both principles in order to ensure better coverage in the detection of occurring attacks. The term hybrid refers to network-and host-based sensor types that are connected to a central management system.

The deficiencies of IDS/IPS systems are the principal of a signature-based approach (e.g., it is possible to detect only patterns or signatures that are known). Such an approach is outdated since due to targeted attacks, the profiling of an attack has changed towards a targeted attack. A target attack is per se not known in advance. Thus, anomaly-based approaches that recognize "abnormal" behavior need to be further developed to be able to react to the "unknown".

Extruder: Data Leakage Prevention

The term data leakage prevention refers to the protection against a suspected, but not measurable and sometimes not even detectable sharing of information to unwanted recipients. Thus, for example all file names that are read by or written to all USB devices are logged so that each change to sensitive data is traceable. Furthermore, with the use of a unique serial number, a USB stick can be assigned to only one specific user. As the stick is encrypted, reading the data on the stick is only possible for colleagues of the department or superiors. In addition, it can also be regulated what kind of applications are allowed to be executed.

4 Cyber Security: A View towards the Future

As seen from the discussion so far, sophisticated and especially trustworthy ICT systems are the "heart" of all critical infrastructures. Securing communication networks and IT systems will be even more demanding in the future. The so-called Future Internet (Internet of Things/Data/Services) envisions a world of extremely powerful mobile devices, billions of "smart" things, new personalized services and applications to dig information, to share opinions and feelings (e.g., social networks as Facebook or Twitter), to monitor our health condition, to allow access to almost infinitive resources in terms of bandwidth, storage and computing power (i.e., cloud computing), accessible from everywhere. Furthermore, we need to face trends such as IPv6, increase of peer-to-peer traffic, encrypted payload.

The way we use the Internet is changing. We are not accessing servers somewhere in the network but rather content and (personalized) services. Sometimes, it is claimed that Future Internet is about content, services and management [33], not about infrastructure.

These trends will change drastically the way of our life, the way we produce and consume energy, transport goods or deal with other people's needs. A trustworthy ICT is certainly the basis for such developments nowadays and in the Future Internet. Privacy of people and security of data and resources are among the most important issues that need to be solved adequately. The changing environment demands the development of secure systems that protect themselves (self-protecting, self-managing) and also others. Furthermore, the content placed somewhere in the network or cloud needs to be protected as well. We see protection on various layers (infrastructure, content, services, or even maybe packets).

Nowadays, security is in most cases an afterthought. Such an approach is not applicable any more. We need security-by-design and management-by-design in order to cope with the new challenges. As depicted in Figure 10, a holistic approach to cyber security that ranges from the real to the virtualized world is needed. The emerging trend of cloud computing, providing IT support as a utility, is the underlying technology for application specific clouds such as the logistics, automotive or energy supplier clouds. Short time-to-market cycles can only be achieved through IT support, adapted to the processes and an efficient infrastructure. Regardless of the type of clouds (e.g., private, public, hybrid or Software as a Service (SaaS), Platform as a Service (PaaS), Infrastructure as a Service (IaaS)), the success of this approach will depend on solving cyber security and privacy issues.

Cyber security nowadays already has to face several threats. The Internet of the Future will add more. New technologies, new services, new application areas where ICT is the underlying technology for critical infrastructures will produce new threats. We are facing them however with "old" concepts such as IP and an endless patching. Research in developing new cyber defense approaches is of mandatory importance towards a trustworthy ICT and critical infrastructures.

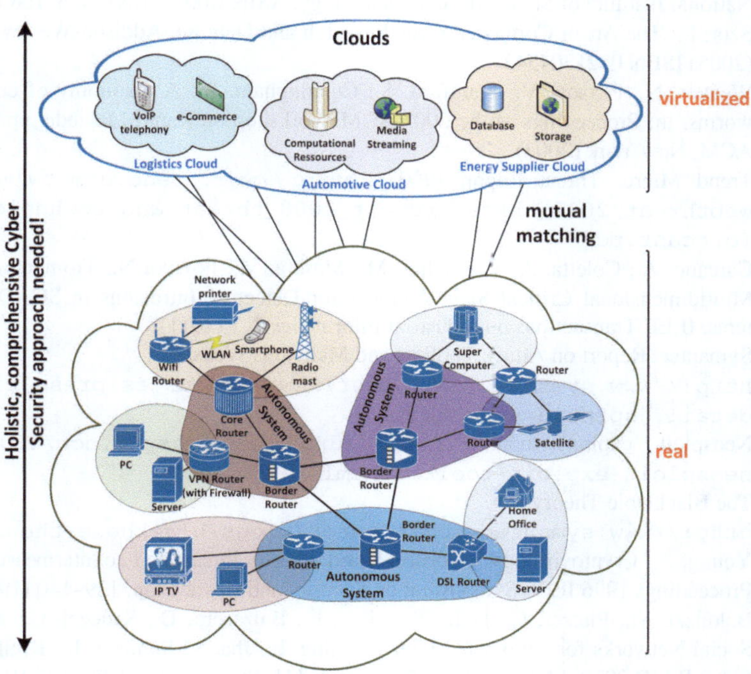

Fig. 10 Holistic view of Cyber Security

References

[1] Khurana, H., Hadley, M., Lu, N., Frincke, D.A.: Smart-Grid Security Issues. IEEE Security & Privacy Magazine 8, 81–85 (2010)

[2] Ray, D., Harnoor, R., Hentea, M.: Smart power grid security: A unified risk management approach. In: 2010 IEEE International Carnahan Conference on Security Technology (ICCST), pp. 276–285. IEEE (2010)

[3] Metke, A.R., Ekl, R.L.: Security technology for smart grid networks. IEEE Transactions on Smart Grid 1, 99–107 (2010)

[4] Kargl, F., Papadimitratos, P., Buttyan, L., Müter, M., Schoch, E., Wiedersheim, B., Thong, T.-V., Calandriello, G., Held, A., Kung, A., Hubaux, J.-P.: Secure vehicular communication systems: implementation, performance, and research challenges. IEEE Communications Magazine 46, 110–118 (2008)

[5] Fischer-Hübner, S.: IT-Security and Privacy: Design and Use of Privacy-Enhancing Security Mechanisms. LNCS, p. 351. Springer (2001) ISBN 9783540421429
[6] Mouratidis, H., Giorgini, P., Manson, G.: Integrating Security and Systems Engineering: Towards the Modelling of Secure Information Systems. In: Eder, J., Missikoff, M. (eds.) CAiSE 2003. LNCS, vol. 2681, pp. 63–78. Springer, Heidelberg (2003)
[7] Swanson, M.: Security self-assessment guide for information technology system, US Department of Commerce, Computer Security Division, Information Technology, National Institute of Standards and Technology, ASIN B002WJINW4, p. 108 (2001)
[8] Szor, P.: The Art of Computer Virus Research and Defense. Addison-Wesley, p. 744 (2005) ISBN 0321304543
[9] Weaver, N., Paxson, V., Staniford, S., Cunningham, R.: A taxonomy of computer worms. In: Proceedings of the 2003 ACM Workshop on Rapid Malcode, pp. 11–18. ACM, New York (2003)
[10] Trend Micro, Threat Report (2007), http://www.trend.com.tw/micro/webthreat/2007_threat_report_2008_threat_and_technology_forecast.pdf
[11] Carcano, A., Coletta, A., Guglielmi, M., Masera, M., Fovino, N., Trombetta, A.: A Multidimensional Critical State Analysis for Detecting Intrusions in SCADA Systems. IEEE Transactions on Industrial Informatics 7, 1 (2011)
[12] Symantec, Report on Attack Toolkits and Malicious Websites, http://www.symantec.com/about/news/resources/press_kits/detail.jsp?pkid=attackkits
[13] Neosploit exploit toolkit, http://dxp2532.blogspot.com/2007/12/neosploit-exploit-toolkit.html
[14] The BlackHole Theory, http://www.symantec.com/connect/blogs/blackhole-theory
[15] Young, A.: Cryptovirology: extortion-based security threats and countermeasures. In: Proceedings 1996 IEEE Symposium on Security and Privacy, pp. 129–140 (1996)
[16] Balduzzi, M., Platzer, C., Holz, T., Kirda, E., Balzarotti, D., Kruegel, C.: Abusing Social Networks for Automated User Profiling. In: Jha, S., Sommer, R., Kreibich, C. (eds.) RAID 2010. LNCS, vol. 6307, pp. 422–441. Springer, Heidelberg (2010)
[17] Ianelli, N., Hackworth, A.: Botnets as a vehicle for online crime. CERT Coordination Center, pp. 1–28 (2005)
[18] Ard, C.: Botnet Analysis. Forensic Computer Science IJoFCS 2, 65 (2007)
[19] Krebs, B.: Shadowy Russian Firm seen as Conduit for Cybercrime. Washington Post A15 (October 13, 2007)
[20] Demchenko, Gommans, L., de Laat, C., Oudenaarde, B. : Web services and grid security vulnerabilities and threats analysis and model. In: Proceedings of the 6th IEEE/ACM International Workshop on Grid Computing, pp. 262–267 (2005)
[21] Targeted Attacks, https://info-point-security.com/images/stories/2010/Symantec_targeted_attacks_2.jpg
[22] Richardson, R.: CSI computer crime and security survey. In: Computer Security Institute, pp. 1–28 (2007)
[23] Anagnostakis, K.G., Sidiroglou, S., Akritidis, P., Xinidis, K., Markatos, E., Keromytis, A.D.: Detecting targeted attacks using shadow honeypots. In: Proceedings of the 14th Conference on USENIX Security Symposium, vol. 14, p. 9 (2005)
[24] Guha, R.K., Furqan, Z., Muhammad, S.: Discovering Man-in-the-Middle Attacks in Authentication Protocols. In: Military Communications Conference, MILCOM 2007, pp. 1–7. IEEE (2007)

[25] Corporate Trust, Studie: Industriespionage - Die Schäden durch Spionage in der deutschen Wirtschaft, Munich pp. 1–56 (2007), http://www.corporate-trust.de/pdf/STUDIE_191107.pdf
[26] Bussmann, K., Krieg, O., Nestler, C., Salvenmoser, S., Schroth, A., Theile, A., Trunk, D.: Wirtschaftskriminalität 2009 Sicherheitslage in deutschen Großunternehmen. In: Martin-Luther-Universität Halle-Wittenberg and PwC AG, pp. 1–65 (2009), http://www.pwc.de/de/risiko-management/.../Studie-Wirtschaftskriminal-09.pdf
[27] Falliere, N., Murchu, L.O., Chien, E.: W32. stuxnet dossier. Symantec Security Response, 1–69 (2011), http://www.symantec.com/content/en/us/enterprise/media/security_response/whitepapers/w32_stuxnet_dossier.pdf
[28] Roesch, M.: Snort-lightweight intrusion detection for networks. In: Proceedings of the 13th USENIX Conference on System Administration, Seattle, Washington, pp. 229–238 (1999)
[29] Noecker, G.: now: Network on Wheels, http://www.net-on-wheels.de
[30] The Secure Vehicle Communication (SeVeCOM) Project, http://www.sevecom.org/
[31] Car 2 Car City Scenario, http://www.car-to-car.org/uploads/pics/BMW_3D_Car2Car_CityScenario.jpg
[32] ENISA, Why is awareness raising, Article, pp. 1–2, http://www.enisa.europa.eu/media/key-documents/fact-sheets/Awareness-1.pdf
[33] Schönwälder, J., Fouquet, M., Dreo Rodosek, G., Hochstatter, I.: Future internet= content+ services+ management. IEEE Communications Magazine 47, 27–33 (2009)

How Logistics Can Create and Support Public Security

Dr. Matthias Witt

LOG GmbH, Managing Director, Adenauerallee 131a,
53113 Bonn, Germany
Matthias.Witt@LOGmbh.de

1 General Development of the World after the Cold War

In the last two decades, the security environment and the attitude toward security both of individuals and of society as a whole has shifted significantly. The end of the East-West conflict and the associated dissolution of the bipolar world order resulting from the fall of the Berlin Wall and the disbanding of the Soviet Union led to a paradigm shift and entirely new conditions, particularly in regard to security policy.

After the historical break resulting from the attack on the Twin Towers of the World Trade Center and the Pentagon – the nerve centers of global financial and military power – on September 11, 2001, security was tightened worldwide. Politics, society and business all had to adapt to these new challenges. This is especially true for the logistics sector.

Since 1990, the world has perceptibly moved together, it has become smaller. Geographic borders are losing importance, and globalization is progressing inexorably. For business, this has created global markets for almost all goods and services, including capital and work, and with all the consequences that entails. The public often perceives these changes more as a threat than as an opportunity.

However, from the perspective of a logistician, globalization offers enormous development potential for opening new markets. It belongs, next to progress in communications and information technology, among the driving forces of the sector. Today, it is not unusual when a product is developed in the USA, produced in Asia and sold in Europe. Logistics must constantly adapt and grow to guarantee efficient solutions to the tasks at hand.

The significance of logistics will increase in the future, since the interdependence of societies will continue to grow. Trends such as the expanding movement of global capital and goods, the advancing exchange of information and knowledge, and intensifying migration will all increase.

The worldwide flow of good and global supply chains already allow us as a society to profit from the positive effects of globalization. But modern industrial societies are also vulnerable to politically motivated terrorists and international criminal organizations, particularly as a result of this tight interconnectedness and its dependency on technical infrastructure.

The eruption of the volcano Eyjafjallajökull on Iceland in 2010 vividly demonstrated how dependent the world already is on a functioning logistics infrastructure, as well as how it can be adversely affected by natural disasters in

M. Essig et al. (Eds.): Supply Chain Safety Management, LNL, pp. 199–209.
springerlink.com

ways that humans are powerless to prevent. The effects on European air traffic caused the travel tourism industry to collapse; many exhibitors were absent from the Hannover Fair because they were stuck in Asia, production lines of large global businesses threatened to grind to a halt and a number of teams were unable to fly to the 2010 European Gymnastics Championships in Birmingham.

Today, the failure of a system – whether transportation, communication or energy supply – shocks almost all areas of life and production and triggers various domino effects, which are virtually impossible to estimate in their complexity. The earthquake and resulting tsunami which flattened the northeastern coast of Japan in March 2011, triggering explosions in the nuclear power plant in Fukushima, are a tragic reminder of this. Stunned by the destruction, the world was forced to watch the Japanese desperately attempt to prevent the worst-case scenario from becoming reality.

Individual nations and international organizations have the difficult task of ensuring public security as efficiently and effectively as possible in this situation with its varying threat perceptions. Here, public security includes the integrity of all material flows, in other words: maintaining public security primarily means that today's complex processes and systems – local, national and transnational – operate smoothly.

Public security has long since left the exclusive province of government: approximately 80 % of all sensitive installations in Germany are now in private hands. A large number of actors now contribute to maintaining security. In the first instance, these are certainly government agencies and organizations charged with security tasks. In Germany, this includes the police and the Constitutional Affairs committees of the federal and state governments, the Federal Intelligence Service, the Military Intelligence Service (MAD) and the THW. But as a result of the many new threat scenarios, additional actors have been added: for example state accredited private aid organizations such as the German Red Cross, Johanniter-Unfall-Hilfe, Malteser Hilfsdienst or Arbeiter-Samariter-Bund.

There is no shortage of experts, but it must be clear where each responsibility lies. For this reason, since 2002, a biannual National Crisis Management Exercise (LÜKEX) has taken place. This exercise simulates different scenarios to prepare as comprehensively as possible to face possible threats. Goals, processes, structures, and the abilities and means of the actors must be networked in such a way that crisis management works.

The two large German logistics companies Deutsche Post and Deutsche Bahn have also been involved in LÜKEX for years. As transport and communications groups, they are an indispensible part of successful crisis management. This demonstrates the central role the logistics sector plays in maintaining public security. It is they who help create security with their solutions by supporting governmental administrations and organizations in various ways.

How they do this and which tasks can be assumed by the logistics sector will now be examined in detail using two selected examples. The first involves securing supply chains, the second concerns supporting the military in its tasks.

2 Supply Chain Security

Global trade has developed at a rapid pace. At the center of global transport is a tightly woven web of sea lanes and the connected ports. It must be assumed that port volumes will continue to increase. This fact makes the transportation network vulnerable to terrorist attacks.

The logistic sector's responsibility for public security within the framework of supply chain security is therefore particularly obvious. Supply chain security continues to grow in global importance. Here, supply chain security is defined as protection from attacks intended to cause damage to the supply chain, i.e. by organized criminality, international terrorism, sabotage or pirates. This must be differentiated from supply chain safety, which covers the aspect of security concerned with maintaining operational performance.

If it were solely up to the USA, today the entire supply chain would be one hundred percent controlled and therefore protected from terrorist attack. As a result of pressure from the USA after the attacks of 9/11, the ISPS Code (International Ship and Port Facility Security-Code) of the IMO (International Maritime Organization) was instituted, which today represents one of the most important foundations for secure ports and trade routes. It governs the loading and unloading of ships, the transport and intermediate storage of goods in ports, dictates security checks and names the measures which must be carried out to avert threats. Because no one wished to be shut out of the global trade, almost all countries committed themselves to implementing the security rules. According to this: along the entire transport route, that is, from the starting point to the border of the port facilities, the transport and logistics service providers and carriers are responsible for the security of the cargo. With entry to the port facilities, the ISPS Code applies to all those involved. The result: since it was instituted, German ports also resemble high-security areas.

Increased entry controls to warehouse and terminals within port facilities and to ships, identity checks at hubs, barriers, modern digital entry control systems, vehicle identification at gates, video surveillance systems with sensors, and electric fences – these can now be found in ports worldwide. This carries with it the risk that ports could be paralyzed, since a higher container turnover always results in more security checks. But understandably, necessary controls may not delay transport processing.

At present, there are over 400 rules and regulations from the USA and the European Union governing the transatlantic transportation of goods. Already in force is the Container Security Initiative (CSI), which states that for transports into the United States, logistics companies are required to report to US security services 24 hours in advance which cargo is to be delivered where. This is intended to make the transport of dangerous goods under an intentionally false declaration more difficult. However, this also includes the "H.R.1 law", under which the USA wishes to scan all containers beginning in October 2012 in order to increase security. It is evident that with over 20 million containers being transported globally, such demands would incur high costs for the logistic companies. The European Commission in Brussels reckoned a few years ago with procurement costs for X-ray and scanner equipment of €430 million and annual operating costs of €200 million.

The EU, too, has driven up logistics costs with some rules. In 2008, the customs reform went into effect, a measure focused on defending against terrorism. It standardizes the EU's security rules for transportation and logistics companies, and proclaims electronic customs procedures as the future. Companies which fulfill the required security standards can apply for the status of 'Authorised Economic Operators' (AEO) and thereby enjoy trade facilitation, which has the positive effects of faster and simplified customs processing. But critics complain both about data retention and the increased financial and bureaucratic burden for transportation, logistics and port operations. Furthermore, the expansion of security technologies and the procurement of new security technology is not free. To recoup these additional costs, increasing prices for end customers can hardly be avoided. But the logistics sector will certainly find a solution to reduce these high costs.

These logistics solutions could be based on GPS technology, for example, using so-called geofencing. With this, the current GPS position is constantly compared with a predefined transport corridor. Deviating from this corridor triggers an alarm. Subsequently, the reaction to the alarm is managed by headquarters. Sensor technology can also be used that measures specific values during container transport and relays these to headquarters. For example, by measuring light, it can be determined if a container has been opened or remained closed. Other sensors can show the effects of force with vibration sensor, or monitor temperature to ensure the cooling of specific objects.

A further option is the use of RFID solutions. These are already used invisibly in passports, for inventory control in shops and in access controls. RFID can help in many other instances to increase the security of citizens and business processes. The technology offers significant development potential. However, the potential security risks must also be considered: due to their widespread use, attackers will have many opportunities to read the RFID tags with commercially available scanners and to misuse the collected data. Therefore, the logistics sector will have to place particular emphasis on the areas of privacy, authentication, authorization and availability. Data collected within the framework of identification must be protected especially well!

The volcanic eruption on Iceland in 2010 showed something else: namely that German companies are well-prepared for the risks of globalization. Only in individual cases did the ash cloud lead to problems; companies were able to deal well with the failures, supply chains held almost everywhere and production was able to continue. The reason for this was risk mitigation by companies and the fact that the logistics sector is used to acting flexibly; it is confronted with closed airports or defective machines almost daily and is used to finding alternative routes and alternate strategies. Although deliveries were delayed by a few days, they still reached their goal. At BMW, plants were only affected by production shutdowns because the strategic decision had been made to maintain the lowest possible stock ("line feeding").

However, German companies should not draw the conclusion from this that they can overcome every "crisis" with flexibility and strength. They should rather further increase their preparations, because similar situations can occur at any time. Many manufacturers of logistic solutions have accepted the challenge to make a contribution to public security and are working hard on technical solutions. Promising approaches can already be seen today.

LOG has also been active in this segment for years, supporting NATO by helping it to track and secure goods deliveries. Using transparent, stabile and secured supply chains, the entire logistics sector will make a contribution to public security; but it cannot do this for free!

Photo 2.1 NATO Consignment Tracking by LOG– the antenna on the building collects RFID-data (Picture: LOG)

3 Assuming Sovereign Military Duties

The logistics sector also supports public security by assuming originally sovereign duties. In almost all Western industrial countries, a trend has developed since the 1990s. The services offered range from supply, catering and cleaning, to logistics and transport, from consulting and training to property and personal security.

As a result of the power vacuum after the end of the Cold War, the private military services sector has thrived. Typical here is that they not only offer a product, but operate it on behalf of the military. They include not only military companies (logistics, repair/operation of weapons systems), but also security firms (protection, guarding), in particular: 'military provider firms', which participate in armed conflict, 'military consulting firms', which offer consulting and training, and 'military support firms', which support logistics, maintenance, transport, reconnaissance or information technology.

In many countries, domestic conflicts developed, civil wars began, nations collapsed, and international conflicts, which were previously hidden behind the façade of the Cold War, flared up. At the same time, the end of the block confrontation led to military cutbacks; today there are approx. 7 million fewer soldiers than in 1989. This saturated the market with ex-soldiers seeking new fields of activity.

Furthermore, at this time the USA felt little inclination to interventionism. If missions were nevertheless undertaken, the number of soldiers was kept to a minimum. The outsourcing of tasks to private suppliers was an obvious next step. The United Nations in particular was identified as a potential client due to its chronic shortage of operational contingents.

The use of private companies by the US army increased, so that in the Gulf War in 1991, there was one 'contractor' for approx. every 50 soldiers. From 1994-2002, the Pentagon signed approx. 3,000 contracts with private firms for missions at various locations.

In Afghanistan and Iraq, the activity of private service providers reached their highpoint. Vinnell, CACI, Blackwater, DynCorp, Custer Battles, Titan Corporation and Aegis Defense Services trained Iraqi and Afghani security forces, guarded bases and the Baghdad airport, the 'Green Zone' and the Afghan President Karzai. As a reference point, for Iraq, approx. 20,000 non-Iraqi employees in private companies are named, representing a ratio of 1:10 to military forces.

Until the new Iraqi army became operational, they were the second-largest fighting force in Iraq, approximately as strong as all of the USA's coalition partners combined. Although the use of armed service providers is in principle not new, through their intensity and huge number a new quality has been achieved.

Uncounted are an additional 50,000 civilian non-Iraqi employees active in the areas of supply and logistics. About 20-30% of the supply capacity in Iraq, so estimates, are delivered privately. Undisputed is their quantitative increase since the 1990s. The revenue of the sector, which was $100 billion in 2003 is estimated to reach $202 billion in 2010. Civilians are directly involved in hostilities today as never before in military history!

There are a number of reasons for this: The end of the Cold War and its effects have already been addressed. Additionally, since the 1990s, a general trend towards privatization in almost all western countries has developed from the desire to save money through outsourcing. The privatization of Deutsche Post and Deutsche Bahn are good examples, and in the federal state of Hessen the first German prison has been partially privatized. The tendency to also transfer military tasks to the private sector is therefore part of a trend.

With globalization, great armies became obsolete, since subnational actors could also exert force of arms through information technology and the close interconnectedness of the world. Thus, the waging of war changed dramatically. Conversely, modern armies now require specialists in all areas, as modern weapons systems require fewer soldiers but more civilian experts. This is clear in training, maintenance, operation, logistics, and in particular in all areas of IT, in the testing of new technologies and for special tasks, for example airborne refueling or systems maintenance. It would be simply unaffordable for the military to train their own personnel and maintain them in the numbers required. It is also hardly possible for sluggish national structures to keep pace with complex technological developments.

In general, a distinction is made today between specialists whose work optimizes the military ('mission enhancing'), and abilities which are not present in the army ('mission essentials'). While a freedom of choice exists for 'mission enhancing' regarding the use of civilians, their use for 'mission essential' tasks has become unavoidable.

Weapons systems manufacturers have gradually changed their business models: where they previously manufactured systems and delivered them to the military, today they offer complete services for the entire product lifecycle; they are maintained from 'factory to foxhole'.

In the long run, the Bundeswehr was also unable to buck this trend, although they began relatively late. Only in 1999 was a framework agreement made between the German Federal Government and approx. 600 companies and trade associations, in which projects in the information sector, service area, vehicle management, training and maintenance operations were identified. In the focus stood the minimization of operating costs and the easing of the burden on the Bundeswehr for tasks which do not belong to the core military responsibility and can be provided at lower cost by the private sector.

To date, diverse tasks have been partly or completely privatized within the framework of outsourcing or public-private partnership projects. The start was made by the federal 'Chemical Weapon and Armament Pollution Disposal Company' (GEKA), which at the end of 1999 began operations in Munster.

In August 2002, the 'Gesellschaft für Entwicklung, Beschaffung und Betrieb' (g.e.b.b.) was founded as a GmbH owned by the German federal Government. Its task was to work entrepreneurially and as a partner in other PPPs of the Federal Ministry of Defence (BMVg), to advise the BMVg and become a driving force in privatization in order to, in the words of the contract, 'achieve a systematically and institutionally safeguarded highest degree of cost effectiveness while meeting the needs and service of the Bundeswehr.' Including the g.e.b.b., 5 organizations with approx. 8,300 former employees of the Bundeswehr were created.

In 2002, the Bundeswehr-Fuhrpark-Service (BwFPS) GmbH vehicle fleet manager in Troisdorf followed, in which the federal government has a 75.1 % stake; the remaining 24.9% is held by the Deutsche Bahn AG. Tasks of the company are the management and operation of the vehicle fleet of the Bundeswehr, primarily domestically. Also in 2002, the LH Bundeswehr Bekleidungsgesellschaft (LHBw) was founded in Cologne, of which the federal government has a 25.1 % stake. The majority of 74.9 % is split among the 'Lion Apparel GmbH', a manufacturer of special clothing and whose sister company supplies the US army with uniforms, and the logistic supplier 'Hellmann Worldwide Logistics GmbH'. Task of the LHBw is supplying soldiers with uniforms more quickly and efficiently. Over the contract duration of 12 years, €718 million are to be saved.

In February 16, 2005, the Heeresinstandsetzungslogistik (HIL) GmbH was founded in Bonn, with a 49% minority stake held by the federal government; other stakeholders are 'Rheinmetall Landsysteme', 'Krauss Maffei Wegmann' and the Diehl daughter 'Industriewerke Saar', with 17% each as industrial holding. Since these companies are also manufacturers of the systems, the hope is for positive effects of the 'lifecycle' processes, similar to 'factory to foxhole' in the USA. HIL assumes the material responsibility for almost all military weapons systems and guarantees a daily availability of 70 %.

In August 2006, the founding in Meckenheim of the BWI Informationstechnik GmbH followed, a consortium of 'Siemens Business Services' and 'IBM'. Goal of this cooperation was the privatization of the IT of the Bundeswehr under the project name 'Herkules'. The contract also provided for the modernization of the datacenters of the Bundeswehr and the introduction of the computer software SAP in the Bundeswehr. This is the largest PPP project in Europe, with a financial volume of approximately €7.1 billion and a contract length of 10 years, and can also be traced back to the innovation contract of 1999. However, it got off to a difficult start, so that negotiations could only be concluded successfully in 2006. Industry is involved in the BWI with 50.1%, the federal government with 49.9%.

Additional projects with private partners are the communications satellites SATCOM, the training support for the transport helicopter NH-90, and training and maintenance of the fighter jet Eurofighter. While the majority of service contracts focus on domestic services, the Bundeswehr also uses civilian services internationally. In strategic air transport, it agreed to the 'Strategic Airlift Interim Solution (SALIS) on January 23, 2006 together with 15 other European nations and Canada. Within this framework, 2 wide-body transporters 'Antonov 124' were stationed in Leipzig by the Ruslan SALIS GmbH, a daughter of Russian 'Volga-Dnepr'. Furthermore, Ruslan SALIS must, in cooperation with the Ukrainian company 'Antonov', provide an additional 4 airplanes within 9 days, when required. The Bundeswehr wants to thus ensure strategic air transport capacity in a timely manner, one of the chronic deficits in the military capabilities of the Europeans.

Photo 3.1 Antonov 124 of Volga Dnjepr in Afghanistan (Picture: Witt)

Model for SALIS is the US concept of the 'Civil Reserve Air Fleet' (CRAF), which has existed since the 1950s but was first used during the buildup on the Persian Gulf in 1990. Within the framework of this contract, civilian US airlines agree to provide the military with transport capacity. CRAF is also in use in current missions. A similar Bundeswehr plan involves strategic sea transport. In the future, the Guaranteed Commercial Strategic Sea Transport (GGSS) will contractually ensure access to commercial sea transport capacity with Danish Ro-Ro ships, using civilian capacity to make up for deficits in military equipment.

Furthermore, there are operator models at over 70 Bundeswehr locations at a total value of more than €80 billion, and an additional 8 cooperation models in the area of material maintenance for the Luftwaffe, simulator training for the helicopter NH-90, and the supply of repair equipment to the military through so-called 'central government warehouses'.

The privatization of the Bundeswehr shows growth potential – with one exception: until now, food services have failed in an ambitious project to privatize mess halls. A pilot project first provided for the privatization of 13 mess halls in southern Germany. Their operation was assumed by 'Dussmann Service' in the summer of 2005 to serve almost 5,000 soldiers daily. In June 2006, the contract was canceled by Dussmann. The primary problems named were the difference in chain of command – the civilian personnel was 'provided' by the Bundeswehr – and economic dissonance.

After this setback, the Bundeswehr decided on an self-optimized solution, which also brought potential for rationalization. By centralizing purchasing, almost 200 positions could be eliminated. After introducing the 'partial kitchen', major portions of the meals were delivered from another mess kitchen and only single components were prepared fresh on location. This lowered the costs for production and storage, kitchen equipment and personnel by 25-30%. Furthermore, future optimization of catering includes a centralization of catering planning in the Bundeswehr.

It must be noted that privatization in the Bundeswehr is advancing. In areas which are not part of the core military capabilities, as a rule civilians are less expensive and of higher quality. Although until recently it was the rule that the Bundeswehr provided logistics and catering itself, in the meantime this monopoly has fallen. In 2011, LOG has been responsible as the principal contractor for catering and the operation of the domestic buildings in Mazar-e-Sharif in Afghanistan, and worked there in a strong network with the companies ES-KO and National Air Cargo.

Photo 3.2 In Mazar-e-Sharif, Afghanistan, in 2011 the soldiers were catered by LOG (Picture: LOG)

Changes are also looming in the area of 'life support'. Where Germany military service providers were generally working as sub-contractors for American service providers, the Bundeswehr administration is currently studying whether the privatization of these diverse tasks seems possible, makes sense and is efficient. A corresponding interest determination procedure has already been performed and is being evaluated.

In the course of the reorientation of the Bundeswehr and due to a lack of funds, the trend towards privatization will continue to grow. The suspension of conscription will also play a part. Overall, the Bundeswehr would be wise to turn more to privatization. In international missions, they work with partners who use civilians widely. As part of multinational staffs, they can even be involved in planning their missions.

Additionally, a change in the 'force mix' to fewer support units and more combat troops can be seen in the Bundeswehr as well. Struggles over contingent and mandate upper limits will increasingly lead to the delegation of more tasks to civilian companies, leading to concealment of contingent strength.

Our project experience from Afghanistan shows that it must be considered in all deliberations: purely logistical and technical challenges are basically solvable, but the future "status" of the company association in a mission must be settled. To what extent are they "integrated" in the contingent? How far do insurance and other protective measures extend? For, without wishing to go into detail, the legal situation is ambivalent, confusing and complicated.

The status of civilian employees as 'non-combatants' with the armed forces is decisive as to whether they represent a valid target for enemy attack and whether they are given prisoner of war status when captured. Although they may accompany armed forces on missions, they must remain away from fighting to not lose their status. Civilian truck drivers delivering consumer goods to the troops open a legal gray zone here. Specialists who maintain military equipment may also cross this line, as do experts for information gathering and intelligence gathering and reconnaissance. This is unsatisfying, since they would then be considered illegal civilian combatants, similar to the 'illegal combatants' imprisoned in Guantanamo. Clarification under international law is required, as is the necessity for the thoughtful use of civilians under consideration of their legal status. For their protection, the subjective impression they make with the enemy must be managed.

It should be noted: military capacities can be supplemented or completely replaced by civilian services. And military-related small businesses are also there for the Bundeswehr as a partner on missions.

4 Conclusion

The logistics sector is globally active and flexible enough to adapt quickly to new conditions at any time. The technical possibilities they can offer and their expertise across diverse fields make them important partners of national governments and organizations in the positive use and development of the possibilities and chances of globalization, while avoiding its negative developments.

Their importance will certainly increase in the future; it will remain a growth sector. New business areas and chances and risks will develop, which until now have either not been served or have only been met to a limited extent. Furthermore, the sector must also always be prepared for extreme scenarios which, although at first glance unlikely, remain conceivable and possible.

The fact seems to be: the future of the market for logistics services will, like the world in general, be determined by the decisions and actions of the actors in this market. The same applies to the future of every single company.

4 Supply Chain Resilience

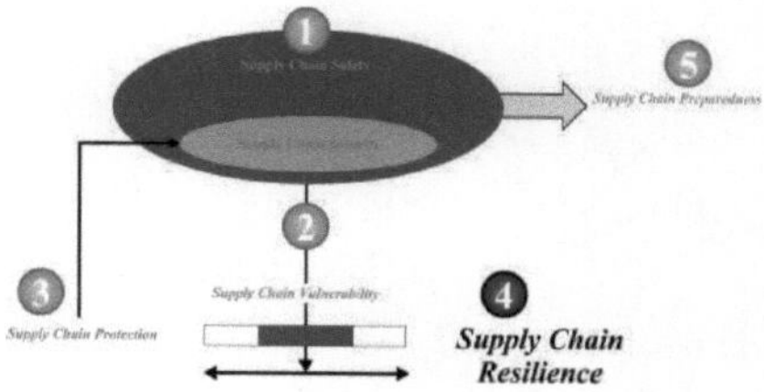

Panama Canal Update
Liliana Rivera and Yossi Sheffi

Self-Healing Supply Networks – A Complex Adaptive Systems Perspective
Philip Cordes and Michael Hülsmann

Supply Chains – How to Support Critical Infrastructures' Safety, Protection, Preparedness, and Resilience
Albrecht Broemme

Supply Chain Event Management – Concept and Use in Business practice
Joerg S. Hofstetter and Wolfgang Stölzle

Adaption-Based Supply Chain Resilience
Dmitry Ivanov, Boris Sokolov, and Joachim Käschel

Panama Canal Update

Liliana Rivera[1,*] and Yossi Sheffi[2,**]

[1] Massachusetts Institute of Technology, PhD candidate at the MIT Center for Transportation and Logistics (CTL), 1 Amherst Street, Building E40-222, Cambridge, MA 02139, USA
mlrivera@mit.edu

[2] Massachusetts Institute of Technology, Elisha Gray II Professor of Engineering Systems, Director of the MIT Center for Transportation and Logistics (CTL), 1 Amherst Street, Building E40-261A, Cambridge, MA 02139, USA
sheffi@mit.edu

Abstract. The Panama Canal extension project is, arguably, the most important current transportation project in the world today. It will allow most Post-Panamax vessels to use the canal and is likely to change transportation flow patterns throughout North and South America, as well as port loads and transportation flows inland in the Americas. This paper gives a short update on the status of the project and its likely impacts.

The Panama Canal Expansion Program is perhaps the most impactful transportation project today. The $5.5 Billion project will enable the Canal to handle up to 12,600 TEU, Post-Panamax vessels, instead of the current maximum of 4,400 TEU, Panamax Vessels. The Panama Canal Expansion Program was launched in September 2007 and is scheduled for completion in 2014, 100 years after the original inauguration of the canal. Many pundits claim that the project will lead to "the biggest shift in the freight business since the 1950s, when oceangoing ships began carrying goods in uniform metal containers." (Severson 2011) Halfway through the program, it is appropriate to examine the progress to date and how are ports outside Panama gearing up to be ready for larger ships.

The Panama Canal Expansion Program has four components: (i) construction of new Post-Panamax locks on the Pacific and Atlantic sides, (ii) excavation of the new Pacific Post-Panamax locks for the channel's north access, (iii) improvements to navigational channels, involving dredging of existing navigation channels, and (iv) improvements to water supply to improve Canal water supply and draft dependability. (ACP 2010) These four components are tackled through six main projects, whose progress as of December 2010 is presented in the accompanying table provided by the ACP (2011).

* PhD Candidate, Center for Transportation and Logistics, MIT.

** Elisha Gray II Professor of Engineering Systems; Director, MIT Center for Transportation and Logistics.

M. Essig et al. (Eds.): Supply Chain Safety Management, LNL, pp. 213–216.
springerlink.com

PROGRAM	PROGRESS
Excavation of the Pacific Access Channel (Phase 1)	100%
Excavation of the Pacific Access Channel (Phase 2)	100%
Excavation of the Pacific Access Channel (Phase 3)	93%
Excavation of the Pacific Access Channel (Phase 4)	21%
Dredging for the the Pacific Entrance Navigational Channel	63%
Dredging for the the Atlantic Entrance Navigational Channel	67%
Dredging for the Deepening and Widening of Gatun Lake and Deepening of Culebra Cut by ACP	40%
Third Set of Locks (Design and built)	8%
Raising Gatun Lake's Maximum Operating Level	2%
TOTAL	19%

According to the ACP, by the end of 2010 19% of the work has been completed. Initial work has been focused on the removal of dirt and dredged materials, and the preparation of the construction of the third set of locks by the Grupo Unidos por el Canal consortium (GUPC). The consortium is led by Spain's Sacyr Vallehermoso and includes Italy's Impregilo, Belgian dredging and marine engineering company Jan de Nul, and Panama's largest construction company Constructora Urbana SA (CUSA). Phases 1 and 2 of the Excavation of the Pacific Access Channel are complete, while phases 3 and 4 are on-going. The excavation work for lock chambers, lock heads and water-saving basins is in progress, as well as drilling and blasting operations for the lock excavations. The Pacific and Atlantic Entrance Navigational Channels' dredging work is moving forward and so is the work in Gatun Lake and the Culebra Cut. To finalize the configuration of the locks' hydraulic system, GUPC contracted Compagnie National du Rhône (CNR), who built a model of the locks in Lyon, allowing for experiments with competing engineering designs. Also, the first floodgates design is almost done and its construction should start in 2011.

The Panama Canal Expansion Program is on schedule, despite a strike and the heavy rains that battered Panama in 2010 and caused the canal's first closure in over 20 years. In 2011 the contractor for phase 4 of the excavation of the Pacific Access Channel is expected to start the construction of the Borinquen 1E damn, the first to be built in the Canal area in the last 75 years. In addition, 2011 should witness the first pouring of concrete for the new set of locks in the Pacific and the Atlantic.

The Panama Canal expansion will impact cargo throughout the Americas, presenting new opportunities, leading to new transportation routes, new distribution patterns and new logistics hubs formation. Today, the fastest and preferred way to send cargo from China to the population centers on the U.S. East Coast is by a combination of ship and rail. It takes about 12-14 days for the ocean voyage from Shanghai to the west coast, and another 7-8 days from the US West Coast to the New York by rail for a total of 19-22 days. Sending the same Shanghai-New York cargo through the Panama Canal takes 25-26 days, while sending it through the Suez Canal takes 27-28 days. Using Panamax vessels the route via the West Coast and overland costs about $600 more than the trip through the canal, yet the economics of Post-Panamax vessels more than compensate for this (Economist 2009). This is the reason that 75% of Asian imports use the West Coast route (With the ports of LA/Long Beach acccunting for 43% of this volume (US DOT 2009)) and only 19% use the Panama Canal (while 6% use the Suez route). The Panama Canal expansion will lower shipping costs by allowing Post-Panamax ships sail directly to the US East Coast. However, it will still take longer than the current ship-rail combination.

Traffic diversion estimate vary widely. Most estimates put the maritime traffic gains through the canal at between 20% and 35% of the current West Coast freight. Naturally, this will also depend on the toll levied by the ACP. Currently the ACP charges each ship $72 per container-capacity – thus a 4,500 TEU vessel pays $324,000 to traverse the canal, whether it is loaded or empty. Given its past practice, however, it is likely that the ACP will continue to segment the market and practice yield management in order to maximize the traffic through the canal and its revenues.

To capture some of the new traffic, almost all large ports in the US east cost and along the Gulf of Mexico have expansion projects on the way involving both harbor deepening and land-side expansion of rail and handling capacities. The Port of Norfolk, Virginia, which is 50 feet deep and currently the only port on the US East Coast that can handle the Post-Panamax ships, has five priority navigation projects that focus on maintaining unrestricted navigation in the Port, allowing for easier access for these vessels. The Port of Charleston, South Carolina is investing in keeping its current depth through a continuous dredging program and working towards deepening the harbor further. At the Port of New York/New Jersey work is underway to increase deep water capacity and there are plans to raise the roadbed of the Bayonne Bridge to allow for the Post-Panamax ships to pass under it. The Port of Tampa has an ongoing terminal expansion to quadruple its size. The port of Miami is establishing an intermodal container rail service and rail and bridge expansions. Savannah Harbor is being dredged. Gulf of Mexico ports in Gulfport, Mississippi and Mobile, Alabama, as well as the Tennessee-Tombigbee Waterway, have signed memoranda of understanding (MOU) with the Panama Canal Authority to encourage increased traffic in the Gulf. Other ports involved in MOU are Ports of Houston, Boston, Miami, New Orleans, Charleston, and Tampa.

The massive investments along the US East Coast and elsewhere in the hemisphere may be overdone. The high estimates of diverted traffic requiring port expansion may not take into account several factors:

1. The competitive response of the existing players. To this end, some of the West Coast ports: Los Angeles, Long Beach, Oakland, Portland, Seattle, and Tacoma have banded together with the Western railroads: Burlington Northern Santa Fe and Union Pacific to form the U.S. West Coast Collaboration (USWCC) to guarantee competitive cost and service options. To ignore the possibility of a competitive response recall the fate of the Groupe Eurotunnel SA which went bankrupt as ferry operators improved their service and reduced their costs in response to the competitive threat of the channel tunnel.

2. As some traffic will start to be diverted to the canal, the efficiency of the West Coast ports will improve as the congestion in ports such as LA/Long Beach, Oakland and Seattle/Tacoma eases up. The resulting improved service will attract shipper and ocean carriers.
3. One of the main objectives of the ACP and the Panamanian government is to increase the rate of transshipments and related logistics operations in Panama. While this may or may not take place in Panama, many other ports in the Caribbean's (including Cuba, if relationships with the US will improve) are set to unload Post Panamax vessels and transfer the container to smaller vessels that can get into any East Coast port.
4. Increased focus on environmental issue and carbon pricing in the future may favor the West Coast route. The reason is that the CO2 emissions per TEU for a large Panamax vessel is only 2/3 of the emissions involved in a trip through the Panama Canal.

References

A Plan to Unlock Prosperity. The Economist (2009)

Autoridad del Canal de Panamá (ACP), El Faro Revista Informativa de la Autoridad del Canal de Panama Enero (2011)

Autoridad del Canal de Panamá (ACP), Panama Canal Expansion Program (2010), http://www.pancanal.com/eng/expansion/informes-de-avance/avance.html

Severson, K.: A Race to Capture a Bounty from Shipping. NY Times (2011)

US DOT (Department of Transportation), U.S. Waterborne Foreign Container Trade by U.S. Custom Ports (2009), http://www.marad.dot.gov/library_landing_page/date_and_statistics/Data_and_Statistics.html

Self-healing Supply Networks: A Complex Adaptive Systems Perspective

Philip Cordes[1] and Michael Hülsmann[2]

[1] Jacobs University Bremen, Research Associate at the Department Systems Management, Campus Ring 1, 28759 Bremen, Germany
p.cordes@jacobs-university.de

[2] Jacobs University Bremen, Associate Professor of Systems Management, International Logistics – School of Engineering and Science, Campus Ring 1, 28759 Bremen, Germany
m.huelsmann@jacobs-university.de

Abstract. This paper aims for a logical deductive literature-based generation of hypotheses regarding the robustness of complex adaptive logistics systems (CALS) based on self-healing processes. Therefore, the increasing necessity for supply networks to gain and maintain robustness in order to ensure a high reliability of their logistics services is shown. Additionally, the concept of CALS is presented in order to deduce the outcomes that result from a technology-based increase of the CALS characteristics. Finally, a set of hypotheses is developed that link the outcomes of CALS with the evolvement of self-healing processes in supply networks in order to deduce implications for their robustness. Hence, a starting point for further empirical and simulation-based research is presented, on which basis an operationalization of the outcomes of CALS and their self-healing abilities can be conducted.

1 Introduction

An increasing amount of interrelations between a rising number of actors involved in logistic processes leads to an increasing sensitivity of logistics systems (e.g. Hülsmann, Scholz-Reiter et al. 2007, Harland, Brenchley et al. 2003). This refers on the one hand to changes in parts of these systems themselves and on the other hand changes in other systems, to which they are related in any way. Consequently, there is an increasing amount of influences that can endanger logistics systems to carry out their functions. Hence, the ability to resist against them is increasingly important for logistics companies as well as whole logistics networks to gain and maintain competitiveness through offering and ensuring a high reliability of logistics services. In other words: Robustness, understood as the system's ability, to restore its operational reliability after being damaged in any way (McKelvey et al. 2008), becomes an increasingly important topic for the associated research.

Since a central management might not be able anymore to handle the rising complexity (Hülsmann, Grapp 2005), an alternative approach could be to trigger self-healing processes in logistics systems. The underlying idea is that the

M. Essig et al. (Eds.): Supply Chain Safety Management, LNL, pp. 217–230.
springerlink.com

source of a logistics system's robustness is not its external control but its internal interaction-based processes of decentralized decision-making of autonomous logistics objects (Windt, Hülsmann 2007). WYCISK ET AL (2008) supply networks that are based on this idea Complex Adaptive Logistics Systems (CALS). Technological developments like RFID-tags or sensor networks are on the way to build the basis for a realization of such a CALS, since they enable logistics objects to interact with each other and to render decisions on their own, without having to ask a central management (McKelvey, Wycisk et al. 2009, Wycisk, Mckelvey et al. 2008). Therewith, self-organizing logistics system structures can be created, which might contribute to trigger self-healing processes and hence, increase the system's robustness.

Consequently, the following question arises: How does the implementation of suchlike technologies and hence the increase of the characteristics of CALS influences the robustness of logistics systems through self-healing processes?

The overarching aim of this article is to generate hypotheses regarding the contributions and limitations of increasing the degree of the characteristics of CALS to the robustness of logistics systems through self-healing processes. Therefore, the paper proceeds as follows: In section 2, the increasing necessity for logistics systems to gain and maintain robustness is deduced in order to clarify the research question and its theoretical and practical relevance. In section 3, the concept of Complex Adaptive Systems and their appliance to logistics – the idea of CALS – are described in order to build the basis for the generation of hypotheses regarding its effects on a supply network's robustness. In section 4, hypotheses will be developed, in order to show both positive and negative potentials of CALS and their associated outcomes for increasing the robustness of logistics networks. Finally, section 5 sums up the findings and presents resulting further research requirements.

2 Robustness of Modern Supply Networks

Modern supply networks are characterized by a high degree of complexity and dynamically evolving relationships (Harland, Brenchley & Walker 2003). One illustrative example is the supply network lying behind the production of Daimler Chryslers automobiles, the first tier of which alone consists of approximately 1500 suppliers (Choi, Hong 2002). Many of these suppliers, in turn, have manifold connections and relationships to other companies. As a result, such supply networks are regarded as "webs" consisting of multitudes of business relationships. In addition to the increasing complexity of products and services, globalization and an increased use of e-business models, are mentioned as key drivers for the underlying complexity of todays supply nets. This inherent complexity results in a tendency towards the outsourcing of business activities (Harland, Brenchley & Walker 2003), which goes hand in hand with typical recommendations on targeted focus on core competencies, in order to survive in hypercompetitive environments (Prahalad, Hamel 1990). The whole picture further contributes to the evolution of large interfirm networks (Sydow, Windeler 1998). Hence, the amount of actors participating in supply chains is increasing, whereas the individual shares of added value are decreasing. Conforming to that, companies usually participate in more than one supply chain, which means that the overlying supply networks consist of manifold supply chains crossing one another.

The management of these supply chains needs to include the planning, steering and controlling of all flows of material, services, money and information. Furthermore, management approaches need to integrate all business activities that are involved in the single value-added steps of the development, creation and exploitation of the respective products or services (Cooper, Lambert & Pagh 1997). Hence, manifold interdependencies occur not only among the actors involved in one supply chain, but also among those that cross several supply chains. This means that the decisions of individual actors influence the decision-making of other actors, which results in the accompanying tendency of increased risks in supply networks in various forms, including strategic, operations, supply, customer or financial risks (Harland, Brenchley & Walker 2003, referring to Simons 1999, Meulbroek 2000). Highly significant losses as well as less significant but often occurring losses can damage a supply network's structure and functionality. It can be said that modern supply networks and involved organizations are getting more sensitive because the perturbations that might endanger them are getting stronger and/or are occurring more frequently. Hence, in order to countervail this development they need to increase their robustness.

The term robustness has a wide range of definitions. CARLSON AND DOYLE (2000) for instance define robust systems as *'systems designed for high performance in an uncertain environment and operated at densities well above a standard critical point'*. THADAKAMALLA ET AL. (2004) follow a more narrow definition that refers to a system's robustness as its ability to *'sustain the loss of some of its structure or functionalities and maintain connectedness under node failures'*. They state robustness to be one essential component, of a system's survivability in addition to responsiveness, flexibility and adaptivity. Following this definition, processes of rebuilding of system's structures and connections between agents would therefore countervail the pressures mentioned above. In other words, self-healing processes in supply networks would increase their robustness.

Hence, robustness becomes an increasingly relevant issue for the management of today's supply networks. As a consequence, novel approaches and associated technologies are needed to enable supply networks to cope with the resulting challenges to increase their robustness. One approach might be to enable the respective system to heal it self after being damaged. In this context, it is referred to self-healing as the process of autonomous recovery after the supply network faces a targeted attack or random damage. Hence, the question arises, what are the technological and organizational pre-conditions for self-healing processes in modern supply networks and how they could contribute to a logistics system's robustness. Therefore, the next section examines the context in which self-healing is a natural occurrence for supply networks: Complex Adaptive Logistic Systems.

3 A Complex Adaptive Systems Perspective on Modern Supply Networks

Complex Adaptive Systems (CAS)

Many kinds of systems, from natural to artificial, can be characterized as complex, such as ecological or social systems (e.g. Surana et al. 2005). The term complexity does not only focus on the number of elements in the system, but particularly on the quantity of relationships between the elements and their environment (e.g. Gell-Mann 2002) as well as the type of their interactions (e.g. Hazy, Goldstein & Lichtenstein 2007). One striking property of complex systems lays in the aspect that understanding the behavior of each element of a system perfectly does not lead to an understanding of the system as a whole (e.g. Miller, Page 2007). Adaptivity is another relevant characteristic of complex adaptive systems, which refers to the system's ability to adapt to changing environments through coevolving interactions among its autonomous elements, often referred to as agents (e.g. Holland 2002, Arthur, Durlauf & Lane 1997).

With recourse to KAUFFMAN (1993) and HOLLAND (2002), WYCISK, MCKELVEY AND HÜLSMANN (2008) identified seven main characteristics of complex adaptive systems (CAS): *Heterogeneity, interaction, ability to learn* and *autonomy of agents*, as well as the system behavior; which exhibits *self-organization, its melting-zone* and *coevolution* properties.

Agents in a CAS are *heterogeneous*, in that they distinguish themselves from each other through different goals, properties, functions and rules (Holland 2002). They are also equipped with heterogeneous resources such as information, which is the basis for their motivation to exchange resources between each other. Therefore, CAS consist of *interacting* agents. In order to interact, the agents have to be able to communicate with each other, which is to say that they react on other agents' actions. Furthermore, agents within a CAS act *autonomously* since their actions are not entirely determined by other control entities. This does not however, mean that there cannot be control elements in a CAS. Partial control can be embedded in system behavior upon design or can evolve on its own. Finally, a CAS is characterized by its agents' *ability to learn*, which refers to agents that may modify their rules as experience accumulates, searching for improvements (Holland 2002).

The behavior of CAS is firstly characterized by *self-organization*, which means that it forms its structure autonomously, without the need of interference from outside the system, using its autonomy and interaction characteristics. Hence, behavior and development is neither preconfigured nor totally chaotic, existing between the so-called edge of order and the edge of chaos. KAUFFMAN (1993) called this the *melting-zone*. According to SURANA ET. AL. (2005) the system's elements exist in quasi-equilibrium and show a combination of regularity and randomness. Finally, CAS *co-evolve* at both micro and macro levels. Individual agents respond to others' actions through changes in their own behavior or decision-rules. The whole CAS will also be an autonomous, learning and self-organizing system that interacts with other systems in its environment.

Although these characteristics are primarily found in natural systems like ecologies, researchers mention their applicability to organizations (e.g. Hazy, Goldstein & Lichtenstein 2007) and economic markets (e.g. Arthur, Durlauf & Lane 1997). WYCISK, MCKELVEY & HÜLSMANN (2008) provide the link to logistics systems as markets consisting of organizations cooperating with and competing against each other, which, in turn, consist of agents like employees and intelligent logistics objects, giving rise to what is termed as Complex Adaptive Logistic Systems (CALS).

Complex Adaptive Logistics Systems (CALS)

The complexity of modern supply networks results from the large amount of involved organizations as well as the relations between these organizations. Rather than being involved in only one supply chain, many companies participate in manifold chains (Lambert, Cooper & Pagh 1998). Therefore, logistics systems can no more be regarded as individual and isolated supply chains, where management focus shifts to integration. Such networks create highly complex and non-linear webs of logistics activities, including the flow of information, products and finances between various suppliers, manufacturers, distributors, retailers and customers (Surana et al. 2005). Several authors stated their view of such modern supply networks to be complex adaptive systems (e.g. Choi, Dooley & Runkttusanatham 2001, Pathak, Dilts & Biswas 2004, Surana et al. 2005, Wycisk, Mckelvey & Hülsmann 2008). These then have been observed to show CAS properties, giving rise to the term *Complex Adaptive Logistics Systems* (CALS) (Wycisk, Mckelvey & Hülsmann 2008).

The analogy between complex adaptive systems and supply networks derive from the following set of properties exhibited by organizations such as logistic providers, retailers, and manufacturers participating in a supply network:

- These organizations transact or cooperate with each other (following the *interaction* property of CAS);
- They benchmark other companies and follow trends and actions of market leaders (following the *learning* property of CAS);
- They have varying competences and financial resources and strive to gain competitive advantages (following the *heterogeneity* property of CAS)
- They are legally independent from each other and follow their own goals (following the *autonomy* property of CAS).

Furthermore,

- such a market's structure forms itself without any top-down orders, leading to *self-organization*,
- although, there is no preconfigured supply network structure the system is subject to domestic and international laws, avoiding chaos and forcing the self-organized system into a *melting zone*,
- it affects and is affected by its environment leading to *co-evolutionary* behavior with a variety of other systems.

Ongoing technological developments and advances lead to the analogy of this perspective on the micro-level. Here, the regarded agents of the CALS are the individual logistics objects operating in organizations that in turn operate in logistics networks as shown above. The central aspect in the discussion of CALS on a micro perspective lies therefore in the technological realization, which refers to the »possible smartness« of logistics entities in the real world. Although there are restrictions in, for instance, the parts' abilities to learn (Wycisk, Mckelvey & Hülsmann 2008), further and ongoing developments in associated communication and information technologies lead to the assumption that these restrictions will more and more disappear in the future. This refers especially to developments in the logistics elements' abilities to process information, to render and to execute decisions on their own. The basis for suchlike artificial decision-making abilities is given by the use of technologies that enable logistic elements to get information about their environment: For the logistics objects' identification RFID (radio frequency identification) can be used, for the positioning GPS (global positioning systems) and for the communication between them UMTS (universal mobile telecommunications system) or WLAN (wireless local area network) (Scholz-Reiter, Windt & Freitag 2004).

One example is the so-called 'Intelligent Container', which monitors perishable goods in transit, and adjusts routes and container environment autonomously (e.g. Jedermann & Lang 2008). Another example on a yet higher level is given by market-based design options for a complex adaptive transportation system, in which the logistics objects (e.g. products) bid on transportation places provided by other logistics objects (e.g. containers) (McKelvey, Wycisk & Hülsmann 2009).

The underlying organizational principle is the idea of autonomous cooperation (e.g.) (Hülsmann, Grapp 2005, Scholz-Reiter, Windt & Freitag 2004, e.g. Windt, Hülsmann 2007). According to WINDT AND HÜLSMANN (2007) autonomous cooperation can be described as *"(...) processes of decentralized decision-making in heterarchical structures. It presumes interacting elements in non-deterministic systems, which possess the capability and possibility to render decisions."* (Windt, Hülsmann 2007, p. 9). Increasing the degree of the resulting constitutive characteristics – autonomy, decentralized decision-making, interaction, heterarchy, non-determinism – through the implementation of technologies like e.g. RFID would directly increase the degree of the CALS characteristics. The underlying objective *"(…) is the achievement of increased robustness and positive emergence of the total system due to distributed and flexible coping with dynamics and complexity."* (Windt, Hülsmann 2007, p. 478). However, whereas positive as well as negative effects on the robustness of logistics systems have already been examined (Hülsmann et al. 2008), the underlying causes and effects that arise through a resulting change of the degrees of the CALS characteristics are still unexplored. The same is true for single autonomous cooperation-enabling technologies such as RFID, for which general results regarding economic impacts of an implementation have not been attained yet (e.g. Ngai et al. 2008). Hence, the question, how much autonomous cooperation should be allowed by the management of logistics companies or networks with respect to the robustness of logistics systems, is still open. In other words, it is not yet clear if a '100 per cent well working CALS' is desirable from a managerial point of view.

With regard to the development costs and implementation risks of technologies, such as RFID (Folinas, Patrikios 2008), which would enable the realization of a true CALS on a micro level, the following question arises: Under which circumstances and to what extend would an increase of CALS characteristics contribute to a logistics system's robustness and hence, would be desirable from an economic point of view? Since a »trial-and-error« approach would be risky and expensive – if an implementation of suchlike technologies does not lead to the desired outcomes – it is reasonable to forecast the respective effects. Hence, it is necessary develop a deeper understanding of a CALS' behavior, in order to be able to evaluate its outcomes later on.

Therefore, the idea of self-healing processes is taken as a bridge between the CALS characteristics and its robustness, in order to generate hypotheses about the causal interrelations between the degree to which a supply network can be regarded as a CALS and its robustness. In other words: How do the CALS characteristics affect the supply network's ability to rebuild its system structures after being damaged and therewith contribute to the system's robustness?

In order to make a first step on answering this question, hypotheses regarding the effects of the outcomes of a CALS on a logistics system's self-healing abilities and therewith its robustness will be generated in the following.

4 Generation of Hypotheses Regarding the Self-healing Abilities of CALS

WYCISK, MCKELVEY AND HÜLSMANN (2008) draw our attention to the following outcomes of CALS: Emergence, Adaptation, Nonlinearity; irreversibility, Butterfly effects, Multi-levels, and Scalability. These outcomes affect the healing abilities of supply networks, and hence their robustness in several ways. In what follows, we deduce exemplary hypotheses for positive as well as negative effects on a supply network's robustness (Table 1).

Table 1 Exemplary positive and negative effects of the outcomes of a CALS on its robustness

Outcome	Positive Effects on Robustness	Negative Effects on Robustness
Emergence	Potential evolution of new supply network structures with enhanced self-healing abilities due to learning of smart parts	Potential evolution of new system structures that are locally optimal but globally sub-optimal
Adaptation	Contribution to the balance between a supply network's flexibility and stability followed by enhanced self-healing abilities	Risk of alterations or deletion of connections between agents that are highly relevant for the system's self-healing abilities.
Non-linearity; Irreversibility	Chances of path dependent-based lock-in situations – in terms of inflexibilities to leave robust system states – Chance of a solidification of self-healing processes in supply networks	Risk of path dependent-based lock-in situations – in terms of inflexibilities to leave non-robust system states – Risk of a solidification of self-destroying processes in supply networks
Butterfly Effects	Chance of triggering extremely forceful self-healing processes by tiny initiating actions of smart parts	Risk of triggering extremely forceful self-destroying processes by tiny initiating actions of smart parts
Multi-levels	Potential evolution of stable but flexible sub-units with adaptive characteristics	Risk of extreme negative events triggered by adaptive dynamics within evolved sub-units as well as risks of non-robust sub-units with negative effects on the robustness of the whole supply network
Scalability	Mentioned positive effects on each level	Mentioned negative effects on each level

Emergence: KAUFFMAN (1993) argues that within the melting zone new structures spontaneously emerge via continuous restructuring processes of and between the heterogeneous agents. Through interaction processes new agent attributes in terms of patterns and behavioral rules are created (Holland 2002). In CALS new supply network orders emerge from the autonomous interactions of the involved companies. One example on the organizational level is the emergence of cooperation among logistics service providers, which then might lead to new market leaders. New system structures could arise continuously without the guidance of an external control entity. Hence, if the new evolving system structure exhibits more self-healing abilities than the old structure, one can deduce that emergence contributes to a supply network's robustness. Here arguments in favour as well as against are conceivable.

Hypothesis (H) 1.1: If the agents in a supply network are able to improve their individual performances via modifications of their behavioral rules in the course of an accumulation of experience, then it is possible that the whole system learns how to reconstitute its functionalities after being damaged, hence learning to self-heal. Then the new evolving structures contribute to the supply network's robustness.

However, since emergent phenomena occur in situations in which 'the behavior of the whole is greater than the sum of its parts' (Wycisk, Mckelvey & Hülsmann 2008) the interplay of the agents does not have to lead to the best results possible. Examples are problems of local optima in which the individual goals of the agents differ from those of the whole system or prisoner dilemma situations in which the agents' expectations about other agents' behaviors lead to sub-optimal results (e.g. Poundstone 1993). Therefore, the second hypothesis states the following:

H1.2: If the agents in a supply network follow their own goals, which are not complimentary to the goal of maximizing the supply network's robustness, and if agents are able to change their behavior in accordance to their own measures of performance, then the risk occurs that autonomous decision-making leads to the evolvement of a new system structure that exhibits less self-healing abilities than the former structure. Then the new evolving structures constrain the supply network's robustness.

Adaptation: WYCISK, MCKELVEY AND HÜLSMANN (2008) refer to adaptation as the process of systemic change, in terms of adding, deleting as well as altering connections between agents, and their attributes and behavioral rules, in order to cope with changes in their environment. Supply networks react to demands from their environment while at the same time creating new environments for other agents in the system (Choi, Dooley & Rungtusanatham 2001). One example is the implementation of new technologies in reaction to changing customer demands, which then might become new industry standards or prove useless to meet these demands. This is closely related to what is called in strategic management literature "strategic adaptivity", which refers to the ability to balance a system's flexibility and its stability, both of which are necessary in order to enable a system to heal itself after being damaged. Flexibility is needed to gather new solutions to problems, and stability is needed in order to prevent changes that endanger the system's functionality, which includes therewith its self-healing abilities. Hence, the following hypothesis can be deduced regarding the robustness of supply networks:

H2.1: If the systemic alterations of a system enable it to be both flexible enough to meet changing environmental demands and stable enough to not loose the system's functionality, then they contribute to a supply network's self-healing abilities and hence, to its robustness.

However, systemic adding, deleting and alterations of connections between agents can exacerbate a system's ability to maintain its stability while being flexible.

H2.2: If the adaptive processes delete or alter connections between agents, which are highly relevant for the system's ability to self-heal after being damaged, then the risk occurs that adaption decreases a supply network's robustness.

Non-linearity; irreversibility: HOLLAND (2002) reasons that non-linear behavior of CAS results in the agents' interactions in non-additive ways. Hence, future states of CAS are unpredictable and irreversible, which applies to modern supply networks, where future states and structures, e.g. market domination, cannot be predicted. Irreversibility results in the fact that 'history matters' (David 1985, 1994, 2001). Irreversibility implies that if something 'went wrong' in the past it cannot be undone. Translating into CALS, economic actions cannot be undone, for example the loss of a logistics service provider's reputation due to unreliable deliveries cannot be redeemed quickly. These characteristics in combination with positive or negative feedback loops can lead to path dependencies, which bear the risk, respectively the chance of an evolvement of lock-in situations (David 1985, 1994, 2001). A lock-in describes an inflexibility to leave a certain system state respectively to change a certain system behavior, whether it is efficient or not, and also whether it is robust or not. Hence two opposing hypothesis can be deduced:

H3.1: If individual agents' actions and resulting supply network behavior imply positive feedback loops in terms of solidifications of agents' behavior that lead to self-healing processes, then non-linearity and irreversibility contribute to a supply network's robustness.

H3.2: If individual agents' actions and resulting supply network behavior imply negative feedback loops in terms of solidifications of agents' behavior that lead to self-destroying processes, then non-linearity and irreversibility decrease a supply network's robustness.

Butterfly effects: LORENZ (1963) introduced the butterfly effect as the possibility that a tiny and alleged event might trigger remarkable dynamics with significant effects. In CALS this phenomenon refers to the evolvement of the so-called 'bullwhip-effect' (Wycisk, Mckelvey & Hülsmann 2008), which describes significant fluctuations of orders despite relative constant demands, resulting from feedback loops along a supply chain (Forrester 1961). The risk of extreme and highly significant events, triggered by tiny but initiating actions, can both endanger and foster a system's robustness.

H4.1: If tiny events trigger extremely forceful self-healing processes, then the butterfly effect contributes to a supply network's robustness.

However, as the example of the bullwhip effect has shown, extreme events can be of a negative kind for CALS.

H4.2: If tiny events trigger extremely forceful self-destroying processes, then the butterfly effect decreases a supply network's robustness.

In combination one can deduce that butterfly effects can lead to both an intensification of perturbations and an intensification to cope with these perturbations.

Multi-levels: SIMON (1962) state that through agents' interactions and interdependencies various forms of sub-systems emerge that lead to multi-level hierarchies. Supply networks consist of multiple supply chains that themselves create sub-units (Wycisk, Mckelvey & Hülsmann 2008) through e.g. co-operations between logistics service providers or transportation contracts between logistics service providers and manufacturers. A stable formation of small assembles of autonomous agents or subunits contribute, according to HOLLAND (2002) to a CALS' adaptivity. Hence, with recourse the hypotheses H2.1 and H2.2 the following hypotheses can be deduced:

H5.1: If the systemic alterations of the system's subunits enable them to be both flexible enough to meet changing environmental demands and stable enough to not loose their functionalities in the supply network then they contribute to the whole network's self-healing abilities and hence, to its robustness.

H5.2: If the adaptive processes of the formed subunits delete or alter connections between agents, which are highly relevant for the subunits ability to self-heal after being damaged, then their and hence the whole supply network's robustness decreases.

Scalability: Authors like MANDELBROT (1961;1983) or KAYE (1989) state that in CAS the same system dynamics can be observed on multiple levels. In CALS, scalability can be observed in terms of CAS characteristics on the level of the whole supply network (interacting organizations like suppliers, manufacturers, etc.) as well as on the level of the organizations themselves (functional, operational or organizational levels). Organizational entities can be either human entities like managers and employees or nun-human entities –such as the smart parts. An occurrence of the same dynamics and hence, the same outcomes (emergence, adaptation, non-linearity/irreversibility, butterfly effects, multi-levels and scalability) on multiple levels include that in a 'well-working' CALS they also occur on the level of the involved organizations. Therefore, scalability implies that if positive effects outweigh the negative ones on the micro level than they would also outweigh them on the macro level, and the other way round. The following hypotheses can be deduced:

H6.1: If positive effects on a supply network's robustness, such as those emanating from one or more of the hypotheses H1.1, H2.1, H3.1, H4.1 and H5.1 – pre-conditioned that one or more of them turn out to be right – outweigh negative effects emanating from one or more of the hypotheses H1.2, H2.2, H3.2, H4.2 and H5.2 – pre-conditioned that one or more of them turn out to be right – then scalability contributes to a supply network's robustness.

H6.2: If negative effects on a supply network's robustness outweigh positive effects than scalability decreases a supply network's robustness.

Consequently, hypotheses 1.1 – 6.2 reflect assumptions, based on existing literature on complex adaptive systems, how an increase of the degree of CALS characteristics could contribute to or hinder self-healing processes in supply networks and therewith, increases respectively decreases their robustness.

5 Conclusions

Since robustness is an increasingly necessary characteristic for logistics systems such as supply networks, in order to enable them to gain and maintain competitiveness, this article deals with its emergence trough self-healing processes. One approach to trigger such processes could be the increase of the degrees to which a supply network can be regarded as a Complex Adaptive Logistics System (CALS): *Heterogeneity, interaction, ability to learn, autonomy, self-organization, melting-zone* and *coevolution*. The outcomes of a CALS include several phenomena, such as emergence, adaptation or butterfly effects. Hence, the analyzed question is whether and how theses outcomes affect the ability of a supply network to heal itself after being damaged in any way.

The associated developed hypotheses show both negative as well as positive effects. One example for a positive effect is the outcome adaptation, which enables supply networks to be both flexible enough to meet changing environmental demands and stable enough to not loose the system's functionality. One example for a negative effect is the butterfly effect, which can trigger extremely forceful self-destroying processes through tiny and at first glance unimportant events.

What can managers of supply networks learn from these hypotheses? First of all, they show that there are potentials to increase the robustness of logistics networks through implementation of technologies that increase the degree of the CALS-characteristics, e.g. RFID. However, there are also potential limitations that have to be considered, which might result in a counterproductive result.

The logical next step after generating the hypotheses is to test them in empirical or simulation-based research. Hence, further research requirements include the operationalization of the degrees of CALS-characteristics, of the outcomes and of self-healing processes. Not until they have been made measurable, it is possible to generate results that allow to reject or to accept hypotheses regarding the interrelations between the degrees to which a supply network can be regarded as a CALS and its self-healing based robustness.

Acknowledgement. This research was supported by the German Research Foundation (DFG) as part of the Collaborative Research Centre 637 'Autonomous Cooperating Logistic Processes – A Paradigm Shift and its Limitations".

References

Arthur, W.B.: Competing Technologies, Increasing Returns, and Lock-in by Historical Events. Economic Journal 99(394), 116–131 (1989)

Arthur, W.B.: Positive Feedbacks in the Economy. Scientific American 262(2), 92–99 (1990)

Arthur, W.B., Durlauf, S.N., Lane, D.A.: The economy as an evolving complex system / Santa Fe Institute2: T1 The economy as an evolving complex system. Addison-Wesley, Reading (1997)

Barbuceanu, M., Fox, M.S., Gruninger, M.: An organisation ontology for enterprise modeling: preliminary concepts for linking structure and behaviour. Computers in Industry 29, 123–134 (1996)

Brintrup, A., McFarlane, D., Owens, K.: Will intelligent assets fly? Towards self-serving aircraft assets. IEEE Intelligent Systems (2009) (in press), `http://doi.ieeecomputersociety.org/10.1109/MIS.2009.89`

Carlson, J.M., Doyle, J.: Highly Optimized Tolerance: Robustness and Design in Complex Systems. Phys. Rev. Lett. 84(11), 2529–2532 (2000)

Choi, T.Y., Dooley, K.J., Rungtusanatham, M.: Supply networks and complex adaptive systems: control versus emergence. Journal of Operations Management 19(3), 351–366 (2001)

Choi, T.Y., Hong, Y.: Unveiling the structure of supply networks: case studies in Honda, Acura, and DaimlerChrysler. Journal of Operations (2002)

Cooper, M.C., Lambert, D.M., Pagh, J.D.: Supply Chain Management: More Than a New Name for Logistics. The International Journal of Logistics Management 8(1), 1–14 (1997)

David, P.A.: Why Are Institutions the 'Carriers of History'?: Path Dependence and the Evolution of Conventions, Organizations and Institutions. Structural Change and Economic Dynamics 5, 205 (1994)

David, P.A.: Clio and the economics of QWERTY. The American Economic Review (75), 332–337 (1985)

David, P.A.: Path dependence, its critics and the quest for 'historical economics'. In: Garrouste, P., Ioannides, S. (eds.) Evolution and Path Dependence in Economic Ideas: Past and Present, Cheltenham, UK and Northampton, USA, pp. 15–40 (2001)

Folinas, D., Patrikios, N.: RFID Implementation Framework in Supply Chain. In: Blecker, T., Huang, G.Q. (eds.) RFID in Operations and Supply Chain Management – Research and Applications, pp. 3–12. Erich Schmidt Verlag, Berlin (2008)

Forrester, J.W.: Industrial dynamics. Productivity Press, Wiley, Cambridge, Mass. [u.a.] (1961)

Frey, D., Woelk, P.O., Stockheim, T., Zimmermann, R.: Integrated multi-agent-based supply chain management. In: Proceedings of Twelfth IEEE International Workshops on Enabling Technologies: Infrastructure for Collaborative Enterprises, WET ICE 2003, June 9-11, pp. 24–29 (2003)

Gell-Mann, M.: What is Complexity? In: Quadrio Curzio, A., Fortis, M. (eds.) Complexity and Industrial Clusters: Dynamics & Models in Theory & Practice, Physica, Heidelberg, pp. 13–24 (2002)

Haken, H.: Synergetics: a workshop; proceedings of the International Workshop on Synergetics at Schloss Elmau, Bavaria, May 2-7. Springer, Berlin (1977)

Harland, C., Brenchley, R., Walker, H.: Risk in supply networks. Journal of Purchasing & Supply Management 9(2), 51 (2003)

Hazy, J.K., Goldstein, J.A., Lichtenstein, B.B.: Complex Systems Leadership Theory: An Introduction. In: Hazy, J.K., Goldstein, J.A., Lichtenstein, B.B. (eds.) Complex Systems Leadership Theory: New Perspectives from Complexity Science on Social and Organizational Effectiveness, pp. 1–16. ISCE Publishing, Mansfield (2007)

Holland, J.H.: Complex adaptive systems and spontaneous emergence. In: Quadrio Curzio, A., Fortis, M. (eds.) Complexity and Industrial Clusters, pp. 25–34. Physica, Heidelberg (2002)

Hülsmann, M., Grapp, J., Li, Y.: Strategic adaptivity in global supply chains—Competitive advantage by autonomous cooperation. International Journal of Production Economics 114(1), 14–26 (2008)

Hülsmann, M., Grapp, J.: Autonomous Cooperation in International-Supply-Networks – The Need for a Shift from Centralized Planning to Decentralized Decision Making in Logistic Processes. In: Pawar, K.S. (ed.) Proceedings of the 10th International Symposium on Logistics (10th ISL), Loughborough, United Kingdom, pp. 243–249 (2005)

Hülsmann, M., Scholz-Reiter, B., deBeer, C., Austerschulte, L.: Effects of Autonomous Cooperation on the Robustness of International Supply Networks – Contributions and Limitations for the Management of External Dynamics in Complex Systems. In: Haasis, H., Kreowski, H.J., Scholz-Reiter, B. (eds.) Dynamics in Logistics – Proceedings of the 1st International Conference on Dynamics in Logistics, Bremen, Germany, August 28-30, Springer, Berlin (2008)

Hülsmann, M., Scholz-Reiter, B., Austerschulte, L., De Beer, C., Grapp, J.: Autonomous Cooperation – A Capable Way to Cope with External Risks in International Supply Networks? In: Pawar, K.S., Lalwani, C.S., De Carvalho, J.C., Muffatto, M. (eds.) Proceedings of the 12th International Symposium on Logistics (12th ISL), Loughborough, United Kingdom, pp. 172–178 (2007)

Jedermann, R., Lang, W.: The Benefits of Embedded Intelligence – Tasks and Applications for Ubiquitous Computing in Logistics. In: Floerkemeier, C., Langheinrich, M., Fleisch, E., Mattern, F., Sarma, S.E. (eds.) IOT 2008. LNCS, vol. 4952, pp. 105–122. Springer, Heidelberg (2008)

Kauffman, S.A.: The Origins of Order: Self-organization and Selection in Evolution. Oxford University Press, New York (1993)

Kaye, B.H.: A random walk through fractal dimensions. VCH, Weinheim (1989)

Lambert, D.M., Cooper, M.C., Pagh, J.D.: Supply Chain Management: Implementation Issues and Research Opportunities. The International Journal of Logistics Management 9(2), 1–20 (1998)

Lorenz, E.N.: Deterministic Nonperiodic Flow. Journal of the Atmospheric Sciences 20(2), 130–141 (1963)

Mandelbrot, B.: Stable Paretian Random Functions and the Multiplicative Variation of Income. Econometrica: Journal of the Econometric Society 29(4) (1961)

Mandelbrot, B.B.: The fractal geometry of nature. Freeman, New York (1983)

McKelvey, B., Wycisk, C., Hülsmann, M.: Designing an electronic auction market for complex 'smart parts' logistics: Options based on LeBaron's computational stock market. International Journal of Production Economics 120(2), 476–494 (2009)

Meulbroek, M.: Total strategies for company-wide risk control, 3rd edn., London. Mastering Risk (2000)

Miller, J.H., Page, S.E.: Complex adaptive systems: an introduction to computational models of social life. Princeton Univ. Press, Princeton (2007)

Ngai, E.W.T., Moon, K.K.L., Riggins, F.J., Yi, C.Y.: RFID research: An academic literature review (1995–2005) and future research directions. International Journal of Production Economics 112(2), 510–520 (2008)

Pathak, S.D., Dilts, D.M., Biswas, G.: Simulating growth dynamics in complex adaptive supply networks. In: WSC 2004: Proceedings of the 36th Conference on Winter Simulation Winter Simulation Conference, p. 774 (2004)

Peitgen, H., Jürgens, H., Saupe, D.: Chaos and fractals: new frontiers of science. Springer, New York (1992)

Poundstone, W.: Prisoner's dilemma: [John von Neumann, game theory, and the puzzle of the bomb]. Anchor Books, New York (1993)

Prahalad, C.K., Hamel, G.: The Core Competence of the Corporation. Harvard Business Review 68(3), 79–91 (1990)

Scholz-Reiter, B., Windt, K., Freitag, M.: Autonomous Logistic Processes: New Demands and First Approaches. In: Monostori, L. (ed.) Proceedings of the 37th CIRP International Seminar on Manufacturing Systems, Budapest, pp. 357–362 (2004)

Simons, R.: How Risky Is Your Company? Harvard Business Review 77(3), 85–94 (1999)

Surana, A., Kumara, S., Greaves, M., Raghavan, U.N.: Supply-chain networks: a complex adaptive systems perspective. International Journal of Production Research 43(20), 4235–4265 (2005)

Thadakamalla, H.P., Raghavan, U.N., Kumara, S., Albert, R.: Survivability of Multiagent-Based Supply Networks: A Topological Perspective. IEEE Intelligent Systems 19, 24–31 (2004)

Simon, H.A.: The Architecture of Complexity. Proceedings of the American Philosophical Society 106(6), 467–482 (1962)

Sydow, J., Windeler, A.: Organizing and Evaluating Interfirm Networks: A Structurationist Perspective on Network Processes and Effectiveness. Organization Science 9(3), 265–284 (1998)

Windt, K., Hülsmann, M.: Changing Paradigms in Logistics – Understanding the Shift from Conventional Control to Autonomous Cooperation and Control. In: Hülsmann, M., Windt, K. (eds.) Understanding Autonomous Cooperation & Control – the Impact of Autonomy on Management, Information, Communication, and Material Flow, pp. 1–16. Springer, Berlin (2007)

Wong, C.Y., McFarlane, D., Ahmad Zaharudin, A., Agarwal, V.: The intelligent product driven supply chain. In: IEEE International Conference on Systems, Man and Cybernetics, October 6-9, vol. 4 (2002)

Wycisk, C., Mckelvey, B., Hülsmann, M.: "Smart parts" supply networks as complex adaptive systems: analysis and implications. International Journal of Physical Distribution & Logistics Management 38(2), 108–125 (2008)

Supply Chains – How to Support Critical Infrastructures Safety, Protection, Preparedness and Resilience

Albrecht Broemme

THW (Federal Agency for Technical Relief), President,
Provinzialstr. 93, 53127 Bonn
Germany
Albrecht.broemme@thw.de

This chapter is based on experiences gained from exercises and disasters. The aim of this chapter is to explain how to make supply chains safer and how to make critical infrastructures more resilient. Efforts for good planning, courage to run exercises and take over experiences by "lessons learned" will help to minimize costs during and after problems with supply chains – whatever the causes might be.

"Hope for the best, prepare for the worst."

1 Supply Chain: Definition

The Council of Supply Chain Management Professionals (CSCMP) defines Supply Chain Management as follows: "Supply Chain Management encompasses the planning and management of all activities involved in sourcing and procurement, conversion, and all logistics management activities. Importantly, it also includes coordination and collaboration with channel partners, which can be suppliers, intermediaries, third party service providers, and customers. In essence, supply chain management integrates supply and demand management within and across companies. Supply Chain Management is an integrating function with primary responsibility for linking major business functions and business processes within and across companies into a cohesive and high performing business model. It includes all of the logistics management activities noted above, as well as manufacturing operations, and it drives coordination of processes and activities with and across marketing, sales, product design, finance and information technology."

The supply chain is an all-integrating and very complex system. Due to these facts, supply chain safety is most important for many kinds of processes: producing "hardware" as well as the marketing of products and supplying people with milk, electricity, money and so on.

Wikipedia, the free encyclopaedia, says: "**Supply Chain Security** refers to efforts to enhance the security of the supply chain: the transport and logistics

M. Essig et al. (Eds.): Supply Chain Safety Management, LNL, pp. 231–244.
springerlink.com

system for the world's cargo. It combines traditional practices of supply chain management with the security requirements of the system, which are driven by threats such as terrorism, piracy, and theft. Some analysts have raised concerns about supply chain security overreach. Typical supply chain security activities include:

- credentialing of participants in the supply chain,
- screening and validating of the contents of cargo being shipped,
- advance notification of the contents to the destination country,
- ensuring the security of cargo while in transit via the use of locks and tamperproof seals,
- inspecting cargo on entry."

The major aspects here are the risks during the transport in containers on ships, trains and trucks – but this is only a part of supply chain security.

In the following chapter of this book it shall be shown, that Supply Chain Safety is a complex part of a successful Supply Chain Management and that Supply Chain Security has become more and more relevant in our society. We are more and more world wide connected and more and more dependent. So, Supply Chain Security as part of Supply Chain Management is nowadays an important aspect of disaster management, civil defence and homeland defence. In general, Supply Chain Security belongs to planning for safety and resilience of critical infrastructures.

2 Critical Infrastructures – In Former Times and Nowadays

Critical infrastructures are rather easy to explain: As long as a critical infrastructure works as it should do, nobody recognises the value of it. But, if there is a break-down, most people get to know, how important a critical infrastructure is. A good example is electricity: most people have no idea, what a break-down really means. For example a study in Germany shows that the majority of our citizens believe that it is not a big problem, if there is no electricity available for about two weeks. Obviously, these men and women are convinced to be able to cook as normal during that time and that the meat in their freezers will not start to rot. Many drivers have no idea how to get their car out of the garage if there is only an electrical door opener. And who has a battery-operated radio at home (including full batteries!) to listen to important news, concerning the situation, caused by the power interruption?

Meanwhile, electricity has been identified as the most important critical infrastructure in our daily life. Without electric power, no computer can be used, most telephones won't ring any longer, the traffic lights are off and the signals for the railway show stop – no train can run. No passengers can travel to work or go home and no goods can be transported by train.

A hundred years ago, that was completely different: My great-grandmother had no telephone, no fridge and no freezer. And for cooking she used a wood- and coal operated stove – which also heated the kitchen. One of the very few electrically powered items was a few lamps. But she also had several petroleum lamps, ready for use at any time. And, my great-grandmother always had some preserved fruit and vegetables in lots of bottling jars. So, there was no need for electricity to live her daily life.

Once a week, my great-grandfather bought a big mug of home brewed beer at one of the restaurants close-by. To get this beer home, no truck was needed. Of course, the brewery needed hops, malt and barley – these components had been stored once or twice a year. And water to brew the beer came from the brewery's own well.

Only fifty years ago, railway signals still had petroleum lamps and had to be operated by a signal man. The signal box was an absolutely safe system, completely independent of external power. The signals had to be set by the muscles of an operator – a man, working in every railway signal box. Nowadays, the signals are electrical ones, they need electricity, and the signal box is dependent on several operating computers. No electricity for the control system means that no train can move, even if the overhead power lines are still working.

Another challenge to our economic system is "just in time delivery". The concept of "Just in Time" is becoming more and more common. Most companies avoid the storage of commodities. This includes the final products as well as the components needed for its production. The main reason for this concept is to increase assets by reducing investment in stored goods. "Just-in-Time" logistics must work with precision: the truck delivering new bottles to a brewery must be unloaded just before the bottles are to be filled. This logistical concept also dictates that no bottles are filled that are not for immediate use. And demand is dependent on the time of year and the weather. If the temperature is higher and if there are no holidays, more beer will be bought than under different circumstances. But, only some of these variables are predictable, others are not. That means: supply chain management has to plan for flexible conditions.

In our everyday life, supply chains work very well. That means that the Supply Chain Management has been well planned and the logistics are working as they should. That means otherwise that there are not many experiences if Supply Chains do not work as they should. This positive circumstance has to been compensated by planning and exercises.

Logistics are part of every supply chain. Logistics means: getting the right things at the right time to the right place – and all that at the best price. There is a correlation between logistics and critical infrastructures: logistics have to deliver things, critical infrastructures have to carry-on everyday commodities. Critical infrastructures are electricity and gas, drinking water and food, communication systems, transportation of people and consumer goods as well as the cash flow. We see: critical infrastructures are part of supply chains, some supply chains are part of critical infrastructures.

"Critical Infrastructures" are defined as organisations and institutions of great importance for the national community. When critical infrastructures break down or are impaired, long-term interruption of supplies, critical disruption of public safety and other dramatic results would occur.

3 How to Make Supply Chains Safe?

Remark: Instead of saying "make safe" it would be better to say "make safer", because it will never be possible to avoid all mistakes and achieve 100 percent safety. The higher the safety level has to be, the more effort (and money of course) has to be invested. And most of these investments are invisible. They are overhead costs, but without them the supply chain might break down. A break down will generate extraordinary high costs depending how long the break takes and what resorts are involved!

As long as the system is running, it is difficult to imagine, that it might stop one day.

Step 1: Work out a Process Analysis

There is a well known saying: "A chain is only as strong as its weakest link." What does this mean for supply chains? To make a system safe, you have to get to know all the links of your supply chain and you have to identify the weak parts of your system. Such an analysis has to show all the internal and external interfaces. In the end, every segment of the supply chain has to been identified.

Most interesting points are the human-machine-interfaces:

a) A person might cause critical situations. E.g. a drunken truck driver causes an accident; all his cargo is lost and will never get to the destination (or at least not just in time).
b) A person might recognise an incoming critical situation and decrease the problem. E.g. a locomotive driver recognises a wrong switch stand and makes an emergency break – to avoid a train accident.
c) A person might recognise an incoming critical situation and respond in such a way as to increase the danger. This happened for example during an experiment in the nuclear power plant in Tschernobyl on April 26th, 1986.

Any part of the supply chain may cause a problem, or a combination of several parts may cause a serious problem. The collapse of an extensively branched system is more likely than the collapse of a simple system. According to experience, the failure of one or more weak parts is the most common reason for technical or logistical problems – never the stronger parts.

Process analysis has to show all parts of a process including the side paths, which are necessary to support the process. It is very important is to identify all interfaces – there are many more than you had imagined. Such analyses had been done in 1999 to avoid the crashing of computer systems by the new millennium.

"**Stress tests**" have to be done, but they need to be done by specialists. Stress tests get becoming more and more common, to avoid major disasters (e.g. the safety of power plants) or to find the best-price solution in relation to high investments (e.g. where to build a new road?).

Anyway, at the end of the process analysis a lot of **checklists** have to been created in order to help organize the next steps. A good check list requires a lot of work. It has to be written for those people who have to use it later on. Many check lists are either too primitive or too complicated, and are therefore of no value.

A good way to improve process analysis is to run exercises. There is no need to run full scale exercises only; table top exercises can do as well. In a table top exercise, a critical situation is given and will be explained. Then the management of the critical infrastructure or of the supply chain has to act and react. Mistakes will occur; wonderful as long as they don't happen in a real situation and as long as they are mentioned in the "lessons learned". In this way, the process analysis will be improved.

To do "lessons learned" means to accept, that mistakes happened. It means to be willing and able to change a process on order to develop a safer process. That means: you need a LEARNING ORGANISATION. If you are clever, you make a mistake only once. If you are intelligent, you mostly learn by mistakes, which happened outside your organisation.

<u>Step 2: Take care of Protection</u>

In many people's minds, protection does not mean preparing themselves against danger because they believe that everything will be taken care of by someone else. A good example of active protection is the smoke detector in private homes as part of a protection chain: people have to buy them, they have to install them and they have to change the battery once a year. Smoke detectors save lives, because they give an early warning signal when a fire breaks out. Than you call the fire brigade and leave the rooms with smoke and fire as soon as possible.

However, protection costs money. E.g.: there are seat belts in every car but most of them will never have an accident. So the seat belts (as well as airbags) will be thrown away at the end of a cars live. If you would be able to order a car without any safety equipment, it would be about 20 percent cheaper! But there is now choice: you have to buy cars with "safety all included".

Protection includes all kinds of active or passive measures to make something safe. To make supply chains safe, you have to know – by process analysis – all the key points for

a) human decisions,
b) human actions,
c) automatically controlled actions,
d) all kind of accidents,
e) inside caused disruptions and
f) from the outside caused disruptions.

Relating to a) Human decisions

Man made decisions should be made by skilled persons only. They have to know their responsibility and what sort of decisions are within their field of responsibility. The field of responsibility has to be clearly limited, and when the limits are reached, there have to be clear instructions what to do in such a case. So, if an employee has a problem outside his field of responsibility – and even within his responsibility – he acts correctly in asking his supervisor for advice or to take over the decision. It would be completely wrong if the employee is asked why "he is not able to decide for himself" or "why he disturbs an instructor with such silly questions".

Whenever the field of responsibility (for each level) is not clearly defined and written down and explained, some decisions will be wrong – and nobody might feel guilty. Guilty are those who have to organize the processes, it might be "liability by mismanagement".

Check lists (see above) will help to find the best solution.

Relating to b): Human actions

Man made actions require well skilled people. They should be better paid than those with no skills. To organise man made actions there is a need for exact descriptions of what to do in each situation. These instructions have to be checked at least once a year to see if they are still meeting requirements.

Useful instructions are difficult to define and checklists are difficult to write. They have to be produced specifically for those who have to understand them. Many instructions are too complicated, so they have no value. To make an instruction easy to understand, it has to be written with simple expressions, short sentences and include some good photographs or – even better – drawings. If an instruction has to be a take away instruction, it has to be printed in pocket size on waterproof paper. It has to be readable even at night without a magnifying glass. Most of the instructions probably don't fulfil all of these requirements.

Some instructions will have to include check lists (see above).

Relating to c): Automatically controlled actions

However, automatically controlled systems may be safer than those, where people have to act ("right" or "wrong"). But every control system is created by a person. Therefore it has to been checked under several different situations.

Meanwhile, there are automatically controlled systems, which "learn" by themselves. Systems which develop themselves are more complicated. And self-made programmes might be accurate or might be wrong. To validate such systems is rather complicated; it needs an audit at a very high level.

Relating to d): All kind of accidents

What kind of accidents happened – it might happen again! There is the need for an overview, what kind of incident occurred. We have to check incidents at least over the last 20 or 30 years – worldwide. The more results we get, the easier is the planning.

It does not really matter if the probability of an accident is low or high – it might occur. And the combination of several accidents is also possible. That means that several events occur more or less simultaneous. These concurrent events might be completely nondependent or dependent. To be realistic: it is impossible to create a master plan for each combination of all kinds of incidents.

Relating to e): Inside caused disruptions

There are many possibilities for inside caused disruptions:

- One element of the supply chain has a problem (e.g. intermediate goods of too low quality or a blackout after fire or floods).
- Several elements are disordered.
- There is a shortness of money.
- The management has internal conflicts.
- The management or the sub management makes wrong decisions.
- Workers are on strike.
- Workers make mistakes.
- Many staff members fall ill.
- The working atmosphere is bad.
- The productivity is not effective.
- The planning was wrong.

Anyway, it is necessary to find out the reasons for internally caused disruptions in order to avoid them next time.

Some of the above points are difficult to change e.g. it will be a long-term process to improve the working atmosphere.

Relating to f): From the outside caused disruptions

If disruptions are caused "from the outside" this might be out of the management's responsibility. These disruptions may cause serious problems in supply chains, as well. As long as companies try to get more and more duties to be outsourced, more and more risks will be outsourced at the same time. It is difficult and it takes a long time to define exactly the different responsibilities on both sides of the border line between "internal" and "external". In case of a disruption, the external supplier will try to deny his debt.

To have an external problem, there are similar steps as they are mentioned on internal problems:

- The sub management makes wrong decisions.
- Workers are on strike.
- Workers make mistakes.
- Many staff members fall ill.
- The working atmosphere is bad.
- The productivity is not effective.
- The planning was wrong.
- Extreme **weather conditions**, as thunder storms, heavy rainfalls, blizzards, floods, hurricanes or long periods of extreme temperatures (heat or frost) can cause serious disruptions. These meteorological situations will cause even more problems, if you never thought about them. Of cause, you cannot avoid a flood. But if you decide where to build a new warehouse you should be sure that it is a "safe place". To check this point, you should not only check the past ten years. Several hundred years will give you a better answer. So, look at old maps. If you find a lake, a river or an old harbour at the location where

you plan to build up a new plant, you should know: one day, the water will be back… And good to know: some endings of names of locations indicate, that this place has been close to a river, a lake or another type of water.

- Seismological caused disruptions, as earthquakes, landslides or tsunamis, are no every-day risks. But there are parts in our world where they might happen any day. In March 2011, a heavy earthquake (9.0 m) caused a serious Tsunami which hid the coastline of Honshu, the main island of Japan. More than 90 percent of the 20,000 killed people had been killed by the tsunami and not by the earthquake itself. The tsunami caused serious damages in nuclear power plants, because the internal emergency power supply had been destroyed as well as the power lines. Why did this happen? In Japan, the nuclear power plants had to be located close to the ocean in order to have enough water for the cooling system. All safety planning based on tsunamis after the year of 1896. According to these results, the maximum high of a tsunami was 7 meters only. But after the JOGAN-Earthquake, the tsunami had been about 16 meter high. But this earthquake happened already in the year of 869 (!), and the dates had not been as reliable as the newer ones. So it had been decided, not to use records that appear unreliable! Meanwhile everybody knows that the tsunami of 2011 came up to 18 meters high. But, the walls between the nuclear power plant "Fukushima" and the ocean had not been built for such a "millennium event"…

To develop an effective prevention, **Group-think-Processes** and **Round Tables** are very usefully, as long as "open mind" is accepted. Open mind means, that negative ideas are welcome as well as recognised problems without a solution already. In every day life, such ideas are not welcome at all, and many contributors do not mention "negative" ideas because they are afraid to get in trouble with the boss.

Clearly defined responsibilities are important for a successful prevention. Never think: "I assume that it will have been done by anyone correctly." Such presumptions are often wrong, but they are part of our daily life. If you have an order it is necessary to address it clearly to a specific person and to demand a short report, as soon as the order has been successful done – as well if there are problems or delays to fulfil the order. This way to communicate is similar to the military, it is the only way to do it right.

According to clear responsibilities, it is better to have few external-driven actions, only. That means: outsourcing is no longer a good solution – as long as you have to minimise the risks. Meanwhile, more and more organisations and companies stop the outsourcing and return to in-sourced solutions. The reason to change such aims is the cognition, that outsourcing has not that "good value" as it should have. At the same time it is a benefit for the safety of supply chains.

Step 3: Take care of Preparedness

After the process analysis and all planning for matters of the prevention, preparedness is the next step. That means to be prepared if the normal situation is sudden disrupted. For such a situation, there is a need of "**Emergency Action Plan**". In many cases, such plans do not exist because people believe that all the efforts for prevention will avoid emergency situations. Preparedness is the competence, to act and to react in critical situations, where the actual situation is very different to normal conditions.

The first and very important point is to recognise the status by realistic situation reports as an unusual or even critical situation. The sooner such situation is recognized, the better will succeed in overcoming the difficulties. Of course, drifting to failure might be fast or very slow – both drifts have to be included in the planning.

The second step is to change over from the daily action plans to an "Accident Management" or even to a "Severe Accident Management". In many cases, it took too long to do this switch because the status report was wrong or because the management did not want to consider that the situation has changed to a critical one. But, if there is a well prepared and well trained accident management planning it should be reversed as soon as necessary.

The third step is to practise a good internal information flow and than an external information flow. Every body has to realise that the daily routine does not work at the moment, but other routines have been planned and will work. Again, checklists will help to do a good job. Of cause, statements of private companies have to be coordinated with those of public agencies. For sure: any kind of mistake or disagreement between the sayings will immediately be identified by the public media. Never forget: the bigger the interest of the media, the more it is necessary to tell **exactly all you really know** (not what you believe!) and what you did and what the next steps will be (not: might be). Never announce e.g. a too short time for the duration of an emergency situation – it is much better if you are able to shorten the time.

The fourth step is to respect during the Emergency Management clearly defined responsibilities. Parts of them are very different to the daily routine. For example in a press conference the CEO himself will have to explain the company's Emergency Management – it would be wrong to do it by the press relations officer only. But this has to be trained by the CEO, because it is not his daily live. So, Emergency Management means a lot of change management on order to keep the situation under control as good as possible.

Step 4: Take care of Resilience

Resilience is all capacity to come back from a critical situation to the normal life as soon and good as possible. And: a fine resilience allows a wider range of uncritical or even critical situations where the no emergency management is necessary. Anyway, there is a need of a "**Contingency Action Plan**", that means planning including checklists what to do in case of different varies of emergencies.

To develop such a "Contingency Action Plan" is rather difficult. To create it, there is a need for a good internal and external **network**. Every company, non-profit organisation or governmental organisation has a lot of experienced specialist; they have to been asked to bring their experience into good planning. Some parts of these planning will have to be done by external specialists, they are expensive and they are no insiders.

There is an important proverb concerning the networking:

"Cooperation should start before we meet abroad in a crisis."

Options to engineer an acceptable risk level for supply chains, in particular for nondurable consumer goods, include:

- a good stock managing,
- to consider alternative sourcing arrangements,
- Business Interruption and Contingency Insurance (also no insurance will be able to cover all risks).

4 THW: Technical and Logistical Support in Desasters

In case of major problems on supply chains or in case of disasters or serious incidents every company has to do the first actions by itself – adequate to the Emergency Action Plans and the Contingency Action Plan. That means e.g. to have take care for the power supply of a store – especially for the computerised control system – by a generator of your own. Of course, this generator has to been tested several times a year **by a real use** of the generator **during several hours**. Than you can be rather sure that the system will keep on running if the normal electrical power supply is down for more than half day.

But, if all planning, all equipment, all emergency tools do not work as it should do, it is good to know that in Germany a federal, governmental organisation, the THW, is able to assist.

THW's History

"THW" is the German abbreviation for "Technisches Hilfswerk", meaning "German Federal Agency for Technical Relief". Nowadays, just "THW" is in use in most languages. The THW can look back on an eventful development and many years of commitment by voluntary and some very few fulltime workers.

Otto Lummitzsch, founder and first director of the THW, 1950 – 1955

Source: *THW*

It was a time of political and cultural changes and renewals in post-war Europe when the federal Minister of Interior, Heinemann, and Lummitzsch met for the first time on August 22, 1950, in Bonn, to discuss the development of a civil protection tool in the Federal Republic of Germany. At that time, there were hardly any structures of civil protection in Germany, so the spoken promise given by the Minister of the Interior on that evening signified a crucial innovation. One month later, Lummitzsch had Heinemann's written assignment in his hand and he began with "the work for the organisation of a civil regulatory service". Lummitzsch became the first director of the new agency. In 1953, the THW became a federal agency, thanks to the constitutional mandate by the Federal Ministry of the Interior.

Right from the start, the guiding principle of THW were the operations with the voluntary staff. To commit oneself on a voluntary basis in order to protect people in distress is a humanitarian idea, which makes THW known, not only at home, but also well beyond the borders of the Federal Republic of Germany and Europe.

THW is a federal Agency, being part of the Federal Ministry of Interior. So, THW's budget comes through the Federal Ministry of the Interior. It is about 180 million EURO per year. THW has all together more than 80,000 volunteers and 800 fulltime, only. There are local sections in nearly 700 cities all over Germany.

THW abroad – the "Blue Angels"

For more than 50 years, the THW has been operating in Germany every day in order to render technical assistance. The help of the "Blue Angels", as the volunteers of the THW were called by the French population after their operation in France in 1999, has ranged from the disasters that moved the nation in the 1960s, such as the flood in Hamburg and the mining accident of Lengede (Germany) to the floods of the Elbe and Oder at the beginning of the new millennium.

The repair work after the storm tide in the Netherlands in 1953 marks the beginning of the missions of the THW abroad. It was followed by humanitarian assistance after drought periods, civil wars and earthquakes in Africa, Europe and South America, as well as in South and South-East Asia after the Tsunami disaster. In 2005, the THW provided technical assistance to the United States for the first time in its history. The earthquake in Haiti is another chapter of humanitarian aid abroad: THW provided the population with drinking water and supported the German Embassy in coordinating the German relief measures.

Through its relief measures in Germany and abroad, THW is making a contribution in the effort to reduce hardship and disasters. Through its operations after disasters and in its long-term reconstruction projects, it implements humanitarian aid worldwide on behalf of the Federal Republic of Germany. On occasion – as after the operation of the THW in Skopje, Yugoslavia in 1963 – this has led to a deepening of the political relations between the countries. Today the THW participates in the worldwide intermeshing of all relief organisations as an internationally active operational organisation. As a partner the THW is assigned a sustaining role by the United Nations as well as in the European Union.

THW' main body – 80,000 Voluntary assistants

THW is a Federal Agency on behalf of the Federal Minister of Interior. THW has more than 80,000 members. But: 99 per cent are voluntary engaged men and women, 800 are paid staff members. That means that all the volunteers have a paid job (or they go to school or university). Voluntaries are not paid at all during their training lessons, exercises or interventions in the THW. The only point is: if there are courses or operations during the working hours, members of THW may interrupt their work without taking holidays. On demand, the employer gets the salary reimbursed.

These volunteers are based in nearly 700 local sections. These buildings of THW have vehicle hall, a court for exercises and a main building with locker rooms, assembly rooms, a kitchen and some offices.

The paid staff is located in 66 district offices, 8 state offices, two federal THW schools and the headquarters.

THW's Operation Options

THW has a lot of technical and logistical options. THW might be explained as companies of engineers or as civil pioneers. THW do not do the same jobs as the fire fighters or paramedics of the Red Cross. In the following you find a short overview on THW's operation options:

Technical Threat Prevention

- Search, rescue and salvage
- Clearing and blasting
- Rescue from water dangers
- Fight against flooding and inundation
- Lighting of operational areas

Technical Support in the Range of Infrastructure

- Electricity supply
- Drinking water supply
- Waste water disposal
- Bridge work

Command and Communication, Logistics

- Establishment and operation of command centres
- Command support
- Creation of temporary telecommunication systems
- Establishment and operation of logistic bases
- Catering and care of operational staff
- Maintenance of material, repair and maintenance work for mission equipment
- Transportation of consumer items for mission demands

Technical Support in the Protection of the Environment

- Fight against oil damage
- Water analysis

Provision of the Population

- Electricity and drinking water provision
- Waste water disposal
- Establishment and equipment of emergency accommodation and collecting points with matching infrastructure

Further Technical Support

- Technical help on traffic routes
- Rescue from heights
- Diving
- Makeshift road works
- Maintenance of civil protection facilities (emergency wells, shelter)

The following two examples show the support by THW in case of disruption of critical infrastructures:

Damage in river of Rhine: THW looks after stranded Mariners

Whether it is water and bread, fruit or vegetables – staple foods are scarce commodities on the vessels which are stranded on the river of Rhine in front of St. Goar (near Koblenz). Upon request of the Water and Shipping Directorate, more than 30 THW volunteers had been active in order to supply the mariners with the bare necessities, as drinking water and gasoline. A lot of vessels had been anchored in the Rhine. By radio, the boat owners informed the river police when, for example, water and staple foods become scarce. THW volunteers then refill the 3,000 litre water tanks, procure food and transport the goods or new crews by THW multi-purpose boats to the vessels.

The reason for this operation was a tank ship with 2,400 tons of sulphuric acid made an overturn, sank and blocked the Rhine. (Meanwhile we know: the ship had been overloaded by 90 tons.) So, all shipping on one of the busiest waterways in Europe had to been stopped. More than 200 vessels had to wait for about four weeks to continue their journey. It took about two weeks to bring the salvage vessels from Rotterdam and Duisburg to St. Goar, another two weeks to salvage the vessel.

THW volunteers look after the mariners who were stranded on the Rhine.

Source: *THW/Michael Walsdorf*

Haiti: Help for better Life Conditions

The building of drainage, the securing of footpaths, the installation of sanitary facilities and the supply of drinking water are tasks which THW fulfils in Haiti to improve the living conditions in the many emergency camps. THW implements the building measures on behalf of the office for Humanitarian Aid and Civil Protection of the European Commission (ECHO).

THW helps in the numerous emergency camps of Haiti, among other things, to set up drainage, to secure footpaths and to install sanitary equipment.

Source: *THW/Oliver Hochedez*

From May 2010 up to June 2011, THW has provided technical help in more than 50 camps in Port-au-Prince. Additionally, since November 2010, THW and Maltese International have fought against the cholera disease in Port-au-Prince. According to official information, more than 220,000 people had been infected with diarrhoea. In many transition camps in the Haitian capital especially, hygienic standards are inadequate.

The German teams cleaned the sewage systems and distribute drinking water canisters, soap as well water cleaning tablets. Sanitary facilities and water tanks had been disinfected with chlorine solution and some parts of their construction improved. With the help of films and other educational measures, the inhabitants are thoroughly trained in the areas of early warning, disinfection and building measures.

Supply Chain Event Management – Concept and Use in Business Practice

Joerg S. Hofstetter[1] and Wolfgang Stölzle[2]

[1] University of St. Gallen, Assistant Professor of Management, Vice Director of the Chair of Logistics Management, Dufourstrasse 40a, 9000 St. Gallen, Switzerland
joerg.hofstetter@unisg.ch

[2] University of St. Gallen
Professor of Logistics Management , Director of the Chair of Logistics Management, Dufourstrasse 40a, 9000 St. Gallen, Switzerland
wolfgang.stoelzle@unisg.ch

Ensuring continuity of supply while maximizing profitability are primary goals of corporate disciplines like procurement, logistics, production, or distribution. Each discipline for itself has established concepts to reduce or cope with inherent hazards that may hinder inbound or outbound product flows as well as shrinkage, all eventually harming continuity of supply.

Ensuring product availability outbound to customers has become a major customer requirement over the past decades. In buyer market economies with limited product or brand loyalty product unavailability immediately result in lower sales. Today's customers are quick to substitute a missing product with a different brand or seek it elsewhere. Consequently, distributors, traders, and original equipment manufacturers (OEM) optimize their operations to ensure product availability to their customers. They also do rely on the delivery performance of their suppliers.

Ensuring product availability inbound from suppliers has become the key prerequisite for today's production concepts. Production plans generally require the simultaneous availability of multiple products to conduct planned work. In particular for customer individual assemblies, like in the European automobile industry, concepts like "just in sequence" are based on 100% product availability. Any missing product requires changing the sequence of other supplies in order to reestablish the right sequence of all parts that need to come together for a customer order. Reduced stock levels have been achieved by highly reliable production planning and by increasing delivery cycles and delivery accuracy from suppliers. As production relies on product availability as planned to keep operations running, penalties apply to suppliers in case of product unavailability. These also apply when sub-suppliers cause material shortages that lead to unavailability.

Past cases of major product unavailability that resulted in retail out-of-stocks or production line stops have pointed to the high complexity of today's supply chains.

M. Essig et al. (Eds.): Supply Chain Safety Management, LNL, pp. 245–265.
springerlink.com

On their way from raw materials to the finished good, many products go through multiple production and logistics steps performed by different companies, often located in various parts of the world. Product shortages and logistics failures still happen frequently. The question for each company to be answered is: Does this impact our delivery performance? And if so, how can we respond?

In order to ensure continuity of supply, companies require hands-on management concepts for supply chain safety - concepts that warn them about and trigger them to respond to potential product shortages, as early as possible. The earlier a company knows about the risk that a certain product may become scarce, the more opportunities this company has to find alternative solutions. Yet, the high complexity of today's supply chains makes this task very demanding. Few companies know the sub-suppliers in their supply chains as they are numerous and may change without any notice. So, as long as external suppliers are involved in a supply chain, a company cannot manage supply chain safety alone, but needs the help of its suppliers and their sub-suppliers as well as logistics and other service providers.

Supply Chain Event Management (SCEM) has been suggested as both a management concept and a tool to improve, or even ensure, supply chain safety. It shall reduce the risk of unavailability through visibility in the supply chain by establishing transparency of workflows and trigger as-soon-as-possible problem solution. The idea is, based on a description of a supply chain, to follow each order comparing the status at specified process steps in the supply chain with the plan, and in case of relevant deviations issue a warning to the subsequent process steps. In practice, SCEM is approached differently by companies.

This chapter of the book addresses the existing heterogeneity of SCEM understandings and SCEM approaches found in business practice and academia today. It takes up the lacking definition regarding the scope and context of SCEM, the missing hands-on measurement for SCEM use, and the darkness about the current use of SCEM in business practice. With Supply Chain Event Management (SCEM) being a sub-concept of Supply Chain Management (SCM), this chapter first specifies the general understanding of SCM. It then describes SCEM, including a hands-on operationalization. Finally, it discusses the current status of SCEM use in business practice.

1 SCM – The Foundation

Supply Chain Management (SCM) is "the management of the network of organizations that are involved, through upstream and downstream linkages, in the different processes and activities that produce value in the form of products and services in the hands of the ultimate consumer" (Christopher, 1992). It is a management concept in which a company's internal and external logistics-related activities are integrated and coordinated (Harland, 2006). SCM covers all activities ranging from procurement, production to distribution and customer service; or from a different view, from the customer order to the after-sales-service. SCM ideally begins with the production of raw materials and ends with the disposal, or recycling, of the final product. SCM plans, executes and controls the material, informational and financial flows, all geared toward the final customer.

The overall objective of SCM is the complete fulfillment of the delivery promised to the customer while minimizing logistical costs. Its formal performance goals comprise of both quality improvement and cost reduction but also the realization of time-advantages and the increase in overall customer satisfaction (Heusler, 2004). The degree to which these SCM goals are achieved is referred to as "Supply Chain Performance" (Karrer, 2006).

Consequently, a supply chain is a process chain covering the entirety of all value adding processes involved in the making of a product. Each individual process comprises of adaptation and execution processes that are oriented on the customer contract. The workflow describes the application of these processes to fulfill a customer order.

Demands on SCM continue to intensify due to increasing cost pressure, increasing customer expectations on logistics quality, decreasing response and delivery time, and increasing clustering of specific industries in few parts of the world (Heusler, 2004). The further advancing segmentation of value creation, triggered by focusing on core competencies and economies of scale, increases the number of independent organizations involved in today's supply chains.

Companies have responded to these demands on one hand by reducing stocks and shortening lead time in their processes. On the other hand, they internationalize their sourcing and production, and transfer more responsibility to suppliers.

The individual organizations involved in a supply chain frequently pursue their own objectives instead of the entire supply chain's objectives. Often, these organizations are part of multiple supply chains. The results are multi-faceted information barriers between the respective actors, giving only insufficient visibility of logistics processes in supply chains and hindering the overall coordination of supply chains (Heusler et al., 2006).

This situation causes three serious effects: First, the probability of an unforeseeable incident in the workflow increases. Second, the degree of impact and depth of the induced work plan changes increases (Heusler et al., 2006). Third, decision processes for those often complex problems require more time, whereas the available reaction time decreases with the increased dynamic (Heusler, 2004).

Highly productive concepts generating high quality information are required in order to make ad hoc decisions in such tense situations. Information must be actively assessed and filtered as well as forwarded according to relevance, which furthermore permits triggering corrective action.

Attention is growing on SCEM as a management concept to respond to those challenges. Although SCEM has been a topic of discussion in the scientific literature for some time (Bretzke et al., 2005) and use in business practice has become more common, the current implementation status of SCEM is unclear.

The University of St.Gallen, in cooperation with its partners BASF, CapGemini, EURO-Log, Hewlett-Packard, PTV and DB Schenker, conducted a survey about the use of supply chain event management (Reiche et al., 2009). The findings are based on the replies of 250 companies from Austria, Germany and Switzerland.

2 SCEM – The Description

Supply Chain Event Management (SCEM) is a management concept that, by monitoring all individual workflows in the entire (interorganizational) supply chain, identifies deviations from the work plan, evaluates their impact on the subsequent process steps, notifies the responsible decision makers for relevant deviations about incidents, develops and suggests corrective action, and controls that corrective action has been taken and achieves the objective to reestablish the work plan.

An event is defined as a specific process step in the workflow. The scientific literature on SCEM is ambiguous on this, as some authors define an event as a recognized deviation (an incident) at a specific process step.

SCEM is part of Supply Chain Management, increasing the visibility of critical logistics processes to ensure order fulfillment as planned. It connects long term planning (supply chain planning) with the operative execution of these plans (supply chain execution) (Heusler, 2004; Stölzle, 2008).

The discussion about SCEM is largely separated into two fields: SCEM as a management concept, and SCEM as a tool. The first is found in the literature on supply chain management and logistics, the second in the literature on business engineering systems and information technology. The second literature stream is substantially larger, yet frequently does not account managerial aspects (i.e., implementation). Conceptual definitions of SCEM in both literature streams are frequently not properly distinguished; neither in the practical nor in the scientific literature of SCEM (Nissen, 2002; Otto, 2003: Bodendorf & Zimmermann, 2005; Sputtek et al., 2008).

SCEM as a tool is a further development of tracking & tracing systems (Stefansson & Tilanus, 2000). These are passive systems, gathering status information about workflows without filtering for relevance. The resulting flood of data limits their efficacy. A central point of criticism of tracking & tracing systems lies in this lack of selectivity for data allocation, which leads to limited applicability of data in regards to decision making. In fact, data needs to be accessed manually, causing high search cost. With increasing complexity of and dynamic in a supply chain's processes the problems intensify. Consequently, management rarely benefits from reduced workload or increased efficiency.

SCEM addresses in particular this problem. Its goal is to optimize the workflow in such a way that it improves the visibility of both the internal but also the interorganizational processes (supply chain visibility). Increased visibility enables early detection of looming turbulences in material and informational flows. The goal is to identify deviations of real workflows from work plans in interorganizational supply chains as early as possible and to initiate corrective action based on pre-defined rules. The selection and evaluation of workflow data as well as simulation and execution of corrective action are meant to contribute to the support of the management function and to the optimization of supply chains itself. SCEM contributes substantially to complexity reduction by focusing only on relevant deviations in workflows.

Practically, SCEM actively surveys interorganizational logistics processes at defined process steps (the so-called events) and reacts towards looming or occurred deviations from work plans. The basis for SCEM is a target work plan defined for each individual order line. This target work plan is created based on standard plan values, enabling automated planning. The workflow of each order is monitored at defined events by measuring defined parameters and comparing them with the plan values. In case a deviation falls below or surpasses the plan value (and its range of tolerance), a notification is issued about the incident, and further alternatives are suggested. Where applicable, this procedure occurs on a self-regulating basis in order to adhere to supply agreements to readopt the work plan.

In order to apprehend the above-mentioned tasks, SCEM combines the following five management elements:

1. Monitor
2. Notify
3. Simulate
4. Control
5. Optimize

The literature is not fully aligned on the definition of these elements, as in particular the last concept is often referred to as "measure". We argue in the following section why "optimize" fits better to the content of this element.

3 SCEM – The Operationalization

This article addresses SCEM as a management concept that includes the required tools. To understand the use of SCEM, it is important to extend our view from the management elements of SCEM to also consider richness and reach of SCEM in practice. The literature addresses those later aspects with four distinct dimensions explained below (Sputtek et al., 2008).

To enable hands-on goal setting and measurement of progress in use of SCEM, measures are defined for each dimension. This operationalization of SCEM has been missing so far in the literature.

The following five dimensions cover the use of SCEM:

1. Management elements
2. Degree of aggregation
3. Degree of differentiation
4. Depth of implementation
5. Breadth of implementation

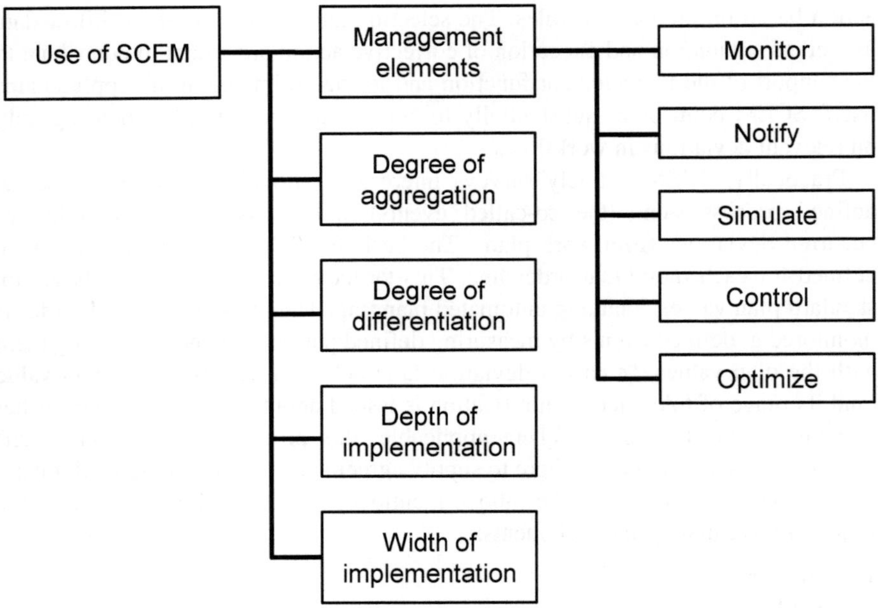

Fig. 1 Operationalization of SCEM

Management elements

Based on the definition of SCEM, five management elements of SCEM characterize its core functions.

The management element "monitor" covers the generation of a work plan with precise plan values and ranges of tolerance, the measurement of the workflow at each event, the comparison of the measured values with plan values, and the evaluation of relevance of identified deviations. The basis of SCEM is the definition of an order specific work plan for the supply chain. For efficiency reasons the creation of work plans should be automated. The measurement and the data gathering of a workflow enable the company to monitor order status in real time regarding time, location, quantity, and condition. The status notifications at each event are neutral. Central to the "monitoring" element is the comparison between the measured values for the workflow and the target values with their ranges of tolerance for the work plans. In case of a deviation beyond the range of tolerance, the information is interpreted as "incident" which triggers the step "reporting".

The management element monitor can be measured by answering the following questions (Bretzke et al., 2002; Heusler et al., 2006; Knickle & Kemmeter, 2002):

1) *Our SCEM automatically generates a work plan for each incoming order.*
2) *This work plan specifies for each object and event:*
   2a) *target times*
   2b) *target locations*

   2c) target quantity
   2d) target condition

3) *Our SCEM captures the workflow for each object and event regarding:*
   3a) real time
   3b) real location
   3c) real quantity
   3d) real condition

The management element "notify" encompasses the creation and transmission of warnings about a detected incident as well as the control over the reaction toward the warnings. As the management element requires a detected incident as trigger for action from the management element monitor, warnings are issued only for relevant deviations. The warning is sent to the person responsible for the process detailing the detected deviation. The inclusion of a due date for corrective action in a warning controls that the recipient has taken action in time. If no corrective action is reported, the warning is escalated to the process supervisor. Ideally, the issuing of warnings and the control for corrective action happens automatically.

The management element notify can be measured by answering the following questions (Bretzke et al., 2002; Heusler et al., 2006; Knickle & Kemmeter, 2002):

4) *Our SCEM automatically identifies deviations between the work plan and the workflow.*
5) *Our SCEM automatically sends warnings to the responsible person, if the detected deviation exceeds the tolerance range.*

The management element "simulate" supports the responsible actor in developing potentials options for corrective action. Simulations may be based on heuristics or linear optimization, and generate results that help in evaluating the different options.

The management element simulate can be measured by answering the following questions (Bretzke et al., 2002; Heusler et al., 2006; Knickle & Kemmeter, 2002):

6) *Our SCEM supports us in simulating potential corrective actions, if the detected deviation exceeds the tolerance range.*
7) *Our SCEM automatically proposes corrective actions, if the detected deviation exceeds the tolerance range.*

The management element "control" consists of the selection and execution of corrective actions, with the objective to bring the workflow back in line with the work plan. The automation of this management element requires a high level of process standardization and a comprehensive classification of potential turbulences.

The management element control can be measured by answering the following questions (Bretzke et al., 2002; Heusler et al., 2006; Knickle & Kemmeter, 2002):

8) *Our SCEM controls if there is a reaction to a warning*
9) *Our SCEM escalates warnings independently to a higher ranging person if no reaction is taken to a warning within a period of time*
10) *Our SCEM automatically takes corrective actions, if the detected deviation exceeds the tolerance range.*

The management element "optimize" abstracts from the individual workflows with their specific incidents by targeting the optimization of the supply chain itself, its processes and standard values. The analysis of incidents allows on one hand to identify problem areas in the supply chain and its processes, and to implement changes that make them less vulnerable. On the other hand, the SCEM can be improved by both defining the most meaningful events, better measurement, and by adjusting the plan values and tolerance ranges. This allows realizing both efficiency gains and reductions in irrelevant warnings.

The management element optimize can be measured by answering the following questions (Bretzke et al., 2002; Heusler et al., 2006; Knickle & Kemmeter, 2002):

11) *We regularly analyze our workflows with the help of our SCEM to identify improvement potentials.*
12) *The results of the workflow analysis are regularly used to optimize our control systems (hardware and configuration).*
13) *The results of the workflow analysis are regularly used to describe our business processes more realistically.*

4 Degree of Aggregation

The degree of aggregation defines the unit or granularity of the monitored objects. For instance, the monitored object may be a complete customer order or an individual order line. The degree of aggregation is described by the number of divisible logistical units, which are monitored by the SCEM. These logistical units may be tangible or intangible. A container is an example for a tangible unit. It can be divided further into pallets or boxes. Intangible units may be a general contract that can be subdivided into multiple orders which can furthermore be divided into order lines. The degree of aggregation indicates the complexity to which extent the workflow status of the object is monitored.

The degree of aggregation can be measured by answering the following questions (Olhager & Selldin, 2003):

14) *Our SCEM monitors the following objects:*
    - 14a) *customer orders (for LLP: transport orders)*
    - 14b) *customer order lines*
    - 14c) *production orders*
    - 14d) *transport orders*
    - 14e) *transport order lines (transport objects, packaging pieces)*
    - 14f) *loading equipment (containers, pallets, boxes)*
    - 14g) *orders from suppliers or service providers*

5 Degree of Differentiation

The degree of differentiation indicates the scope and the parameters that monitor the processed objects. This may include data about timeliness, location, condition, or completeness. The more status information is collected, the more transparent is the supply chain process.

The degree of differentiation can be measured by answering the following questions (Bretzke et al., 2002; Heusler et al., 2006; Knickle & Kemmeter, 2002):

15) *Our SCEM controls:*
  15a) *on-schedule clearing of events*
  15b) *the location of the controlled object*
  15c) *the condition of the controlled object*
  15d) *the completeness of the controlled object*
  15e) *the physical condition of the loading equipment*
  15f) *the timely reaction to warnings of the SCEM*

6 Depth of Implementation

The depth of implementation describes how far up and down the supply chain the SCEM is implemented. Ideally, SCEM covers all process steps along the entire supply chain from the raw material to the end customer. This may include multiple sites and organizations. A profound depth of implementation means that not only suppliers and customers but also the company's different business locations and service providers are considered. Furthermore, implementation depth describes the quality and quantity of information exchange in SCEM.

The depth of implementation can be measured by answering the following questions (Olhager & Selldin, 2003):

16) *All of our company's business locations that were targeted to be included in our SCEM are covered today.*
17) *All of our customers that were targeted to be included in our SCEM are covered today.*
18) *All of our suppliers that were targeted to be included in our SCEM are covered today.*
19) *All of our service providers that were targeted to be included in our SCEM are covered today.*

7 Width of Implementation

The width of implementation describes the amount and kinds of different logistics processes along the supply chain that are controlled by SCEM. Ideally, the process spectrum covered by SCEM stretches across the entirety of logistical processes from the order receipt and workflow to distribution.

The width of implementation can be measured by answering the following questions (Pfohl, 2010; Pfohl, 2004; Supply Chain Coucil, 2007):

20) The following business processes are integrated in our SCEM:
  20a) transport processes
  20b) storage processes
  20c) handling processes
  20d) order fulfillment processes

The possible answers for all questions 1-20 may range on a 7 point Likert scale from "not at all" (1) to "to the fullest extend" (7), whereby the latter indicated full use of this SCEM dimension.

The quantitative assessment of the degree of SCEM use in business practice allows to reliably measure and compare real use against targets, over time, within industries or geographies. It allows further to visualize the degree of SCEM use.

As all dimensions are equally important to describe SCEM use, what also applies to the sub-dimensions, the calculation of the median represents a company's degree of SCEM use.

8 SCEM – The Prerequisites

SCEM as a management concept that includes the required tools, draws from both technological as well as managerial resources. This suggests that SCEM use requires technical and organizational prerequisites.

The technical prerequisites include the control of supply chains via IT-systems, the degree of automation of data exchange, the integration of external partners such as suppliers, customers and service providers into the data exchange, and the actuality of data. The literature generally recommends that storage of data should occur by as few systems as possible, ideally by one single system. Fully automated data exchange and data generation in real time require substantial investments but generally contribute to improved efficiency and forecasting accuracy (Teuteberg & Schreber, 2003). The integration of all business partners into the SCEM also requires greater efforts in preparation and larger investments, yet is a pre-requisite for the implementation along the entire supply chain.

The technical prerequisites can be measured by answering the following questions:

21) We control our supply chain with one integrated IT-system.
22) Data exchange with our external partners happens automatically (e.g. EDI)
23) Data exchange with our external partners happens manually (e.g. Web applications)
24) Our electronic data exchange includes:
  24a) our customer
  24b) our suppliers
  24c) our service providers
25) The exchange of SCEM relevant data happens in real time.

26) *The actuality of the data in our IT-systems is appropriate to monitor the workflows.*
27) *The actuality of the data in our IT-systems is appropriate to identify relevant deviations of the workflow from the work plan.*
28) *Our systems apply the following technologies to monitor objects:*
    - 28a) *RFID*
    - 28b) *barcode*
    - 28c) *optical character recognition*
    - 28d) *GPS*
    - 28e) *GSM (location based services)*
    - 28f) *decentral mobile computing (e.g., on-board computer)*
29) *Our SCEM uses the latest traffic information.*
30) *Relevant traffic alerts trigger our SCEM to issue a warning.*

The organizational prerequisites include the support by top management, the mapping and optimization of the existing business processes, and a definition of process interfaces with external business partners. The involvement of top management underlines the strategic dimension of SCEM. The organizational prerequisites primarily address aspects that need to be provided by each company's management.

The organizational prerequisites can be measured by answering the following questions:

31) *Our top management backs our SCEM.*
32) *Our top management provides sufficient resources for our SCEM.*
33) *Our top management feels responsible for our SCEM.*
34) *Our company has a dedicated organizational entity, responsible for our SCEM.*
35) *Before implementing SCEM our company had:*
    - 35a) *mapped our transport processes*
    - 35b) *mapped our handling processes*
    - 35c) *mapped our storage processes*
    - 35d) *mapped our order fulfillment processes*
    - 35e) *optimized those processes*
    - 35f) *defined the interfaces to our suppliers*
    - 35g) *defined the interfaces to our customers*
    - 35h) *defined the interfaces to our service providers*
36) *Our company has established a structure that provides resources (off normal operations) to identify improvement opportunities from our SCEM.*
37) *We established clear personal responsibilities for dealing with warnings issued by our SCEM.*
38) *All employees involved in our SCEM have participated in substantial training.*
39) *We have developed an incentive system that supports our SCEM.*
40) *For SCEM we use standardized terminology in our company.*
41) *The definition of objects monitored by out SCEM allows full visibility of our processes.*

42) *We have defined the objectives of our SCEM in writing.*
43) *The partners (suppliers, customers, service providers) connected to our SCEM are honest among each other.*
44) *Occurring problems among partners (suppliers, customers, service providers) connected to our SCEM are raised openly.*
45) *The partners (suppliers, customers, service providers) connected to our SCEM fulfill their duties even if they are not controlled.*
46) *Our SCEM controls the following criteria:*
    - 46a) *constitution of transport (i.e. completeness)*
    - 46b) *reliability of transport (i.e. punctuality)*
    - 46c) *transport loss ratio*
    - 46d) *delivery service (i.e. reliability)*
    - 46e) *supply readiness*
    - 46f) *number of relevant deviations of workflows from work plans*
    - 46g) *lead time*

The possible answers for all questions 21-46 may range on a 7 point Likert scale from "not at all" (1) to "to the fullest extend" (7), whereby the latter indicates full availability of this dimension.

9 SCEM – The Targeted Objectives

Supply chain performance, as the targeted outcome of SCM, is comprised of multi-dimensional, future-oriented and interconnected dimensions (Karrer, 2006). In the context of logistics processes along multi-company supply chains, this is in particular the fulfillment of the customer expectations. The literature discusses further aspects characterizing supply chain performance (Pfohl, 2010). Companies report benefits of SCEM on the following performance criteria.

Customer benefits derive directly from the key SCM objectives, i.e. the optimization of costs, quality and time. Further aspects relevant to customers are the responsiveness to inquiries about and the quality of order status information.

Process optimization targeted by SCEM refers primarily to the improvement of logistics processes such as transport, handling, storage, and order fulfillment.

Efficiency applies especially to the degree of utilization of warehouses and transport equipment.

Time optimization applies to the time needed for transportation, delivery, and order fulfillment.

Flexibility refers to flexibility of workflows, transport, and delivery.

Reliability encompasses the constitution of transport (i.e. completeness), the reliability of transport (i.e. punctuality), the transport loss ratio, the delivery service (i.e. reliability), the supply readiness, the number of relevant deviations of workflows from work plans, and the lead time.

Internal cost criteria concern accumulated stock keeping costs, safety stock, extra tours and extra production runs.

10 SCEM – The Current Use in Business Practice

The average degree of SCEM use in business practice is "partial" only (4,14 on a seven point scale from 1 (not at all) to 7 (to the fullest extend)). This shows that SCEM is currently only partially used as a management concept in business practice. The degree of SCEM use per company is distributed fairly normally, with a median value that is almost exactly in the center of the scale.

Management elements

As the management elements constitute the core of SCEM, their analysis is of particular interest in the evaluation of SCEM use.

The initial creation of a work plan per object is a mandatory requirement to apply SCEM. At most companies, this creation is only partly generated automatically.

For the management element ***monitor*** companies responded which criteria they define and which of those they actually monitor. The criteria of time, location and inventory dominate the monitoring of workflows. Conditions are less frequently monitored. Most likely, this is caused by differences in required efforts. The automatic measuring of conditions is far more tedious than the acquisition of classical logistical parameters such as location, time and quantity. Generally, it appears that the original tracking & tracing functions still dominate on an operative level.

The management element ***notify*** consists of two sub-divisions, the identification of incidents (relevant deviations) and their transmission. The automatic identification is more strongly applied than the following step of transmitting. Companies are somewhat reluctant to allow automated issuing of warnings, as they fear overreaction.

The management element ***simulate*** is fairly used in practice. While some companies use simulations to develop and evaluate reaction options, the selection is in most cases done manually.

The management element ***control*** is little used in practice. The monitoring of reaction toward a warning is the most prominently implemented element. Automated escalation if reaction is missing is still an exception today. The automatic seizing of corrective action is hardly implemented. This may be explained by the high degree of automation and the technological demands associated with it. An automatic seizing of corrective action would, in this respect, tackle the entire operative aspect of SCEM.

The management element ***optimize*** is more frequently implemented in business practice today. This matches the expectation as optimize is the element that reconfigures SCEM. The configuration of SCEM is challenging as only realistic process descriptions and realistic plan values allow differentiating the relevant incidents from deviations that have little impact on the subsequent workflow of an object. Running this management element does not require major automation. Rather, it takes up descriptive statistics of the management elements monitor and notify, often just referring to the data of the established tracking & tracing systems. Optimization often occurs periodically at discrete points in time for which no specific degree of automated optimization is necessary. Overall, companies only make use of a part of the large potential offered by workflow analysis to the optimization of control systems or business processes.

Overall, the degree of implementation of the different management elements varies substantially. This variation is connected to the degrees of automation and standardization required for the different management elements. SCEM requires the definition of processes and values that well match reality. It further requires the development of categories for the many different kinds of incidents, processes and corrective actions. On this basis, automation can be applied. All this indicates the substantial investments companies face implementing SCEM.

Degree of Aggregation

Among the different possible monitoring objects, customer orders and customer order lines are most frequently used. Loading equipment is given the least attention, what is surprising, as tracking & tracing systems often monitor unit loads in particular. The tracking of customer orders is therefore more prominently characterized by the position of the order as opposed to the characteristics of the order.

Degree of Differentiation

The monitoring of objects is concentrated on location, time and quantity of the objects. For those factors, companies check for deviations from the work plan. The object's condition is checked rarely, and so is the identification of deviations. Also, few companies control the timely reaction to warnings. The reason might be, that these factors require a high degree of automation of the control systems which companies are not yet ready for.

Depth of Implementation

SCEM is far from being comprehensively implemented along the entirety of the supply chain. The majority of the targeted implementation today is found internally within companies. Among the targeted implementation of external organizations, the integration of customers is more advanced than the integration of suppliers or service providers.

Width of Implementation

SCEM in business practice integrates primarily the order fulfillment and storage processes. Transport and handling processes are less frequently integrated. On one hand, this indicates the strong focus of today's SCEM on customer orders. On the other hand, this reflects the high complexity and low level of implementation of systems controlling handling and transportation.

Technical Prerequisites

The survey results highlight the general trend of consolidation of internal IT-systems into one single system. External partners are incorporated, or given the opportunity to integrate, into this system. The electronic data exchange occurs primarily automatically. It is remarkable that companies integrate customers more strongly than suppliers and service providers. This reflects the strong, and for the SCM focal, customer emphasis for the business. The data transfer for the control of the customer process generally occurs in a timely manner, and data actuality can be regarded as appropriate.

One can generally deduct from the responses that the technical prerequisites, in particular within companies, are generally established. The trend for standardization and automation are clearly recognizable as well.

However, it is also noteworthy that the barcode is the most commonly used identification while RFID and GSM are clearly the exception. Also, traffic information is hardly used for SCEM.

Organizational Prerequisites

Top management generally backs SCEM and takes responsibility but hesitates providing resources. The definition of processes and process related interfaces is a frequent prerequisite for the use of SCEM; whereas the client-oriented interfaces again play the major role.

Companies have established structures and competencies in order to support SCEM in the same way as they use SCEM. Interestingly, companies have hardly adjusted incentive systems to motivate people towards an increased use of SCEM.

Companies make use of SCEM data to control performance to the extent they use SCEM. Transport loss ratio and the number of relevant deviations of workflows from work plans are somewhat less frequently applied.

Dimension		Mean	Std. deviation
Management elements: Monitor			
1)	Our SCEM automatically generates a work plan for each incoming order.	4.1	2.1
2)	This work plan specifies for each object and event:		
2a)	target times	4.8	2.1
2b)	target locations	4.7	2.1
2c)	target quantity	4.6	2.1
2d)	target condition	4.1	2.2
3)	Our SCEM captures the workflow for each object and event regarding:		
3a)	real time	4.8	2.0
3b)	real location	4.7	2.0
3c)	real quantity	4.8	2.1
3d)	real condition	4.2	2.1
Management elements: Notify			
4)	Our SCEM automatically identifies deviations between the work plan and the workflow.	4.0	2.0

Fig. 2 Current use of SCEM in business practice

5)	Our SCEM automatically sends warnings to the responsible person, if the detected deviation exceeds the tolerance range.	3.6	2.1
Management elements: Simulate			
6)	Our SCEM supports us in simulating potential corrective actions, if the detected deviation exceeds the tolerance range.	2.5	1.6
7)	Our SCEM automatically proposes corrective actions, if the detected deviation exceeds the tolerance range.	2.2	1.4
Management elements: Control			
8)	Our SCEM controls if there is a reaction to a warning	2.8	1.9
9)	Our SCEM escalates warnings independently to a higher ranging person if no reaction is taken to a warning within a period of time	2.4	1.7
10)	Our SCEM automatically takes corrective actions, if the detected deviation exceeds the tolerance range.	1.9	1.4
Management elements: Optimize			
11)	We regularly analyze our workflows with the help of our SCEM to identify improvement potentials.	3.9	1.8
12)	The results of the workflow analysis are regularly used to optimize our control systems (hardware and configuration).	3.7	1.8
13)	The results of the workflow analysis are regularly used to describe our business processes more realistically.	3.8	1.8
Degree of aggregation			
14)	Our SCEM monitors the following objects:		
14a)	customer orders (for LLP: transport orders)	5.4	1.8
14b)	customer order lines	5.1	2.0
14c)	production orders	4.6	2.2
14d)	transport orders	4.6	2.0
14e)	transport order lines (transport objects, packaging pieces)	4.5	2.0
14f)	loading equipment (containers, pallets, boxes)	4.1	2.1
14g)	orders from suppliers or service providers	4.5	2.0
Degree of differentiation			
15)	Our SCEM controls:		
15a)	on-schedule clearing of events	4.9	1.8
15b)	the location of the controlled object	4.4	2.0

Fig. 2 *(continued)*

15c)	the condition of the controlled object	3.9	2.1
15d)	the completeness of the controlled object	4.4	2.0
15e)	the physical condition of the loading equipment	3.4	2.0
15f)	the timely reaction to warnings of the SCEM	3.4	2.0
Depth of implementation			
16)	All of our company's business locations that were targeted to be included in our SCEM are covered today.	4.9	1.9
17)	All of our customers that were targeted to be included in our SCEM are covered today.	4.0	1.8
18)	All of our suppliers that were targeted to be included in our SCEM are covered today.	3.6	1.7
19)	All of our service providers that were targeted to be included in our SCEM are covered today.	3.7	1.8
Width of implementation			
20)	The following business processes are integrated in our SCEM:		
20a)	transport processes	4.5	2.1
20b)	storage processes	4.9	2.0
20c)	handling processes	4.4	2.1
20d)	order fulfillment processes	5.4	1.6
Technical prerequisites of SCEM use			
21)	We control our supply chain with one integrated IT-system.	4.9	2.0
22)	Data exchange with our external partners happens automatically (e.g. EDI)	4.6	1.7
23)	Data exchange with our external partners happens manually (e.g. Web applications)	3.9	1.6
24)	Our electronic data exchange includes:		
24a)	our customer	4.6	1.9
24b)	our suppliers	4.0	1.9
24c)	our service providers	3.8	2.0
25)	The exchange of SCEM relevant data happens in real time.	5.3	1.8
26)	The actuality of the data in our IT-systems is appropriate to monitor the workflows.	5.1	1.7
27)	The actuality of the data in our IT-systems is appropriate to identify relevant deviations of the workflow from the work plan.	4.8	1.7

Fig. 2 *(continued)*

28)	Our systems apply the following technologies to monitor objects:		
28a)	RFID	1.9	1.6
28b)	barcode	5.6	1.8
28c)	optical character recognition	3.2	2.4
28d)	GPS	2.3	2.0
28e)	GSM (location based services)	2.0	1.9
28f)	decentral mobile computing (e.g., on-board computer)	3.2	2.2
29)	Our SCEM uses the latest traffic information.	1.9	1.6
30)	Relevant traffic alerts trigger our SCEM to issue a warning.	1.7	1.4
Organizational prerequisites of SCEM use			
31)	Our top management backs our SCEM.	5.3	1.6
32)	Our top management provides sufficient resources for our SCEM.	4.5	1.6
33)	Our top management feels responsible for our SCEM.	4.6	1.7
34)	Our company has a dedicated organizational entity, responsible for our SCEM.	4.4	1.9
35)	Before implementing SCEM our company had:		
35a)	mapped our transport processes	4.4	1.8
35b)	mapped our handling processes	4.4	1.8
35c)	mapped our storage processes	4.9	1.7
35d)	mapped our order fulfillment processes	5.2	1.5
35e)	optimized those processes	4.4	1.5
35f)	defined the interfaces to our suppliers	4.4	1.8
35g)	defined the interfaces to our customers	4.9	1.8
35h)	defined the interfaces to our service providers	4.3	1.9
36)	Our company has established a structure that provides resources (off normal operations) to identify improvement opportunities from our SCEM.	3.8	2.0
37)	We established clear personal responsibilities for dealing with warnings issued by our SCEM.	4.4	1.9
38)	All employees involved in our SCEM have participated in substantial training.	4.4	1.7
39)	We have developed an incentive system that supports our SCEM.	2.9	1.8
40)	For SCEM we use standardized terminology in our company.	4.1	1.8

Fig. 2 *(continued)*

41)	The definition of objects monitored by out SCEM allows full visibility of our processes.	4.1	1.8
42)	We have defined the objectives of our SCEM in writing.	3.9	1.9
43)	The partners (suppliers, customers, service providers) connected to our SCEM are honest among each other.	4.4	1.5
44)	Occurring problems among partners (suppliers, customers, service providers) connected to our SCEM are raised openly.	4.9	1.4
45)	The partners (suppliers, customers, service providers) connected to our SCEM fulfill their duties even if they are not controlled.	4.4	1.4
46)	Our SCEM controls the following criteria:		
46a)	constitution of transport (i.e. completeness)	4.7	2.1
46b)	reliability of transport (i.e. punctuality)	4.8	2.0
46c)	transport loss ratio	3.9	2.2
46d)	delivery service (i.e. reliability)	4.8	2.0
46e)	supply readiness	4.5	2.1
46f)	number of relevant deviations of workflows from work plans	3.9	2.0
46g)	lead time	4.6	2.0

The possible answers for all questions may range on a 7 point Likert scale from "not at all" (1) to "to the fullest extend" (7), whereby the latter indicates full availability of this dimension.

Fig. 2 *(continued)*

Discussion

SCEM is in use, even if only rudimentarily, by most businesses. However, the extent of use varies, offering substantial opportunities for improvement. Due to a merely partial implementation of individual dimensions and underdeveloped links amongst these dimensions, the vast potential offered by SCEM is still to be realized.

Tracking & tracing of spatial and time related status information of the monitored objects dominates in the use of SCEM. Companies initially create a work plan for each object, being a mandatory requirement to apply SCEM. However, the degree of automation varies remarkably.

Increasing the degree of automation offers substantial potentials. The results of the study show that all dimensions of SCEM, which require low automation, are more implemented than those, which require high automation. As many companies agree on the effectiveness of SCEM, a higher degree of automation could directly lead to an improvement in supply chain performance. However, this does not mean that the full automation of all processes along the supply chain necessarily leads to success. Companies need to understand what degree of automation makes sense for their specific business case.

Further opportunities are offered by SCEM implementation along the entire supply chain. The study clearly showed that the implementation of SCEM within companies is further advanced than the interorganizational implementation. The implementation into the upstream supply chain, the core idea of SCEM, is particularly low. However, only a complete and tailor-made implementation of all dimensions of SCEM will give way to the complete realization of potential.

Technical prerequisites are important for the implementation of SCEM and are, in regards to the current degree of implementation, often sufficiently available. Organizational prerequisites are less provided. A possible reason for this is a hesitant allocation of financial resources. Top management still seems to be unclear about the effectiveness of the investment.

The study shows that, although its benefit for companies has been recognized, the implementation still lacks behind its potentials. A higher use in the future would lead to higher transparency and visibility in logistical processes. A continuous expansion along the entire supply chain enables the quicker identification of emerging problems and more rapid responses with corrective action.

SCEM is a management system with responsibilities and processes supported by IT systems. Broad support of management, structures with clearly defined responsibilities, and training of people applying the system are just as important as the fully automation of data acquisition and processing as well as of simulation and decision-making.

SCEM is a concept that is applicable across different industries. Its promised benefits of more complete order fulfillment and improved quality, cost and efficiency potentially interest companies independently from their industry.

References

Bodendorf, F., Zimmermann, R.: Proactive Supply-Chain Event Management with Agent Technology. International Journal of Electronic Commerce 9(4), 57–89 (2005)

Bretzke, W.-R., Stölzle, W., Karrer, M., Ploenes, P.: Vom Tracking & Tracing zum Supply Chain Event Management: aktueller Stand und Trends. Düsseldorf (2002)

Christopher, M.G.: Logistics and Suppry Chain Management, London (1992)

Harland, C.M.: Supply Chain Management: Relationships, Chains and Networks. British Journal of Management 7(1), S63–S80 (1996)

Heusler, K.-F.: Implementierung von Supply Chain Management: Kompetenzorientierte Analyse aus der Perspektive eines Netzwerkakteurs. Wiesbaden (2004)

Heusler, K.-F., Stölzle, W., Bachmann, H.: Supply Chain Event Management: Grundlagen, Funktionen und potentielle Akteure. WiSt – Wirtschaftswissenschaftliches Studium 35(1), 19–24 (2006)

Karrer, M.: Supply Chain Performance Management: Entwicklung und Ausgestaltung einer unternehmensübergreifenden Steuerungskonzeption, Wiesbaden (2006)

Knickle, K., Kemmeter, J.: Supply Chain Event Management in the Field: Success With Visibility, Boston (2002)

Nissen, V.: Supply Chain Event Management. Wirtschaftsinformatik 44(5), 477–480 (2002)

Olhager, J., Selldin, E.: Enterprise resource planning survey of Swedish manufacturing firms. European Journal of Operational Research 146(2), 365–373 (2003)

Otto, A.: Supply chain event management: three perspectives. The International Journal of Logistics Management 14(2), 1–13 (2003)

Pfohl, H.-C.: Logistiksysteme: Betriebswirtschaftliche Grundlagen. 8th ed., Berlin (2010)

Pfohl, H.-C.: Logistikmanagement: Konzeption und Funktionen. 2nd ed., Berlin (2004)

Reiche, F., Hofstetter, J.S., Stölzle, W.: Ereignisorientierte Steuerung von Lieferketten: Nutzen, aktueller Stand der Nutzung und Potenziale, Göttingen (2009)

Sputtek, R., Hofstetter, J.S., Stölzle, W., Kirst, P.: Developing a Measurement Instrument for Supply Chain Event Management-Adoption. In: Kreowski, H.-J., Scholz-Reiter, B., Haasis, H.-D. (eds.) Proceedings of Dynamics in Logistics: First International Conference, LDIC 2007, Bremen, Germany, Heidelberg, August 2007, pp. 391–404 (2008)

Stölzle, W.: Supply Chain Event Management. In: Klaus, P., Krieger, W. (eds.) Gabler Lexikon Logistik: Management Logistischer Netzwerke und Flüsse, 4th edn., Wiesbaden, pp. 541–546 (2008)

Supply Chain Council, Supply-Chain Operations Reference Model (2007)

Stefansson, G., Tilanus, B.: Tracking and tracing: Principles and practice. International Journal of Technology Management 20(3/4), 252–272 (2000)

Teuteberg, F., Schreber, D.: Mobile computing and auto-ID technologies in supply chain event management - an agent-based approach. In: Proceedings of ECIS, vol. (45) (2003)

Dr. Joerg S. Hofstetter is Assistant Professor of Logistics Management at the University of St.Gallen, Switzerland, and Vice Director of the university's Chair of Logistics Management.

Dr. Wolfgang Stölzle is Professor of Logistics Management at the University of St.Gallen, Switzerland, and Director of the university's Chair of Logistics Management.

Adaptation-Based Supply Chain Resilience

Dmitry Ivanov[1], Boris Sokolov[2], and Joachim Käschel[3]

[1] Hochschule für Wirtschaft und Recht Berlin, Inhaber der Professur für International Supply Chain Management, Campus Schöneberg, Badensche Straße 52, 10825 Berlin, Germany
dmitry.ivanov@hwr-berlin.de

[2] St. Petersburg Institute for Informatics and Automation of the Russian Academy of Sciences (SPIIRAS), Deputy Director for Research, Russia
sokol@iias.spb.su

[3] Technische Universität Chemnitz, Inhaber der Professur für Produktionswirtschaft und Industriebetriebslehre, Thüringer Weg 7, 09126 Chemnitz, Germany
joachim.kaeschel@wirtschaft.tu-chemnitz.de

Abstract. In this paper, we develop an adaptation-based supply chain resilience framework based on the control theoretic perspective for supply chain planning domain regarding the agility and disruption-resistance to achieve maximal economic performance and stability in supply chains. We propose a detailed analysis of supply chain resilience based on a mutual classification of flexibility and reliability elements. Subsequently, an algorithm of decision-making on supply chain planning regarding ensuring both supply chain reliability and flexibility is presented. The quantitative approaches and formal tools are based on the modern control theory in combination with operations research techniques, global stability, controlled adaptation and the use of attainable sets.

The developed framework and tools which are supporting it allow us to find this balance with regard to supply chain protection and adaptability. Moreover, it makes it possible to take into account individual risk perceptions of managers, different strategies with regard to risk management, and to consider not only supply chain economic performance but also supply chain stability, which is becoming more and more important in ongoing economic transformation.

The developed framework contributes both to the methodical part of SCM and to its practical part where the developed methodical guidelines can be localized for concrete application issues. The framework allows us to approach the issues of mitigating uncertainty and increasing resilience of supply chains from the control theoretic perspective. From the practice point of view, the gained insights provide decision-makers with the possibility to balance reliability and flexibility to achieve maximal economic performance and resilience in supply chains.

Keywords: Supply chain, resilience, reliability, flexibility, stability, performance.

1 Introduction

Ensuring agility along with disruption resistance and resilience are crucial issues in supply chain (SC) planning. To answer these challenges, supply chains are to

M. Essig et al. (Eds.): Supply Chain Safety Management, LNL, pp. 267–287.
springerlink.com

be designed more complex subject to different redundancies subject to reliability and flexibility that can be considered as two crucial elements of SC resilience. A balance of reliability and flexibility with regard to uncertainty is one of the most important conditions for supply chain economic performance and resilience.

There is considerable variation in the definitions of terms related to SC uncertainty, resilience, robustness, and performance (Klibi et al. 2010). Basically, there are three main properties of an SC which can be analyzed regarding uncertainty. These are: (1) the ability to cope with volatility and continue plan execution once perturbed, (2) the ability to continue plan execution and to achieve the planned performance in the presence of disturbances, and (3) the ability to maintain, execute and recover (adapt) the planned execution along with the achievement of the planned (or adapted, but yet still acceptable) performance. In the systems and control theories, these properties are analyzed as stability (property 1), robustness (property 2), and resilience or disaster-tolerance (property 3). This paper focuses on the case (3), i.e. on the *resilience analysis*.

The objective of SC as a controlled system is to ensure performance over time. The achievement of the planned performance is subject to maintaining SC behavior and its changes as an adaptation to changes in environment. Recently, the research community has begun to shift to a paradigm that the performance of SCs is to consider *adaptable, stable, and crisis-resistant processes* to compete in a real perturbed execution environment (Sheffi 2005, Ponomarov and Holcomb 2009, Ivanov and Sokolov 2010). In this setting, a number of research advancements can be indicated.

Responsiveness, agility, and flexibility shape enterprise competitiveness and contribute to increasing supply chain (SC) performance (Christopher and Towill, 2001; Christopher, 2005; Kouvelis et al., 2006; Mangan et al., 2007; Gunasekaran and Ngai; 2008; Swafford et al., 2008). On the other hand, achievement of the planned (potential) SC goals can be inhabited by perturbation impacts in a real execution environment (Hendricks and Singhal 2005; Kleindorfer and Saad, 2005).

In line with ongoing economic transformations, SCs are being designed more complex, subject to different redundancies to ensure reliability and flexibility. The real SC performance is based on the maintaining planned execution trajectories (reliability) and responsive cost-efficient agile changes or recovering after being disturbed (flexibility) (Ivanov and Sokolov, 2010). The profit losses through non-purposeful (e.g., demand fluctuations) and purposeful (e.g., terrorism or thefts) perturbation impacts can amount to 30% of the annual turnover (Kleindorfer and Saad, 2005, Williams et al., 2008). Profit of a SC can decrease if a disruption in SC becomes public (Hendricks and Singhal, 2005; Craighead et al. 2007).

To answer these challenges, SCs become to be designed more complex subject to different redundancies with regards to reliability and flexibility. The reliability is mostly seen as a cost-driven property while flexibility is considered from the profitability point of view (Sheffi and Rice, 2005; Christopher, 2005). At the same time, both reliability and flexibility create a certain robustness reserve that inevitably causes additional costs (Bertsimas and Sim 2003).

Although the problem of mutual relations between reliability and flexibility has been widely discussed in business literature, there is a lack of explicit qualitative frameworks and formal quantitative models. In this paper, we develop an explicit statement of adaptation-based SC resilience for the SC planning domain regarding the agility and disruption resistance to achieve maximal economic performance and stability in SCs. The framework contributes to the investigation of resilience influence on the SC performance, both from the costs and income points of view in a real perturbed execution environment. Furthermore, we will develop tools for decision-making on a balance of SC reliability and flexibility with the composite objective of maximizing both the SC resilience and the SC performance to answer the practical question about the investments in reliability and flexibility redundancy and the real increase or decrease in performance.

The rest of this paper is organized as follows. We start the paper with the state-of-the-art analysis. Section 3 discusses the methodical basics of the proposed approach and presents the conceptual framework. Sections 5 and 6 describe the developed decision-making tool and analyse conceptual and experimental results. We conclude the paper by summarizing the main findings.

2 State of the Art

The issue of how to design and plan SCs with regard to responsiveness, flexibility, and disruption resistance is very important – first, for SC agility itself and, secondly, for designing robust SCs (Van Landeghem and Vanmaele, 2002) and re-designing SCs for new products (Graves and Willems, 2005), new order penetration points, and a variety of disruptive factors (Ivanov and Sokolov, 2010).

In practice, different redundancy policies to mitigate uncertainty are applied. E.g., Dell combines the transport and inventory strategies by storing cheap components in Europe and order-driven replenishing of expensive components in Asia. On the contrary, Cisco stores expensive components in the USA and produces cheap components in Asia. Another example is the German automotive supplier MTU Aero Engines Ltd which is able to run the production three weeks based on the safety stocks.

Recent literature has identified different methods to strengthen SCs to mitigate uncertainty impacts. *First,* different reliability reserves (material inventory, capacities buffers, etc.) can be referred to. For this issue, valuable approaches and models for SC design and planning under uncertainty were elaborated, widely presented in Tayur et al. (1999) and de Kok and Graves (2004). Fisher et al. (1997) emphasized the necessity to consider costs of uncertainty reduce while configuring SCs. Kleindorfer and Saad (2005), Khan and Burnes (2007), Peck (2007) provided conceptual frameworks that reflect the joint activities of risk assessment and risk mitigation that are fundamental to disruption risk management and resilience in SCs. Hendricks and Singhal (2003) provided the evidence that the investments in the increasing SC reliability can be seen as insurance for possible economic losses. Similar ideas were discussed by Chopra and Sohdi (2004) with regard to investments in SC reliability and the covered risks. Graves and

Willems (2005) developed a dynamic program with two state variables to solve the SC configuration problem for SCs that are modelled as spanning trees, and applied it to optimizing the SC configuration for new products. Meepetchdee and Shah (2007) developed a framework of logistical network design with robustness and complexity considerations.

Second, new strategies such as leagile, agile, and responsive SCs as well as structural-functional reserves (like a pool of alternative suppliers from the virtual enterprise concept) can be applied to make SCs more flexible in a wider sense of the word (Christopher and Towill, 2001; Ivanov et al., 2007; Gunasekaran and Ngai, 2008; Ivanov et al., 2010). Vickery et al. (1999) and Swafford et al. (2008) paid particular attention to SC flexibility. Beamon (1999) and Naim et al. (2006) considered quantity, supply, product, transport, and innovation flexibility. Gunasekaran and Ngai (2009) considered a framework of SC responsiveness. Tachizawa and Thomsen (2007) empirically investigated the aspects of flexibility related to the upstream SC. Coronado and Lyons (2007) investigated the implications of operations flexibility in industrial SCs and the effect it has on supporting initiatives designed for build-to-order (BTO) manufacturing. Wadhwa et al. (2007) presented a study on the role of different flexibility options (i.e. no flexibility, partial flexibility and full flexibility) in a dynamic SC model based on some key parameters and performance measures. Swafford et al. (2008) showed that that IT integration enables a firm to tap its SC flexibility which in turn results in higher SC agility and ultimately higher competitive business performance. Reichhart and Holweg (2007) synthesize the existing contributions to manufacturing and SC flexibility and responsiveness, and draw on various related bodies of literature that affect a SC's responsiveness, such as the discussion of product architecture and modularization. Ozbayrak et al. (2006), Jang (2006), Knoppen and Christiaanse (2007), Ivanov (2009, 2010) and Ivanov and Sokolov (2010, 2011) showed that flexibility and robustness in the SC (unlike in manufacturing systems) is primarily interrelated with adaptation through managerial actions.

The *third* research direction is related to better coordination in SCs. A recent AMR Research study shows the great importance of demand-oriented SC coordination: demand-forecast accuracy creates high responsiveness and cuts costs right through the SC (Friscia et al., 2004). According to the study, the companies that are best at demand forecasting maintain on average 15 percent less inventory, 17 percent stronger perfect-order fulfillment and 35 percent shorter cash-to-cash lifecycles. Different concepts of *coordination* have been developed over the last two decades, such as efficient consumer response (ECR), collaborative planning, forecasting, and replenishment (CPFR) in retail as well as just-in-time (JIT) and vendor-managed inventory (VMI) in industries. Enablers of the coordination are information technologies, such as enterprise resource planning (ERP), advanced planning systems (APS), electronic data interchange (EDI), and radio frequency identification (RFID). Collin and Lorenzin (2006) emphasize that, in practice, coordination determines the postponement strategy and the position of the order penetration point or decoupling point in the SC. The coordination has become a key factor in mitigating the *bullwhip effect* and in overcoming information asymmetry (Lee et al. 1997, Chen et al. 2000,). Moreover, due to Internet technologies, it has become possible to integrate customers into SC considerations, resulting in the development of the build-to-order SCM (Gunasekaran and Ngai, 2005; Sharif et al., 2007).

Fourth, a set of postponed decisions (product postponement, rolling/adaptive planning) can be used (van Hoek, 2001; Ivanov et al., 2010). The agility and coordination of SCs is greatly linked to manufacturing and logistics postponement strategies. Ernst and Kamrad (2000) introduced a conceptual framework for evaluating different SC structures in the context of modularization and postponement. The study provides examples of efficient postponement and modularization combining by HP, Suzuki, and Benetton. Additional case examples include Dell Computers, Nike, IBM, and General Motors and are given by Tully (1993) and Gunasekaran et al. (2008). A crucial issue in postponement and modularization is the determination of an order penetration point. By efficient coordination in relation to these two decoupling points, a powerful opportunity for agile response can be created (Christopher and Towill, 2000). The integration of lean (upstream of the OPP) and agile (downstream of the OPP) SCs was extensively discussed in Mason-Jones et al. (2000) and Christopher and Towill (2000). Recent study by Wang et al. (2009) reports on a three-dimensional concept based on the integration of product, engineering and production activities to define customer order decoupling points.

All these approaches can be named as SC excessiveness or redundancy. The above-described redundancies generally serve two problem areas (Fisher, 1997). First, they intend to protect SC against perturbation impacts based on certain reserves (de Kok and Graves, 2004; SCOR, 2008). This issue is related to the SC reliability. Secondly, redundancies are created to amplify the fork variety of SC paths to react quickly and flexibly to changes of a real execution environment. This issue is related to SC flexibility (Vickery, 1999; Swafford et al., 2008; SCOR, 2008).

Finally, we would like to draw attention to the fact that, in contrast to technical systems, managers in a SC do not strive for a 100% guarantee of the result; they consciously tend to take risks. The studies by Sokolov and Yusupov (2006) and Peck (2007) point out the problem of contradiction between objective risk (determined by experts, applying quantitative scientific means) and perceived risk (perception of managers). Actually, the objective risk treatment is rooted in technical science where 100% reliability is mandatory. In socio-economic systems, like SCs, a value of 95% reliability as an orientation for SCs is empirically suggested (e.g., Sheffi, 2005). In business, risk is some kind of *economic incentive*. Moreover, recent literature shows that different managers perceive risk to different extents, and these perceptions can change in the same manager due to changes in his environment (Sokolov and Yusupov, 2006). That is why the constellations of SC reliability and flexibility should not strive for a unique optimal solution, but rather allow a number of alternative solutions to be formed with different degrees of economic performance and risk.

3 Conceptual Framework

In general, all the means to mitigate uncertainty and increase stability in SCs can be divided into reservations of different resources and adaptation with regard to changes in their usage (or newly attracted or alternative) resources. In any case, all these means are intended to protect SC whether with a direct use of reserved resources (e.g., safety inventories) or an indirect use of the available (or additionally reserved) resources by means of their replanning or reforming on the adaptation basis. We will refer the first group as SC reliability and the second as SC flexibility. As both flexibility and reliability are actually excessiveness in the SC to mitigate uncertainty, they should be analysed mutually and with regard to economic performance and stability of SCs.

To investigate the interrelations between reliability and flexibility, we propose these properties to be analyzed from the system-modelling and complexity-management point of view (Ivanov and Soklov, 2010). From these perspectives, the problem of a system under control and uncertainty is related to an area under control (system space) and an area under uncertainty (uncertainty space) (see Fig. 1).

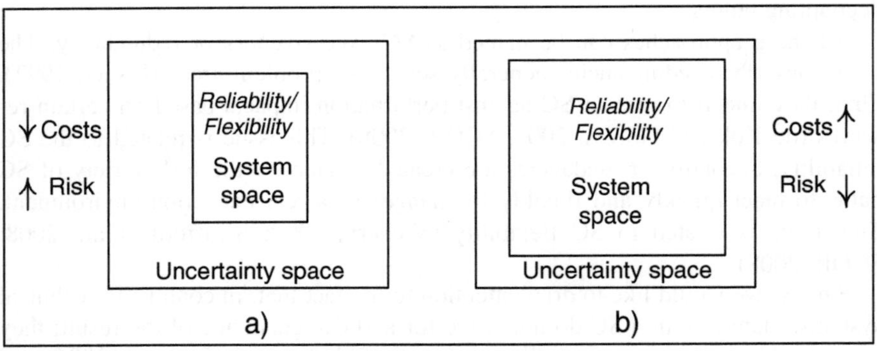

Fig. 1 Reliability, flexibility and uncertainty

By broadening the system (control area) and narrowing the uncertainty area, the system control can be adapted. Hence, the mutual relations between the reliability and flexibility fall under the categories of amplification of control variety and attenuation of environmental variety. Thus, amplifying the variety of our control area and reducing the area of uncertainty, (i) a balance of control and perturbed impacts as well as (ii) maintenance of planed execution processes (reliability) and a quick cost-efficient agility or process recovering once disturbed (flexibility) can be reached. In considering these properties from a system-cybernetic and formal level, we can conclude that reliability and flexibility do not compete but they are very close and complementary in their nature.

Let us define SC reliability and flexibility as well as their interrelations. We derive SC redundancy elements with regards to the reliability and flexibility. Finally, interrelations between the reliability and flexibility are analyzed from the complexity management point of view.

The reliability of SCs is a complex characteristic of a non-failure operation, durability, recoverability, and the maintaining of SC processes and a SC as a whole; this is connected with the creation of a reserves system (the introduction of resource excessiveness) for the prevention of failures and deviations in SC processes.

The flexibility of SCs is a property of a SC concerning its ability to change itself quickly structurally and functionally depending on the current execution state and reaching SCM goals by a change in SC structures and behaviour.

The resilience of SCs is the ability to maintain, execute and recover (adapt) the planned execution along with the achievement of the planned (or adapted, but yet still acceptable) performance. This is connected with the creation of an *adaptation system* (with regard to operations and resources) for the prevention, improvement, or acquisition of new characteristics for the achievement of goals under the current environmental conditions varying in time.

In Figure 2, the conceptual framework of SC performance and resilience is presented.

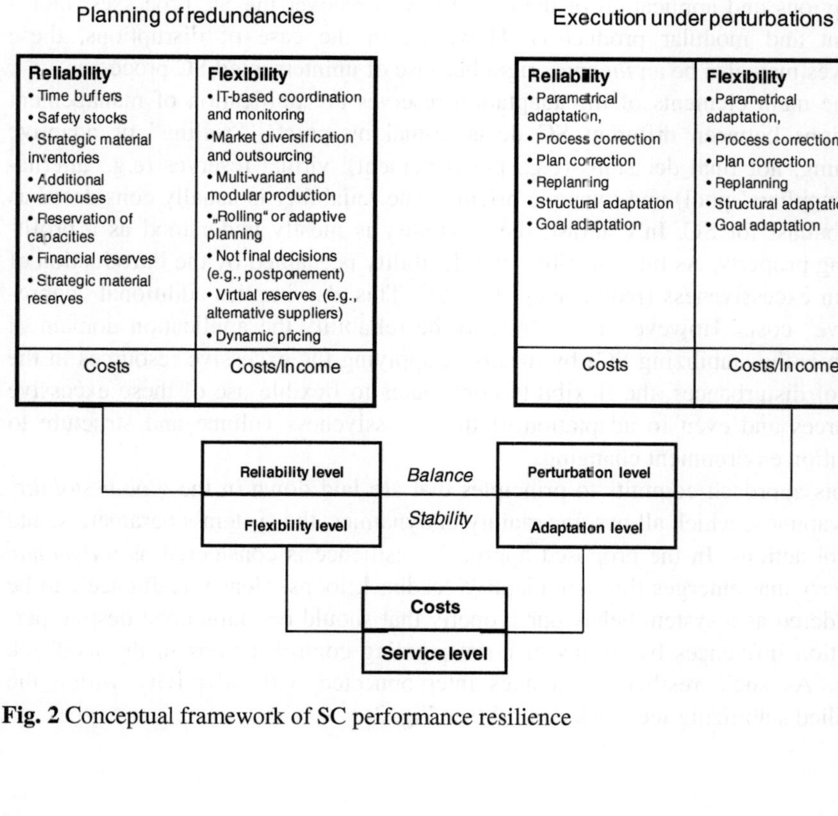

Fig. 2 Conceptual framework of SC performance resilience

A large variety of control influences can be applied to increase agility and to compensate for the disturbances, a large variety of control influences can be applied, i.e. (Christopher, 2005; Ivanov and Sokolov, 2010):

- SC security management;
- asset reserves;
- strategic material inventories;
- market diversification and outsourcing;
- product lines' flexibility and modularity;
- safety stocks and time buffers;
- reserves of SC capacities;
- SC coordination, monitoring and event management.

All these measures are actually the adaptation reserves of SCs. They are characterized by different degrees of operativeness (i.e., using safety stocks or market diversification). The reserves in the SC may be referred to the robustness and adaptability. The main elements of the *reliability reserves* are time buffers, safety stocks, and additional facilities, reservation of capacities, and IT-based coordination and monitoring. These elements cause certain *costs* for the creation of the reliability reserves, their maintaining, and recovery handlings in the case of disruptions and application of these reserves to recover the SC processes, multivariant and modular production. However, in the case of disruptions, these reserves may also be an *income* origin because of uninterrupted SC processes.

The main elements of the adaptation reserves are unification of management functions between different SC decision-making points, "rolling" or adaptive planning, not final decisions (e.g., postponement), virtual reserves (e.g., alternative suppliers' pool) and dynamic pricing. The reliability is usually considered as cost ballast for SC. In contrast, the flexibility is mostly understood as a profit-driving property. As the reliability, the flexibility is enabled by the introduction of certain excessiveness (redundancy) in a SC. This also implies additional "unproductive" costs. However, in contrast to the reliability the application domain of which is the stabilizing SCs by means of applying the excessive resources in the case of disturbances, the flexibility contributes to flexible use of these excessive resources and even to adaptation of the excessiveness volume and structure to execution environment changing.

This approach commits to principles that are laid down in the *global stability* by Lyapunov, which allows uncertainty in dynamics, the system's parameters, and control actions. In the proposed approach, resilience is considered as a *dynamic property* that emerges through adaptive feedback loops. Hence, resilience can be considered as a system behaviour property that should be maintained despite perturbation influences by means of corresponding control actions in the feedback loops. As such, resilience becomes interconnected with adaptivity within the so-called stabilizing feedback control (see Fig. 3).

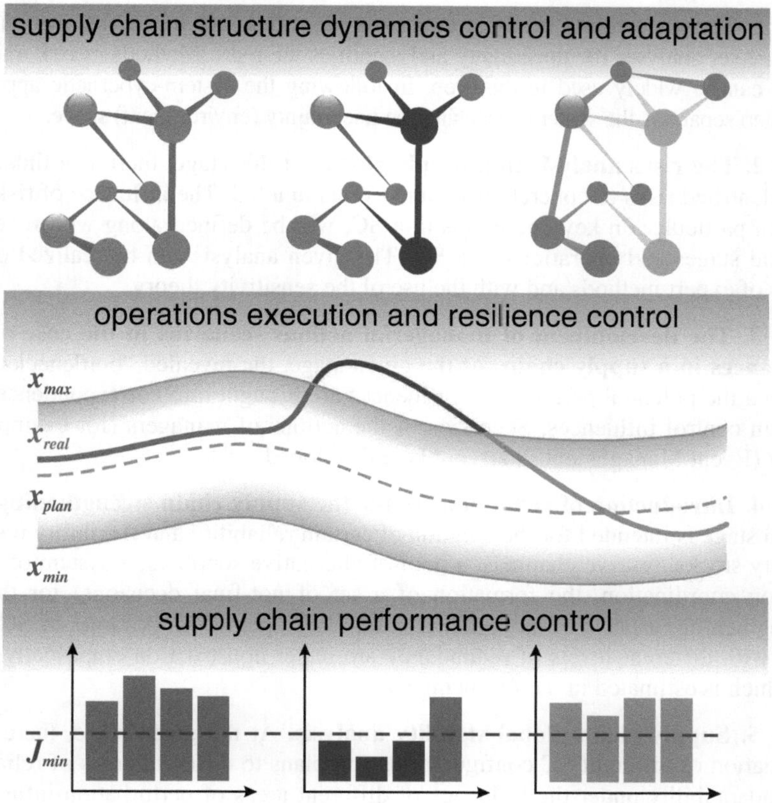

Fig. 3 Perturbations and adaptive recovery of SC execution, resilience and output performance

In Figure 3, the influence of perturbation impacts on SC execution and output performance is analysed. SC execution trajectory $x(t)$ is constructed at the planning stage and strengthened by the elements of reliability and flexibility. This results in an interval $[x_{min}; x_{max}]$ remaining within of which SC achieves its output performance goals J. In the case of global stability (resilience) with controlled adaptability, perturbation impact causes the decrease in SC operations execution but the output performance is achieved by means of the adaptive recovery as the SC structure is controlled.

4 Decision-Support System

4.1 *General Algorithm of Planning Supply Chain Reliability and Flexibility*

Let us consider the general logic scheme of decision-making on the choice of a SC configuration or plans, taking into account SC reliability and flexibility. It consists of 10 steps, which will be considered below.

Step 1. The uncertainty analysis and risks identification. In the first step, a decision-maker analyses the uncertainty and identifies the risks. Methods of risk management can be widely used in this step. In following the system-cybernetic approach, this step separates the system space and the uncertainty (environment) space.

Step 2. The risks analysis in a supply chain. At this stage, there is a linkage of the identified risks to concrete parts and events in a SC. The influence of risk on a SC, in particular on key operations in a SC, will be defined along with revealing critical stages and operations in a SC. The given analysis can be realized on the basis of expert methods and with the use of the sensitivity theory.

Step 3. The development of managerial actions scenarios in the case of disturbances in a supply chain. At the given stage, the revealed "bottlenecks" in a SC and the potential perturbation influences are brought into correspondence with certain control influences. Scenarios of the actions of managers (for example, as EMP (Event Management Plan)) are here developed.

Step 4. Introduction of redundancies for the supply chain strengthening. The given stage is intended for the creation of certain reliability and flexibility reserves (safety stocks, reserve channels, a pool of alternative suppliers, a system of information coordination, the formation of a set of not-final decisions) for the SC strengthening, especially its bottlenecks and key operations. A range of SC variants with different levels of redundancy are under construction empirically, each of which is estimated in the following stage.

Step 5. Supply chain global stability analysis. At the given stage, there is an estimation of different SC configurations and plans to different areas of reliability and adaptability under the influence of different areas of perturbation influences. This analysis is based on the stability of a SC (in the wide interpretation, the SC global stability) that is understood as a complex property of a SC, characterizing the ability of a SC to maintain, realize, and restore goal-oriented functioning in ever-changing execution environment under influence of perturbation factors (Ivanov and Sokolov, 2010).

Step 6. An estimation of costs for the redundancies and disturbances elimination. At the given stage, there is a cost estimation of various measures for the strengthening and adaptation of SCs. The block of costs for the maintenance of SC reliability and flexibility along with the "productive" costs (e.g., total costs of ownership), makes the base for an estimation of cumulative costs in a SC as a result. On the basis of expert methods, there is an elimination of a part of the alternative plans considered in stage 5 (for example, owing to unrealistic costs for the reliability and flexibility maintenance or plans with an unacceptably low level of stability).

Step 7. The formation of a set of alternative supply chains. At the given stage, the set of alternative SCs received after the performance of step 6 is formed.

Step 8. The final analysis of the supply chain stability. The same as in step 5, but on the narrowed set of alternatives and taking into account costs for SC reliability, flexibility, and adaptation.

Step 9. The results calculation and analysis according to the supply chain economic performance and stability. The given stage consists of the analysis of alternative SCs generated in step 8 concerning the level of SC economic performance and stability. The decision-makers can analyse different constellations of SC efficiency and stability, and select the most preferable one from a number of alternative in accordance with the individual risk perception.

Step 10. The final choice of a supply chain configuration or plan. The final choice of a SC plan occurs on the basis of managerial individual preferences and the risk perception.

4.2 Quantitative Modelling Approach

In this section, we describe the principles of the quantitative approach to decision-making on a balance of SC reliability and flexibility with the composite objective of maximizing both the SC stability and the SC performance to answer the practical question about the investments in reliability and flexibility redundancy and the real performance increase.

4.2.1 Simulation Model Concept

The assessment of the compliance of the SC excessiveness and a current execution environment is performed in a special elaborated simulation model. Giving as an input certain environment specifications (e.g., demand fluctuations, technological failure, etc.), one or several state trajectories can be calculated to which SC performance is mostly attracted.

The analysis of these trajectories may provide a number of useful observations, e.g., if the trajectories mostly contain states with a high consumption of safety stocks, SC plans should be reconsidered and information coordination should be enhanced. If the states with delivery delays are frequently encountered, the capacity buffers should be introduced. Analogously, SC bottlenecks can be systematically revealed, analysed, and strengthened. This can be very useful for SC service level and flow capacity analysis. Usually, numbers of bottlenecks (i.e., the states with permanent delivery delays) cause the service level and flow capacity to decrease drastically. Also, the costs of the SC reliability, flexibility, and adaptation can be analysed, e.g., if the trajectories seldom contain the states with high reliability and flexibility maintenance costs, these redundancies can be eliminated. On the other hand, if the transitions between states in trajectories are mostly connected with high adaptation costs, a detailed analysis of the excessiveness in the neighbouring states is needed. Hence, such a model offers constructive methods for the quantitatively grounded analysis of SCs with regard to balancing SC reliability and flexibility with different execution environments. The last point is the last important precondition of the SC global stability. In the progress of this section, a SC stability assessment model will be presented to quantify the balance between SC stability and performance.

4.2.2 Application of Global Stability Analysis to Find an Optimal Redundancy Space and Constellation of Supply Chain Reliability and Flexibility

The main idea of the applying the SC global stability in this study is to find an optimal (from a decision-maker's point of view) constellation of the redundancy space in the SC (with regard to both reliability and flexibility) and the uncertainty space. The SC global stability analysis model addresses the problem of the direct connection of SC stability and economic performance analysis with regard to the areas of uncertainty and control (Ivanov and Sokolov, 2010).

This approach commits to principles that are laid down in the *global asymptotic stability* by Lyapunov, which allows uncertainty in dynamics, the system's parameters, and control actions (Casti, 1979). In this approach, stability is considered as a *dynamic property* that emerges through feedback loops. Hence, stability can be considered as a system behaviour property that should be maintained despite perturbation influences by means of corresponding control actions in the feedback loops. As such, stability becomes interconnected with adaptivity within the so-called stabilizing feedback control.

The idea of this approach to stability is to a certain extent similar to the issue of SC performance and risk management (Ritchie and Brindley, 2007). In both cases, the problem consists of analysing potential SC goals and uncertainty that may negatively affect the achievement of these goals in a real perturbed execution environment. However, while risk management tries to estimate this constellation from the prospective point of view (from a certain planning instant of time up to the end of the planning horizon considered), the global stability approaches the issues from a *dynamic* point of view. Besides, such an approach where stability is achieved by the controlled adaptation reflects very well the nature of SCs that are managed by managers and their control influences (and not by automatic devices like technical systems; that is why we prefer to use the global stability with controlled adaptation rather than conventional BIBO – bounded-in-bounded-out stability).

The model is based on the dynamic interpretation of the SC's functioning process and uses the method of AS. A detailed discussion of the model would be extremely technical and assumes expert knowledge of techniques of OR and control theory. That is why we will provide here the general step algorithm of the work of the model. Mathematics can be seen in (Ivanov and Sokolov, 2010).

General description of the approach

The SC planner can analyse different alternative SC plans, fill these plans with reliability and flexibility elements to different extents, and then investigate how these changes influence the key performance indicators (so-called potential economic performance) and stability (the characteristic describing the chances to attain this potential performance in real perturbed execution environment) of SC. This analysis can be performed with regard to different execution (uncertainty) scenarios and different areas of control (manageability). The model allows multi-criteria estimation and the analysis of the SC's stability to be carried out, considering the combined variants of initial data about possible perturbation influences (the determined, indistinct, stochastic, interval data, and their combinations). The model allows (i) to analyse the stability of various alternative SCs plans to be carried out concerning various kinds and scales of perturbation and control influences, (ii) to calculate for each of the plans and possible scenarios a stability index, having given the decision-maker the possibility of a choice of that plan that corresponds to his/her individual risk perception.

Constructing attainable sets and stability index calculating

After different alternative plans with different degrees of protection against uncertainty (that is, with different control areas subject to different scale and scope of the reliability and flexibility elements) and, correspondingly, different values of key performance indicators (KPI) have been generated, they should be analysed with regard to uncertainty (different areas of perturbation influences) to find the balance of control area (in the generated plans) and perturbation area (different execution scenarios).

The essence of a stability index calculation is based on the construction and comparison of two attainable sets (the area of admissible values of SC goal indicators and the approximated area of SC attainability under the influence of perturbation factors). The stability index is expressed as the area of intersection of these two rectangles.

The constructing attainable set is composed of the following:

- Area of admissible disturbances is defined with the help of vector functions for minimal and maximal disturbances. We propose this area to be named as the AS of the SC under disturbances.
- This AS corresponds to the indicators values, which assess the SC economic performance
- To make the further material more comprehensive, we will examine only two components of vector index. These components correspond to the indicators of SC service level and SC profit. In this case, while geometrically describing the attainable set, it becomes possible to use Descartes' coordinate system.
- Admissible limits of oscillations are defined. They construct special area in the indices space. We are not interested in all the attainable sets but only the subarea, in which the SC is capable of carrying out the planned processes. To perform the stability analysis, internal and external approximations of AS should be constructed.
- Within the framework of a SC plan stability assessment, the task is to find a point that lies on the border of the AS under uncertainty and a line which is a tangent for the given set and includes the point the point that is searched for. Hence, we obtain the external approximation of the AS.
- This AS approximation is a geometrical figure. This approximation is a rectangular area constructed as a result of four lines crossing.
- A stability assessment of the SC plans comes down to the calculation and analysis of a general index of the SC stability for fixed disturbance scenarios and control influences within each generated SC plan $\boldsymbol{u}_i(t)$ ($i = 1,\ldots,n$; where n is a number of alternative plans).

- The essence of a stability index calculation is based on the construction and comparison of two sets (the area of admissible values of SC performance metrics and the approximated area of SC attainability under the influence of perturbation factors). The stability index is expressed as the area of intersection of these two rectangles: P_J (j= 1,…,m; where m is a number of SC performance indicators) – the area of admissible values SC performance indicators and the attainability area under disturbances.
- The selection of the most stable SC plan is carried out according to the condition:

$$S_i^*(\boldsymbol{u}_i(t)) = \max_{1 \le i \le n} \min_{1 \le j \le m} S_i(\boldsymbol{u}_i(t)), \tag{1}$$

where $S_i(\boldsymbol{u}_i(t))$ is the area of intersection; n – the total amount of analysed SC plans; and m – the total amount of disturbance scenarios at the stage of SC plan realization. The square $S_i(\boldsymbol{u}_i(t))$ of the intersection reflects the desirable result declared at the beginning of this subsection – the general index of the SC stability. The smaller the square, the more stable is the SC. The larger the square, the less stable is the SC.

If the SC stability index is equal to zero, this means that the space of SC control (reliability and flexibility) and the space of uncertainty are balanced. If the stability index is >0, additional redundancies should be introduced in the SC, another constellation of reliability and flexibility should be considered, or possible risks should be taken into account.

5 Experiments

For experimental calculations on the basis of the models presented above, a software package has been developed on the basis of C++ and XML. Let us provide an example. The problem consists of finding a SC execution plan with the best balance of reliability and flexibility elements with regard to service level and costs; that is, to find the most stable plan subject to given (e.g., by top management) goals reflected in performance indicators (i.e., minimal level of service level and maximal level of costs).

The example consists of the analysis of three alternative SC configurations (or plans) concerning three variants of perturbation influences $\boldsymbol{\xi}(t)$ at the given area of perturbation influences Ξ ($\boldsymbol{x}$ (t), t). The SC plans (structures) are characterized by the different areas of the planned admissible control influences Q ($\boldsymbol{x}$ (t), t) with regards to the reliability and the given area of general admissible control influences V ($\boldsymbol{x}$ (t), t) with regard to the flexibility. The structure 1 is characterized by a lower profitability, a higher service level, and higher excessiveness costs (due to the higher reliability and flexibility) compared with the structures 2 and 3. In structure 1, the area Q, V ($\boldsymbol{x}$ (t), t) is balanced with the area Ξ ($\boldsymbol{x}$ (t), t). In structures 2 and 3, the area Q, V ($\boldsymbol{x}$ (t), t) is smaller than the area Ξ ($\boldsymbol{x}$ (t), t).

In the given example, the following perturbation impacts have been considered: a decrease in the availability of a resources of 30% (scenario 1), a resource productivity decline of 5% (scenario 2), and the cumulative influence of these two perturbation impacts (scenario 3) have been considered. The experimental results are presented in Figure 4.

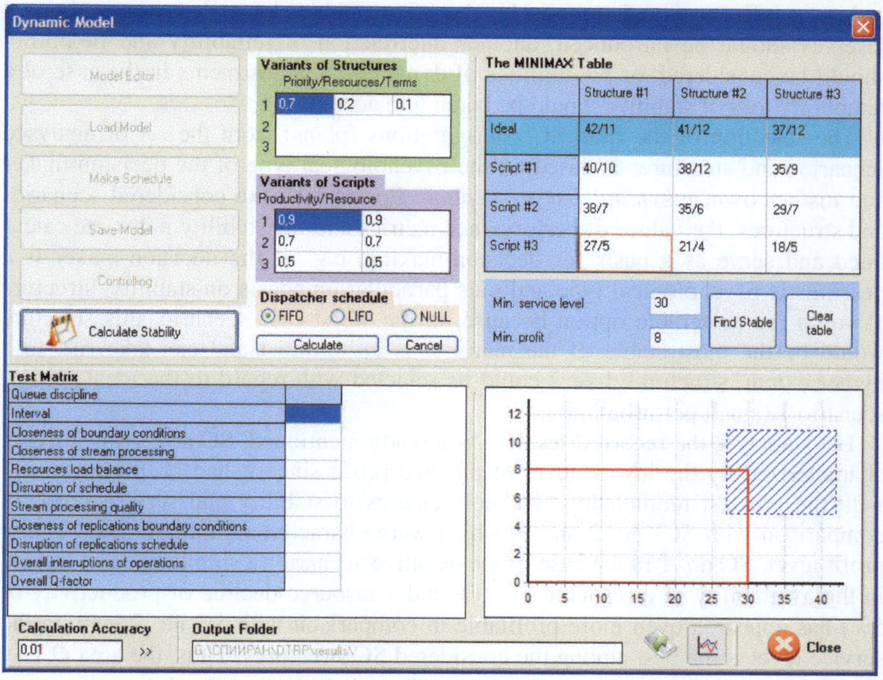

Fig. 4 Simultaneous analysis of SC economic performance and stability

With regard to the considered structures and execution scenarios, the values of the performance metrics are calculated. In the upper-right-hand part of the interface, corresponding values of economic performance to different SC plans and execution scenarios are presented (structure is equal to plan; script is equal to scenario; e.g., for scenario 3 to structure 1, the economic performance indicators are amounted to 27/5, this means: service level is equal to 27 thousand euros, and profit is equal to 5 thousand euros).

The stability index is defined on the basis of the area of intersection of two attainable sets. In the worst scenario case (scenario 3), for case 1 (SC structure no. 1), it is equal to 9 (this case is presented on the interface in Figure 2), for case *b* (SC structure no. 2) – 36, and for case *c* (SC structure no. 3) – 36. As such, SC structure no. 1 is the most stable, i.e. the SC remains stable even in the case of the occurrence of the considered perturbation influences.

The developed geometrical interpretation of the results where the stability index is expressed as the area of intersection of these two rectangles allows managers to very conveniently compare the stability of different plans.The smaller the square, the more stable is the SC. The larger the square, the less stable is the SC.

Cases having zero value of the stability index mean that the *excessiveness* of the corresponding SC structures is enough to protect the SC from the given perturbation influences and to consider the SC as stable. In other cases, additional reserves should be introduced, another interrelation of reliability and flexibility should be considered, or possibilities of dynamic SC adjustments in the case of a temporary loss of stability should be taken into account.

The selection of the final SC configurations (plans) from the set of analysed scenarios and structures is based on the psychological type of the decision-maker and his/her own *individual risk perception*. For each of the considered scenarios and structures, the values of performance metrics and the stability index are calculated and serve as a basis for decision-making, e.g., if the decision-maker is a pessimistic psychological type and puts particular emphasis on stability, structure 1 would the preferable option because, in the worst-case scenario, this structure would be the most stable. If the manager is an optimist and can take risk to a higher extent, structures 1 or 2 could be selected with regard to the ideal case or scenario 1 (small perturbations).

Let's analyse the received results. As already mentioned, SC no. 1 was initially characterized by the lowest level of planned profit since it had been constructed with the greatest redundancy and requirements to stability and service level. In comparison with SCs no. 2 and 3, which were characterized initially by a higher profit level, SC no. 1 in the case of the negative scenario (a simultaneous decrease in the availability of a resource of 30% and a resource decline of productivity of 5%) has appeared even more profitable in comparison with SC no. 2, and on the service level is the best among the considered SC structures. Thus, the area $Q, V(\boldsymbol{x}(t), t)$ is balanced with the area $\Xi(\boldsymbol{x}(t), t)$. The reliability and flexibility excessiveness can be considered as to be balanced with the execution environment.

In the programme, the decision-makers have a wide range of additional analytical possibilities with regard to the different SC structures and execution scenarios, i.e., they can change the admissible intervals of the goal parameters' oscillations and scope and the scale of the perturbation impacts. Additionally, the priorities of the SC goal metrics can be changed. There is also a possibility for a detailed analysis of the order dynamics, operation dynamics, enterprise activities dynamics, and "bottlenecks". The developed prototype implements the above-described theoretical models and makes a step towards designing stable and profitable SCs.

6 Results Discussion

With the presented developments, the study contributes to implementing the decision-making support to balance reliability and flexibility in order to achieve maximal economic performance and stability in SCs. We proposed a detailed analysis and classification of flexibility and reliability elements. Subsequently, an algorithm of decision-making on SC planning regarding the ensuring of both SC reliability and flexibility is presented. The quantitative approaches and formal tools are based on the modern control theory in combination with operations research techniques, global stability, controlled adaptation and the use of attainable sets. Finally, the developed tools are presented to implement decision-making support and SC managers' training.

Reliability and flexibility do not compete but they are very close and complementary in their nature. Both of them influence the SC performance both from the costs and income points of view. A balance of reliability and flexibility with regard to uncertainty is one of the most important conditions for SC stability. It is evident that through the adaptation, SC flexibility and reliability are interrelated. From the dynamics point of view, the reliability elements can also be considered as flexibility elements and the flexibility elements can also be considered as reliability drivers. This is quite natural because both the reliability and flexibility serve as an "uncertainty cushion" of a SC. Balancing the elements of flexibility and reliability, different constellations of service level, costs and stability can be analysed in a methodically well founded manner and with regard to a risk-covering strategy and an SCM strategy.

The presented decision-support tools have been developed to provide decision-makers with the possibility to analyse their SCs with regard to reliability, flexibility and uncertainty. The simulation results allow different interrelations of SC redundancy space (with regard to reliability and flexibility) and the space of uncertainty (with regard to agility and disruptions) to be analysed. Based on this analysis, decision-makers can prove whether the *excessiveness* of the corresponding SC structures is enough to prevent SC against the given perturbation influences and to consider the SC as stable. In other cases, additional reserves should be introduced, another constellation of reliability and flexibility should be considered, or possibilities of dynamic SC adjustments in the case of a temporary loss of stability should be taken into account.

The developed dynamic interpretation of stability is one of the main bases of the ReFlex framework. This dynamic interpretation consists of the following components:

- establishing concrete links to concrete processes, operation and parameters of SC execution dynamics through explicit interconnecting with the structure and operations dynamics models;
- considering not only the likelihood of a project's success but also the concrete adjustment steps for handling and adapting SCs in the case of disruptions based on the explicit interconnection of the stability, reliability and adaptability; and
- explicit interlinking of the synthesis and analysis models.

With the development of the decision-supporting frameworks and tools for SC reliability and flexibility regarding the agility and disruption-resistance, this study contributes to development of SC theoretical knowledge. From the practice point of view, the gained insights provide decision-makers with the possibility to balance reliability and flexibility to achieve maximal economic performance and stability in SCs.

Let us discuss the limitation of this study. First, since the framework as considered in this paper is of a generic nature, we considered in the numerical model only the strategic performance indicators of service level and costs without a deeper analysis with regard to the performance indicators of the second and third level of SCOR. Secondly, we considered the situation of plan analysis with given SC plans. However, the models and algorithms of planning remain beyond the scope this paper. With these models and algorithms, some features of the proposed approach would be clear for readers. Here, we refer to our other publications. Third, in the present state, the developed models and tools support decisions on balancing reliability and flexibility with regard to maintaining SC stability. However, a detailed analysis would be needed if different degrees of this maintenance should be considered (e.g., SCM could take a very risky strategy with partial risk insurance). Fourth, the developed framework, as presented in this paper, is of a generic methodological nature and needs case-specific localization for concrete applications.

For future research, we can distinguish the following areas. First, an in-depth analysis of SC costs and profit-centre models is required to conduct detailed process and operations research. Second, reference SC models with regard to different reliability and flexibility content for different environmental specifications can be elaborated. Third, different SC strategies could be investigated with regard to different reliability and flexibility content. These could be a strategy of maximum reliability maintenance, a strategy of maximum flexibility maintenance, or a strategy of risk financing (insurance contracts to cover the possible losses caused by disruptions).

References

Beamon, B.: Measuring supply chain performance. IJ of Operations and Production Management 19(3), 275–292 (1999)

Bertsimas, D., Sim, M.: The price of robustness. Operations Research 52, 135–153 (2003)

Casti, J.L.: Connectivity, complexity and catastrophe in large-scale systems. Wiley-Interscience, New York and London (1979)

Chen, F., Drezner, Z., Ryan, J.K., Simchi-Levi, D.: Quantifying the bullwhip effect in a simple supply chain: the impact of forecasting, lead times, and information. Management Science 46(3), 436–443 (2000)

Chernousko, F.L., Zak, V.L.: On Differential Games of Evasion from Many Pursuers. Journal of Optimization Theory and Applications 46(4), 461–470 (1985)

Chopra, S., Sohdi, M.S.: Managing risks to avoid supply chain – breakdown. MIT Sloan Management Review (Fall 2004)

Christopher, M.: Logistics and Supply Chain Management: Creating Value-Adding Networks. Financial Times. Prentice Hall, Dorchester (2005)

Christopher, M., Towill, D.R.: Supply chain migration from lean and functional to agile and customized. International Journal of Supply Chain Management 5(4), 206–213 (2000)

Christopher, M., Towill, D.R.: An integrated model for the design of agile supply chains. International Journal of Physical Distribution and Operations Management 31, 235–244 (2001)

Clarke, F.H., Ledyaev, Y.S., Stern, R.J., Wolenskii, P.R.: Qualitative properties of trajectories of control systems: a survey. Journal of Dynamic Control Systems 1(1), 1–48 (1995)

Collin, J., Lorenzin, D.: Plan for supply chain agility at Nokia. International Journal of Physical Distribution & Logistics Management 36(6), 418–430 (2006)

Coronado, M., Lyons, A.E.: Evaluating operations flexibility in industrial supply chains to support build-to-order initiatives. Business Process Management Journal 13(4), 572–587 (2007)

Craighead, C., Blackhurst, J., Rungtusanatham, M., Handfield, R.: The Severity of Supply Chain Disruptions: Design Characteristics and Mitigation Capabilities Decision Sciences 38(1), 131–156 (2007)

Ernst, R., Kamrad, B.: Evaluation of supply chain structures through modularization and postponement. European Journal of Operational Research 124, 495–510 (2000)

Fisher, M., Hammond, J., Obermeyer, W., Raman, A.: Configuring a Supply Chain to Reduce the Cost of Demand Uncertainty. Production and Operations Management 6(3), 211–225 (1997)

Fisher, M.L.: What is the right supply chain for your product? Harvard Business Review, 105–116 (March-April 1997)

Friscia, T., O'Marah, K., Souza, J.: The AMR research supply chain top 25 and the new trillion-dollar opportunity. AMR Research Report (November 2004)

Graves, S.C., Willems, S.P.: Optimizing the supply chain configuration for new products. Management Science 51(8), 1165–1180 (2005)

Gunasekaran, A., Kee-Hung, L., Cheng, T.C.E.: Responsive supply chain: a competitive strategy in a networked economy. Omega 36(4), 549–564 (2008)

Gunasekaran, A., Ngai, N.W.T.: Build-to-order supply chain management: literature review and framework for development. Journal of Operation Management 23(5), 423–451 (2005)

Gunasekaran, A., Ngai, N.W.T.: Modeling and analysis of build-to-order supply chains. European Journal of Operational Research 195(2), 319–334 (2009)

Guseinov, K.G.: Approximation of the attainable sets of the nonlinear control systems with integral constraint on controls. Nonlinear Analysis 71, 622–645 (2009)

Hendricks, K.B., Singhal, V.R.: The effect of supply chain glitches on shareholder wealth. Journal of Operations Management 21, 501–522 (2003)

Hendricks, K.B., Singhal, V.R.: Association Between Supply Chain Glitches and Operating Performance. Management Science 51(5), 695–711 (2005)

Ivanov, D., Sokolov, B.: Adaptive Supply Chain Management. Springer, London (2010)

Ivanov, D., Sokolov, B., Kaeschel, J.: A multi-structural framework for adaptive supply chain planning and operations with structure dynamics considerations. European Journal of Operational Research 200(2), 409–420 (2010)

Ivanov, D.: DIMA – A Research Methodology for Comprehensive Multi-Disciplinary Modelling of Production and Logistics Networks. International Journal of Production Research 47(5), 1133–1155 (2009)

Ivanov, D.: Adaptive aligning of planning decisions on supply chain strategy, design, tactics, and operations. Int. J. Prod. Res. 48(13), 3999–4017 (2010)

Ivanov, D., Sokolov, B.: Dynamic supply chain schedulin. Forthcoming in Journal of Scheduling (2011), doi:10.1007/s10951-010-0189-6

Jang, P.Y.: A flexible and adaptive control architecture for the efficient supply chain management (SCM). WSEAS Transactions on Communications 5(6), 1015–1025 (2006)

Kleindorfer, P.R., Saad, G.H.: Managing Disruption Risks in Supply Chains. Production and Operations Management 14(1), 53–68 (2005)

Klibi, W., Martel, A., Guitouni, A.: The design of robust value-creating supply chain networks: A critical review. European Journal of Operational Research 203(2), 283–293 (2010)

Knoppen, D., Christiaanse, E.: Interorganizational adaptation in supply chains: a behavioral perspective. International Journal of Logistics Management 18(2), 217–237 (2007)

de Kok, A.G., Graves, S.C.: Supply Chain Management: Design, Coordination and Operation. Elsevier, Amsterdam (2004)

Kouvelis, P., Chambers, C., Wang, H.: Supply Chain Management Research and Production and Operations Management: Review, Trends, and Opportunities. Production and Operations Management 15(3), 449–469 (2006)

Lee, H.L., Padmanabhan, V., Whang, S.: Information distortion in a supply chain: the bullwhip effect. Management Science 43(4), 546–558 (1997)

Mason-Jones, R., Naylor, B., Towill, D.R.: Engineering the leagile supply chain. International Journal of Agile Management Systems 2(1), 54–61 (2000)

Meepetchdee, Y., Shah, N.: Logistical network design with robustness and complexity considerations. International Journal of Operations and Production Management 37(3), 201–222 (2007)

Naim, M.M., Potter, A.T., Mason, R.J., Bateman, N.: The role of transport flexibility in logistics provision. International Journal of Logistics Management 17(3), 297–311 (2006)

Okhtilev, M., Sokolov, B., Yusupov, R.: Intelligent technologies of monitoring and control, Nauka, Moscow (2006) (in Russian)

Ozbayrak, M., Papadopoulou, T.C., Samaras, E.A.: Flexible and adaptable planning and control system for an MTO supply chain system. Robotics and Computer-Integrated Manufacturing 22(5-6), 557–565 (2006)

Peck, H.: Supply Chain Vulnerability, Risk, Robustness, and Resilience. In: Mangan, J., Lalwani, C., Butcher, T. (eds.) Global Logistics and Supply Chain Management, pp. 229–248. John Wiley and Sons, New York (2007)

Ponomarov, S.Y., Holcomb, M.C.: Understanding the concept of supply chain resilience. International Journal of Logistic Management 20(1), 124–143 (2009)

Reichhart, A., Holweg, M.: Creating the customer-responsive supply chains: a reconciliation of concepts. International Journal of Operations & Production Management 27(12), 1144–1172 (2007)

Ritchie, B., Brindley, C.: An emergent framework for supply chain risk management and performance measurement. Journal of the Operational Research Society 58, 1398–1411 (2007)

Sharif, A., Irani, Z., Lloyd, D.: Information technology and performance management for build-to-order supply chains. International Journal of Operations & Production Management 27(12), 1235–1253 (2007)

Sheffi, Y.: The resilient enterprise. MIT Press, Massachusetts (2005)

Sheffi, Y., Rice Jr., J.B.: A supply chain view of the resilient enterprise. MITSloan Management Review 47(1), 45–47 (2005)

Sokolov, B., Yusupov, R.: Risk integrated modelling for control in complex organizational-technical systems. Journal of Control and Informatics 1, 1–22 (2006)

Swafford, P.M., Ghosh, S., Murthy, N.: Achieving supply chain agility through IT integration and flexibility. International Journal of Production Economics 116(2), 288–297 (2008)

Tachizawa, E.M., Thomsen, C.G.: Drivers and sources of supply flexibility: an exploratory study. International Journal of Operations & Production Management 27(10), 1115–1136 (2007)

Tayur, S., Ganeshan, R., Magazine, M.: Quantitative models for supply chain management. Kluwer Academic Publishers, Boston (1999)

Tully, S.: The modular corporation. Fortune, 52–56 (1993)

Van Hoek, R.I.: The rediscovery of postponement a literature review and directions for research. Journal of Operations Management 19, 161–184 (2001)

Vickery, S., Calantone, S., Droge, C.: Supply Chain Flexibility: an Empirical Study. Journal of Supply Chain Management 35(3), 16–27 (1999)

Wadhwa, S., Saxena, A., Chan, F.T.S.: Framework for flexibility in dynamic supply chain management. International Journal of Production Research 46(6), 1373–1404 (2008)

Williams, Z., Lueg, J.E., LeMay, S.A.: Supply chain security: an overview and research agenda. International Journal of Logistics Management 19(2), 254–281 (2008)

5 Supply Chain Preparedness

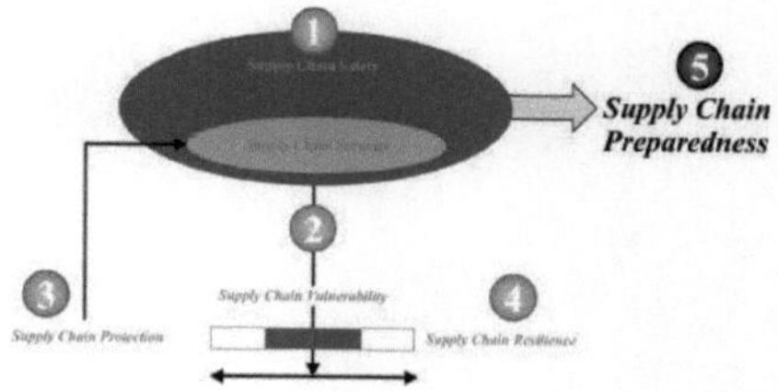

Monitoring and Certification of Supply Chain Safety

Axel Stepken

TÜV SÜD AG, Chairman and Chief Executive Officer (CEO),
Westendstraße 199, 80686 München / Munich, Germany
axel.stepken@tuev-sued.de

As global trade continues to grow, global players will need effective supply and production chains to ensure global sourcing at cost-effective rates and place their products on international target markets. End-to-end monitoring of supply and production chains and certification of products and processes are critical for ensuring that the requirements of different target markets are complied with and the expectations of target groups from different cultural backgrounds are fulfilled.

In 2010, global trade grew by over 13 per cent. According to the World Trade Organisation (WTO), the first half of 2010 saw unprecedented growth. This trend will continue and the global exchange of goods will increase further, in spite of some uncertainties and the possibility of short-term setbacks. Another notable development is the shift in trade flows. While a few decades ago goods were exchanged primarily between Europe and North America, whirlwind economic progress in the Asia-Pacific region has established a global trade triangle and caused a massive shift in the flow of goods. In the course of this development, China has taken Germany's place as the largest export nation in the world and in 2011 most products imported into Germany will be made in China. The majority of these products are not, as is frequently assumed, low-grade goods, but primarily technical products including electrical appliances and machines.

Other countries in the Asia Pacific region have followed the example and become members of the group of global market leaders. South Korea, for instance, has evolved into one of the leading suppliers of systems and system components for conventional and nuclear power stations, while countries such as Bangladesh or Vietnam have acquired a leading position in the production and export of textiles, clothing and shoes.

Factors accompanying and further advancing the rise in global trade and the shift in trade flows include the globalization of financial markets, the global networking and interaction between communication and information media, the ongoing removal of trade barriers – for example, by establishing regional economic areas such as the Single European Market (SEM) or free trade zones such as that created by the North American Free Trade Agreement (NAFTA) and by concluding bilateral and multilateral agreements between individual countries – and the increasing replacement of existing and established national rules and regulations by internationally accepted codes and standards.

M. Essig et al. (Eds.): Supply Chain Safety Management, LNL, pp. 291–297.
springerlink.com

In contrast to the extremely dynamic development of financial markets and communication and information media, the process of removing trade barriers and developing international or harmonized codes and standards is proving to be rather slow and will continue for some time. In the meantime, specific national and regional laws, directives and standards apply in many countries and regions. Furthermore, buyers and consumers from various markets and cultural backgrounds have different expectations and demands. If companies want to launch and successfully sell their products – from food processors to coal power stations and from hybrid vehicles to pacemakers – on international markets, they need to give timely and sufficient consideration to all influencing factors.

Most problems arising during the development and market launch of new products, for example, are caused by late or insufficient consideration of the various connections and interfaces in the global supply and production chains. This complex process commences as early as the product specification and extends throughout pre-testing, type or prototype testing and quality assurance during pilot and mass production to inspections along the logistics chain and after-sales support in the country of destination. All these tasks require specialist know-how in the individual sectors of industry and product groups, and – in view of the globalization of the supply and production chains – a comprehensive global network with strong regional or local presence for conducting the required factory inspections. While multinationals may have the resources to fulfil these tasks, other companies rely on specialist providers of testing and certification services which offer the appropriate expertise and the required network of experts and testing laboratories.

The following case studies describe how testing and inspection along the global supply and production chains work in practice, and the role of product and process certification in this context.

1 Case Study: Chemical Group

An international chemical group headquartered in Germany planned to erect a new production facility in China. The plant was made up of many components produced by suppliers from different countries.

The core business of the chemical group is the production of special plastics, which it supplies primarily to the auto industry and to building services systems. China is well on the way to becoming the world's leading automotive manufacturer. The construction sector in the People's Republic is likewise showing double-digit growth. Following market trends, the chemical group planned to establish a new production plant for special plastics in China. However, in this context the company was focusing not only on the Chinese market. Due to global supply chains, the plastics produced in China would be used by processing companies around the world and therefore had to meet the requirements of the individual markets.

For production to be cost-effective, the new production facility needed to function safely and reliably. To meet this end, an integrated approach focusing on this target and the interaction between the individual plant components had to be adopted in the design and construction of the plant. This process extended from the planning of the plant to the selection and monitoring of suppliers across the world, quality assurance during the assembly of the individual components on-site at the suppliers' factories and the placing into service of the completed production facility. In this context, steps had to be taken right from the outset to ensure compliance with both the national and international standards for this type of production facility and the chemical group's own global standards.

TÜV SÜD accompanied the development of this production facility from the start, planning and managing the quality assurance process. For this project, TÜV SÜD assembled an interdisciplinary team comprising experts from Germany, the USA, India and China. This team not only designed a scheme for monitoring the entire design process, but also ensured target-performance comparison based on ongoing testing and verification. The comparison and alignment of national and international standards, including the China Code, the ASME Code and the chemical group's own quality standards, proved particularly challenging. The experts then had to ensure that suppliers from a host of different countries with different industrial cultures met the requirements for the individual plant components. For this purpose, the key steps in the production process were verified in on-site factory inspections at the suppliers' premises and compared with the existing plans.

The plant has now been successfully placed into service. TÜV SÜD also contributes significantly to the safe, reliable and cost-effective functioning of the plant in the operational phase, conducting in-service and turnaround inspections. The inspections and the inspection results, i.e. detailed knowledge of the plant's state of repair, have proven indispensable for the time- and cost-effective planning and design of the servicing and maintenance process by the chemical group.

2 Case Study: Discounter – Non-food Products

In addition to groceries, a German discounter offers a wide assortment of 'non-food items' from household appliances to toys, shoes and clothing. The discounter has no in-house expertise in the non-food sector and purchases these products across the world.

The discounter's core competence is the purchase and sale of goods. However, the discounter cannot cover the whole scope of professional expertise needed to select and evaluate the entire assortment of products it offers – from eggs to electric drills. In the non-food sector especially, the company faces the challenge of having to purchase products at cost-effective rates across the world while guaranteeing the safety and quality of these products. After all, product-related problems, such as product safety recalls, result in direct material damage, disappoint customer expectations and have a negative impact on the brand image.

These adverse events can only be avoided by ensuring uninterrupted end-to-end testing and inspection along the supply and production chain. This complex process starts with the product specification and type testing, i.e. the testing of a prototype, extending to quality assurance during mass production, inspections throughout the logistics chain and after-sales support. For this task, the discounter relies on a specialist service provider – in this case TÜV SÜD – which has the required know-how and experience and offers a global network of experts and testing laboratories for an array of products.

In this specific example, at a trade show in Shanghai one of the discounter's purchasing agents found a vacuum cleaner which might be of interest for the discounter. The vacuum cleaner appeared to be solidly made and was produced by the same production plant in the Chinese province of Shenzhen from which the discounter had already purchased, and successfully sold, hedge clippers some months previously. The purchasing agent sent a sample vacuum cleaner to the competent TÜV SÜD laboratory in Munich. There the vacuum cleaner was subjected to a 'pre-test' which gave a first impression of its function, safety and quality. A pre-test may reveal, for example, that a product must be modified to comply with the legal requirements in Germany or to meet customer expectations.

In the next step of this process, the purchasing agent then commissions the supplier to manufacture a modified type or prototype and submit it for detailed testing and analysis to the TÜV SÜD laboratory in Munich. In the case at hand, the discounter's priorities were to ensure not only that the vacuum cleaner complied with the legal requirements for mechanical, electrical and chemical safety, but also that the experts verified additional quality characteristics of the product such as usability – including the instruction manual – and durability, and confirmed these characteristics by affixing a certification mark. These certification marks, including safety marks such as the GS mark – widely known in Germany – or more far-reaching quality labels of testing and certification organizations, act as visible proof that a product was tested by an impartial third party against a clearly defined list of criteria and actually offers the promised level of safety and/or quality to the discounter's customers.

Another essential step in the purchasing process is the transition from prototype or pilot production to mass production. In this phase, the TÜV SÜD experts must verify that the appliances made in mass production correspond to the examined type and that a consistent level of quality is assured in production. To do so, they test random samples taken from the production line and audit the supplier's entire quality assurance process or quality management system.

However, end-to-end testing and inspection does not conclude with the start of mass production, but also covers the entire logistics chain, from the manufacturer's premises and pre- and post-shipment inspections (PSI) at the ports of origin and destination right up to the discounter's shelves. Last but not least, reviewing of customer complaints, inspection of returned goods and the assessment of complaints management make up services in after-sales support. Based on this comprehensive assessment, the discounter in our case study was able to rest assured that the vacuum cleaner offered in its branches corresponded to the product it had originally tested, found to be good and ordered.

To ensure end-to-end control is effective in practice, the entire process, and the interfaces in particular, must be clearly defined. After all, this supply and production chain is also governed by the principle that one weak link in the chain may cause it to break, and result in an unusable or even dangerous product being placed on the market.

3 Case Study: Supermarket Chain – textiles

The past 15 years have witnessed a shift in the global production of textiles towards Asia Pacific. Three-quarters of the world's textile production facilities were already installed in this region by 2008. China, India, Pakistan and other Asian countries have evolved into the leading producers of textiles and clothing. Before these products can be placed on the key sales markets in Europe and North America, they must fulfil the relevant safety and quality standards of these markets. Over recent years, another criterion has been added: compliance with international social standards in production.

A British supermarket chain offers a broad product range, which includes textiles. Most of these textiles are garments purchased from a host of different producers in the Asia Pacific region. While the supermarket chain has the necessary resources to control and monitor logistics, it does not have any expertise in product testing or in the certification of social standards. The supermarket chain outsourced this task, i.e. one part of the entire supply and production chain, to an external service provider – TÜV SÜD. Over recent years, TÜV SÜD has invested massively in extending and improving its infrastructure for testing textiles and social standards, focusing particularly on the Asia Pacific region, and has succeeded in establishing a competent network of specialists and laboratories in the key production countries.

The extensive testing and certification process covers laboratory testing of raw materials, factory inspections and on-site audits including control of product packaging and labelling, and certification in accordance with social standards. Laboratory testing focuses on the quality of the fibres used and the chemical safety of the fabrics. Testing for harmful substances is guided by the provisions and limits set forth in the applicable codes and standards, including the EU chemicals regulation REACH in the European Union, the Consumer Product Safety Improvement Act (CPSIA) in the USA and the international Restricted Substance List (RSL) adopted by manufacturers and retailers. Testing of raw materials is complemented by the analysis of random samples taken from the production line.

The audits carried out at the production facilities focus on the quality management system based on the well-known ISO 9001 standard and on verifying compliance with social standards. In addition to the internationally recognized SA 8000 standard of Social Accountability International, further standards play an important role, particularly on regional markets. They include the BSCI standard (Business Social Compliance Initiative) in the European Union, the WRAP program (Worldwide Responsible Accredited Production) in the USA and the ETI standard (Ethical Trade Initiative) in the UK.

The certification of quality management systems has become accepted across the world. The certificate acts as proof that a company has established and maintained effective quality assurance and continuous improvement processes. In many sectors of industry, certificates are imperative for participation in invitations to tender and contract-awarding procedures. A similar trend is observed in social standards. Increasing numbers of manufacturers, retailers and importers are striving for certification in accordance with various social standards. As demonstrated by the countries of origin of the certified companies, this applies particularly to developing countries. In 2009, most certifications as per the WRAP and BSCI standards were carried out in China, Bangladesh, India, Pakistan and Vietnam.

Reports about inhuman conditions in production facilities may permanently damage the image of a company or a brand. This also applies if violations of social standards are caused by downstream suppliers. Given this, compliance with social standards must be verified and controlled across the entire supply and production chain.

4 Case Study: Manufacturer of Electronic Systems (Jabil)

For many companies, the establishment and certification of a quality management system is standard procedure. The continuous improvement of processes is indispensable for cost-effectiveness and thus a key factor for the competitive strength of a company. QM certificates also provide guidance in the search for partners and suppliers in the international exchange of goods and services throughout global supply and production chains. Given rising environmental awareness, certification of environmental management systems may play a similar role in the future. The certificate documents that a company adopts a conscientious approach to environmental issues and prioritizes sustainable operations. Generally, the certification process identifies potential for savings, such as energy savings, and results in a reduction of production costs.

A USA-based international leading developer and manufacturer of electronic systems which it supplies to electronics and technology companies operates in a fiercely competitive environment. Key factors for the company to maintain, or even improve, its competitive strength are continuous improvement of its internal processes, assurance of a high level of quality and exploitation of the potential for efficiency improvement and savings. Given this, the company opted to have its existing quality system certified in accordance with ISO 9001 and to establish an environmental management system as per ISO 14001.

TÜV SÜD was commissioned to conduct the certification of the quality management system and the environmental management system at the company's individual locations throughout the world. To save time and money, the company decided on a one-stop provider of certification services. For this task, TÜV SÜD assembled a project team of 22 auditors located in China, France, Italy, Mexico, Poland, Singapore, Taiwan, the UK and the USA. This team developed, coordinated and agreed the specific audit provisions and requirements to ensure the comparability of the audit results obtained at the different locations.

The audits at the individual sites and the continuous alignment with the strategic goals of the company within the scope of the audits resulted in improved processes in the fields of quality and environmental management. Harmonization of the audit procedures ensured that the processes at the various locations were comparable and best practice examples existing in the company were identified.

On the basis of the assessments and analyses, the TÜV SÜD auditor team also recommended actions to further improve the quality and environmental management systems and their specific processes.

This example shows that quality and environmental management certificates may contribute significantly to improving a company's performance and competitive strength. The certification process also offers the possibility of screening existing processes for their potential to reach the strategic objectives defined by the company, pointing out best practice examples in the company and thus advancing the company's continuous improvement process.

In the global supply and production chain, certificates for quality management systems in accordance with ISO 9001 and environmental management systems in accordance with ISO 14001 provide important guidance when looking for partners and suppliers, as they document that the company has established effective processes and lives up to its promises in the fields of quality and environmental protection.

5 Conclusion

Increasing economic globalization and rising competitive pressure represent a challenge to companies, which have to ensure cost-effective purchasing in global sourcing and offer products on various international sales markets. Globalization has brought in its wake a rise in potential partners and suppliers from different cultural backgrounds. In an increasingly complex environment, companies must ensure end-to-end testing and inspection across their entire supply and production chains while delivering the safety and quality they promised to their customers. Certificates offer important guidance in selecting partners and suppliers. For example, they provide evidence that the company maintains an effective quality and/or environmental management system and complies with social standards. End-to-end safety management across supply and production chains in conjunction with process and product certification is imperative for the long-term success and continued growth of global trade. Manufacturers, retailers and importers will increasingly outsource this responsibility to specialist service providers with the appropriate expertise, the necessary experience and a global network of experts and testing infrastructure. By outsourcing this task, companies can focus on their core competencies, minimize the risks of global sourcing and exploit the opportunities offered by new markets.

Compliance and Supply Chain Safety

Dr. Josef Mauermair

Steria Mummert Consulting AG, Senior Manager,
Hans-Henny-Jahnn-Weg 29, 22085 Hamburg, Germany
Josef.Mauermair@steria-mummert.de

1 Introduction

This article is about the relevance and influence of an effective compliance system on the general framework of supply chain preparedness and its primary goal of ensuring the continuity of supply. This article is first of all the result of my personal project practices and experiences in compliance management and controlling systems in public and logistic sectors over the past two decades.

Ensuring end-to-end supply chain compliance is a critical success factor for all elements and relations, material movements and information flows within and along a global supply chain. In global supply chains and in some business sectors (for example in sectors with military or dual use products) the requirement of complying with complex laws, regulations and standards[1] may also become mandatory. All producers, vendors, suppliers and retailers will have to prove and improve their compliance system continuously and to deal with the relevant laws, business and administration regulations and ethical codes. It's a "To be or not to be"-question for all business partners of the chain, to build up and run an effective and efficient compliance system. Fraud, gaps, failures, incompetence, environmental and many other factors will cause cost, revenue, rentability and reputation disruptions in all dimensions of business affairs.

Section 1 reflects the term "compliance" in the sources, objectives and effects of compliance with a special focus on the relations between compliance and supply chain continuity. Based on this fundamental work I describe the specific structures of rule building and monitoring and disruption aspects.

Section 2 deals with the practical approach of a "rule life cycle model" for the benefit of a better understanding of the possibilities and limits of planning, running and controlling rules as the core objects of the compliance system. I describe the different phases of the life-cycle from a theoretical and a supplier's perspective.

---

[1] ISO 28000, EU Regulation 178, Customs-Trade Partnership Against Terrorism (C-TPAT), Zugelassener Wirtschaftbeteiligter (AEO), Bioterrorism Act der FDA (Food and Drug Administration), Sarbanes Oxley Act der SEC (Securities and Exchange Commission), Basel II, Third Party Liabilities, Technology Asset Protection Association (TAPA), International Ship and Port Facility Security Code (ISPS-Code), EG-Dual-Use-Verordnung, KrWaffKontrG, EAR Export Administration Regulations, ITAR International Traffic in Arms Regulations, 9/11 Act.

M. Essig et al. (Eds.): Supply Chain Safety Management, LNL, pp. 299–308.
springerlink.com

2 Fundamental Terms

2.1 Compliance and Supply Chain Continuity

Compliance. The term 'Compliance' is related to a lot of different context-dimensions like medicine, psychology and regulation-systems. In the regulation context compliance is used according to the following definition in wikipedia: "*In general, compliance means conforming to a rule, such as a specification, policy, standard or law. Regulatory compliance describes the goal that corporations or public agencies aspire to in their efforts to ensure that personnel are aware of and take steps to comply with relevant laws and regulations.*" Therefore compliance means conforming to stated requirements defined for example in laws, regulations, contracts and ethical codes

The world of compliance relates to one of the three core disciplines of governance, risk management, compliance as the integrated framework for ensuring correct business and administration affairs in all corporate segments of a specific company (cf. Hauschka 2008, Wecker/van Laak 2008, Wieland/Steinmeyer/Grüninger 2010). According to the importance of this discipline compliance-affairs are part of top management responsibilities and activities.

Compliance Management System. A Compliance management system is a set of approaches, structures, processes and applications for building, running and improving correct and compliant corporate business and administration affairs. In this context it's absolute necessary to search for the playgrounds of incompliant actions and events and their causes and functional and economic effects with a deep understanding and experience of individual and group behaviour, statistics (especially sample-, test- and network-theory) and legal operations.

Management and fulfilment of business and administration affairs means that all involved roles have to deal with compliance sources, especially with laws, standards, certifications, directions, contracts, documentation, publication, insurance, authentication, authorization and ethical codes and a well-experienced culture of confidence.

The core objects of all these sources are 'rules' and the main compliance objectives are:

- Ensuring correctness.
- Ensuring business and administration continuity.
- Creating reputation.

The main effects are:

- Avoiding costs, penalties and sanctions.
- Ensuring and increasing rentability.
- Ensuring and increasing performance.

Compliance and supply chain continuity. In relation to the structures and processes of supply chain the successful management of compliance affairs have a significant effect to the risk and chance dimension of chain-performance (cf. Weber/Wallenburg 2010, Huth/Lohre 2009, Moder 2008, Teuteberg/Friedemann 2007, Ziegenbein 2007, Kersten/Blecker 2006).

As shown in the figure below the set of defined compliance sources and objectives has significant relevance and influence on all elements and relations of a global supply chain and their involved partners.

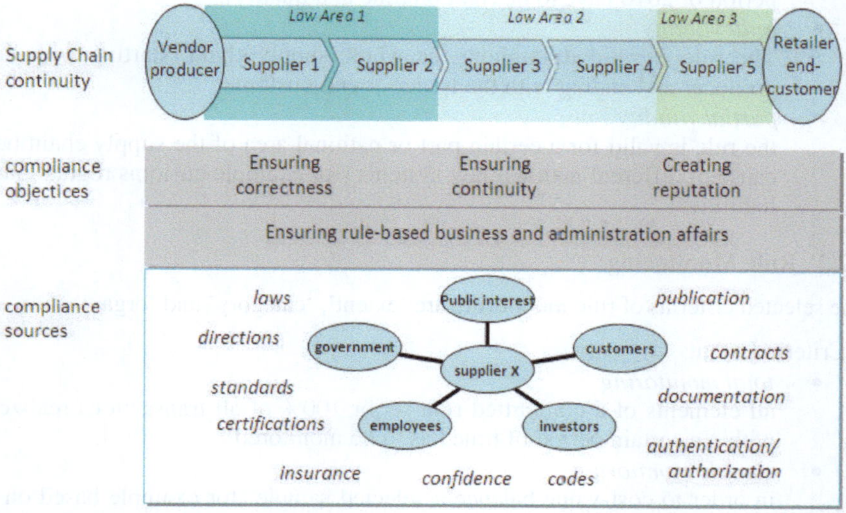

Fig. 1 Relation of compliance-objectives and -sources and supply chain continuity (own illustration)

2.1 *Structures of Rules*

To understand, detect and respond the various forms of possible disruptions in the practical run of a global supply chain, it's very important, to give an short insight in some structural aspects of rules as the core objects of all compliance sources.

2.2.1 Rule Building

The main criterias of rule building are 'transparency' and 'validity'.

- Criteria Transparency:
  - *explicit rule*
    the rule is published and continously available for each involved partner of the supply chain.
  - *implicit rule*
    the rule is unwritten and unpublished but part of the well-known assumptions of a single person or a certain community of persons outside of any audit and control procedures. Implicit rules are very dangerous, there is an significant relation between implicit rules, personal dependencies and disruptions in business and administration affairs.

- Criteria Validity:
  - *absolute validity*
    the rule has to be complied for 100% (for example: general prohibition of children's work for all UN-members).
  - *relative validity*
    the rule has to be complied in accordance to a committed level (for example: 99% of all logistic and movement transactions in the contract period of 2010 have to be right on time and quality).
  - *total validity*
    The rule is valid throughout the whole supply chain, starting with the producer and ending with the trader or end-customer.
  - *partial validity*
    the rule is valid for a certain part or national area of the supply chain because of different national law-systems (for example customs regulations, liabilities)

2.2.2 Rule Monitoring

The selected criterias of rule monitoring are 'extent', 'category' and 'organization'.

- Criteria Extent:
  - *total monitoring*
    all elements of a committed rule-set or 100% of all transactions realized within a certain period of time has to be monitored
  - *sample monitoring*
    in order to cost-value balance a selected sample (for example based on a ABC-analysis) of the rule-set or of the transactions delivered has to be monitored
- Criteria Category:
  - *continuous monitoring*
    monitoring rules or transactions in every moment of applying or realization
  - *periodical monitoring*
    monitoring rules or transactions in defined time-intervalls of applying or realization
  - *ad-hoc monitoring*
    monitoring rules or transactions as a result of certain events or analyses
- Criteria Organization:
  - *internal monitoring*
    planning, running and controlling of the monitoring activities by internal units and resources
  - *external monitoring*
    planning, running and controlling of the monitoring activities by external service providers and their resources.

2.2.3 Structures of Rule Disruptions

There are many forms of causing disruptions in supply chains as shown in the figure below. The key question for structuring the world of disruptions is the following: "Is the disruption result caused by intent/negligence by incompetence or by accident/chance?"

by intent or negligence		by incompetence	by accident or chance	
causing section	phenomenon		causing section	phenomenon
supplier X producer, trader other suppliers	theft fraud spying sabotage manipulation fault gap		environmental factors	natural desaster strike accident terroristic act

Fig. 2 Overview of structures of rule disruptions (own illustration)

3 Life-Cycle Model of Rules for Ensuring SC-Preparedness

A very useful and practical approach for creating and sustaining effective and well-proved rules as shown in the figure below is a life-cycle model of rules with the consistent succession of building, running, monitoring, acting and evaluating stages.

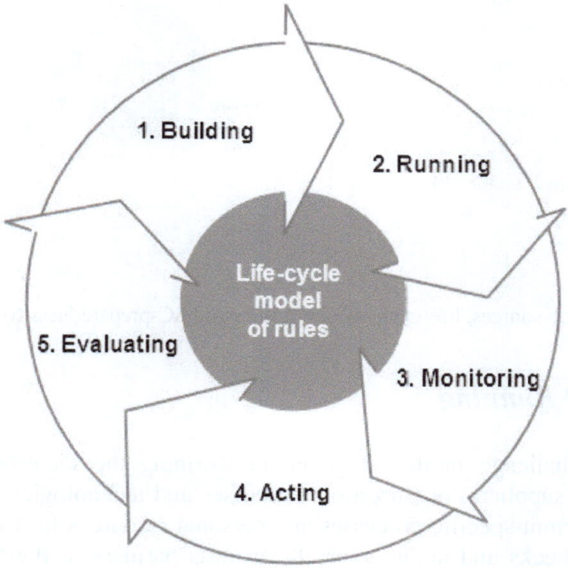

Fig. 3 Life-cycle model of rules (own illustration)

From the perspective of an involved partner of a supply-chain-construction this model is relevant for all sources of compliance without laws.

3.1 Stage 1 Building

The situation of a global supply chain is often characterized by a) a set of supply chain partners (producers, suppliers and traders) with specific business interests,

b) different national law and regulation areas and c) very complex products with a lot of commissioning, movement and storage conditions. Building and setting effective rules will – in best case - create sustainable win-win-constellations for all partners within a global supply chain and therefore it's valuable to invest time and resources in searching for compliable and acceptable rules.

The various compliance sources as shown in the figure below will offer a high level of support for the planning, mitigation, detection, response and recovery stages of the supply chain preparedness if the life-cycle-model will be consequently applied.

Nr	Compliance sources	Relevance for the stages of supply chain preparedness				
		Planning	Mitigation	Detection	Response	Recovery
1	laws					
2	codes					
3	standards					
4	certifications					
5	directions					
6	contracts					
7	documentation					
8	publication					
9	insurance					
10	authentication					
11	authorization					
12	confidence					

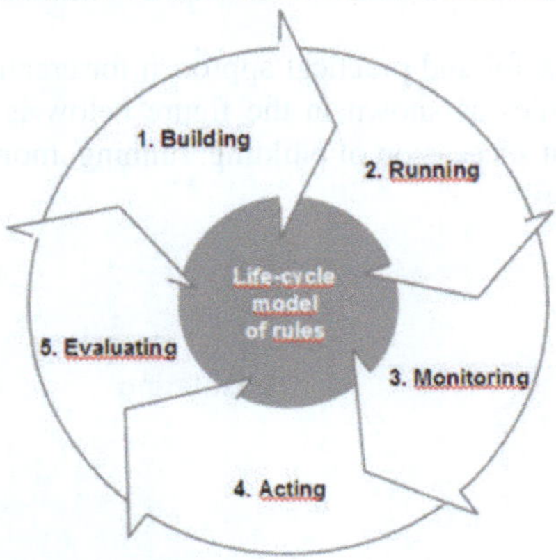

Fig. 4 Compliance-sources, life-cycle model of rules and SC-preparedness (own illustration)

3.2 *Stage 2 Running*

The critical challenge in this stage is transforming the identified compliance sources in the suppliers' organization, processes and technologies. It's critical, to define and commit specific processes and personal resources for running the relevant control checks and audits over all business partners of the SC. You never should commit you and your SC-Partners to a set of operating rules if you cannot fulfil your end of the contract.

Organization. From this point of view there are different solutions possible for running the compliance affairs in day to day business and administration:

- the linkage of all compliance affairs (policies, strategy, concept, responsibility) to top management and/or installation of compliance-departments (for example department of internal control/revision)
- and/or institutionalization of boards (for example anti fraud board)
- and/or decentralized compliance-functions
- corporate and center directions and checklists.

In this context the supplier has to prepare and decide an adequate sourcing-policy (total insourcing, coorporation with service partners, outsourcing of single function and modules to specialized service corporations).

Processes. The Compliance sources are relevant for all business and administration processes, especially for

- Recruiting and resource development
- Auditing and monitoring of business partners
- Contracting and legal services
- Process- and quality management
- Accounting, billing and treasury
- Communication and training

Technologies. For effective and efficient running and high level standards of performance the compliance system will be supported by DWH-/BI-technologies (Brauer/Steffen/Biermann/Schuler 2009), especially for measuring, analyzing, visualization and reporting of heterogeneous, dynamic mass data pools (for example consignment and billing transactions). Further on standard statistical programmes and simulation algorithms in OLAP-applications are very useful. The market of less or more integrated Enterprise Compliance and Risk Management is characterized by high diversity and maturity.

Early warning systems however are from lower relevance, because the objects of monitoring are first of all facts (events and transactions) as we see in the following stage.

3.3 Stage 3 Monitoring

Based on my practical experience the key to successful monitoring processes, results and effects is to develop and implement an integrated controlling system with the combination of monitoring-areas as we see in the figure below.

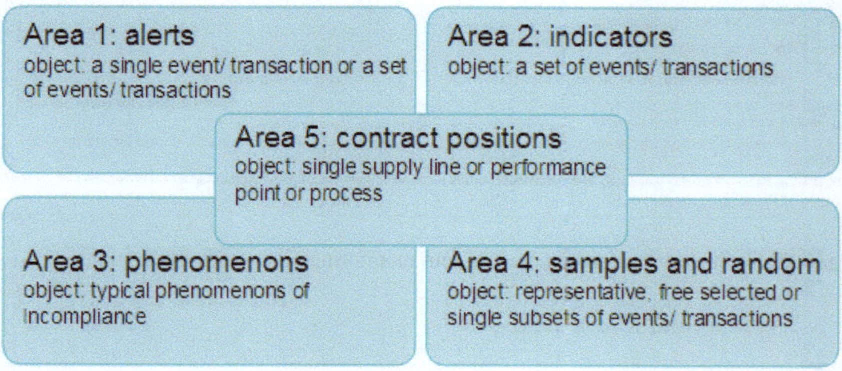

Fig. 5 Integration of different monitoring-areas (own illustration)

The reason for combining these areas in a powerful monitoring strategy is the understanding of the following stress statements:

- Statement 1: "*It is not profitable to build up and run a compliance system in a world of global supply chains with enormous complex and heterogeneous rules and an extraordinary rate of change.*"
- Statement 2: "*The deeper the criminal intentions of one or a few partners in a global supply chain, the more probable is the achievement of advantages and earnings from incompliant activities.*"
- Statement 3: "*If a compliance strategy and system is published, incompliant activities are easy to develop and calculate referring to chances and risks.*"
- Statement 4: "*A Compliance system is a cost-intensive bureaucracy world from outerspace with un-measurable and doubtful benefits keeping off managers and employees from their core business.*"

Nevertheless an enormous variety of proved analytical instruments, reports and activities is available for supporting the five monitoring-areas. The most important challenge is to select, combine and focus a powerful specific set of instruments and linking their results and reports with management planning and controlling processes.

Nr	instrument	espacially for monitoring area ...				
		alerts	KCI[4]	pheno-menons	samples, random	contracts
1	producer and supplier auditing and monitoring-programmes, histories and reports	X	X	X		
2	Product and production spezifications, internal analytical test data	X		X		
3	External analytical test and monitoring data[1]	X		X	X	X
4	Statistical analysis (esp. Correlation, Benford, trendanalysis, special sample- and test-models)	X	X	X	X	X
5	Visualization tools	X	X	X		
6	Compliance performance boards	X	X	X		
7	Standard and ad-hoc reporting		X	X	X	X
8	Transaction operation profiles			X	X	X
9	Committed control-programmes			X		X
10	Special exploration	X	X	X	X	
11	Technical monitoring systems[2,3]	X	X	X	X	X

1 Organizations for delivering test data and monitoring contracted producers and suppliers in developing countries, that they are complying with global labour and environmental standards

2 RFID for delivering transaction, product and production informations, Sealing, Labeling, EPC Electronic Product Code, EDI Electronic data interface

3 Tracking and tracing for status monitoring, alert based delay-communication and customer clearance, recall indication and management

4 KCI = Key Compliance Indicator

Fig. 6 Monitoring instruments in the different monitoring-areas (own illustration)

3.4 Stage 4 Acting

In case of indication or detection of disruptions along the supply chain with offences against valid law or with influence on the suppliers own business interests there are a lot of possible actions which can be considered, decided and executed.

They are characterized by a certain intensity of escalation and a certain level of execution-expense as shown for some examples in the figure below.

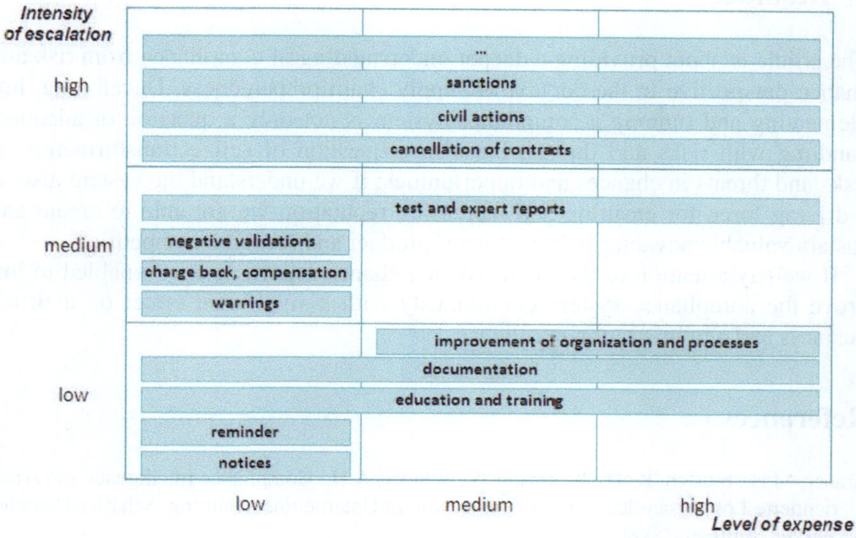

Fig. 7 Action-Portfolio of dealing with rule-disruptions in supply-chains (own illustration)

3.5 Stage 5 Evaluation

From the supplier's perspective all different sources of compliance (without laws) should be evaluated either periodically or in consequence of monitoring results. In my experience it's effective to choose an objective-based approach of evaluation with the purpose of relating outcomes to the key compliance objectives "ensuring correctness", "ensuring continuity" and "creating reputation"; this specific approach allows judgements about their level of attainment.

The very factors that lead to a supplier's excellent compliance system are

- Clarification: "*Are the compliance rules transparent and clearly to understand corporate-wide and for all relating partners?*"
- Consistence: "*Are the rules within a defined rule-set complementary and are there no concurrent effects because of their direct or indirect dependencies?*"
- Measurement: "*Is it possible to measure the results and/or to explore the effects of rules either by indicators, phenomenon-analysis, hypothesis-testing, surveys or other methods in a confident way with adequate cost-benefit balance?*"

- Influence: "*Are there any destructive changes (for example resistance, counter attacks and ignoration patterns) identifiable in the behaviour of managers, employees and business partners as a result of rule setting actions?*"
- Fit for change: "*Is the compliance system fit for expected or probable changes in the future inside the firm and relating to the main external conditions and influences?*"

4 Resumee

The article is about providing a deeper understanding of compliance from risk and chance perspective in the context of supply chain preparedness. Developing, implementing and running a compliance system is not only a question of adequate handling with risks and threats, but also a question of active transformation of risks and threats in chances and opportunities; if we understand the system also as a driving force for ensuring and increasing reputation we are able to create and sustain valuable advantages in customer, product and resource competition.

If we pay attention to the main risk and chance aspects, we are enabled to improve the compliance system continuously with a significant effect on a firm's business and administration excellence.

References

Brauer, M.H., Steffen, K.-D., Biermann, S., Schuler, A.H.: Compliance Intelligence - Praxisorientierte Lösungsansätze für die risikobewusste Unternehmensführung. Schäffer-Poeschel Verlag, Stuttgart (2009)

Hauschka, C.E.: Corporate Compliance – Handbuch der Haftungsvermeidung im Unternehmen. Verlag C.H. Beck, München (2007)

Huth, M., Lohre, D.: Risikomanagement in der Speditions- und Logistikbranche: Bestandsaufnahme zu Verbreitung und Reifegrad. Speditions- und Logistikverband Hessen/Rheinland e.V., Frankfurt am Main (2009)

Kersten, W., Blecker, T.: Managing Risks in Supply Chains: How to Build Reliable Collaboration in Logistics. Erich Schmidt Verlag, Berlin (2006)

Moder, M.: Supply Frühwarnsysteme. Die Identifikation und Analyse von Risiken in Einkauf und Supply Management, Wiesbaden (2008)

Teuteberg, F., Friedemann, J.: Controlling und Risikomanagement in Wertschöpfungsnetzwerken. Research Report, Osnabrück (2007)

Weber, J., Wallenburg, C.M.: Logistik- und Supply Chain Controlling, 6. Aufl. Schaeffer-Poeschel Verlag, Stuttgart (2010)

Wecker, G., van Laak, H.: Compliance in der Unternehmenspraxis - Grundlagen, Organisation und Umsetzung. Gabler Verlag Wiesbaden (2008)

Wieland, J., Steinmeyer, R., Grüninger, S.: Handbuch Compliance-Management. Konzeptionelle Grundlagen, praktische Erfolgsfaktoren, globale Herausforderungen. Erich Schmidt Verlag, Berlin (2010)

Ziegenbein, A.: Supply Chain Risiken. Identifikation, Bewertung und Steuerung. VDF Hochschulverlag, Zürich (2007)

Supply Chain Innovation and Risk Assessment (SCIRA) Model

Stephan Klein-Schmeink[1] and Thomas Peisl[2]

[1] Gesellschaft für Entwicklung, Beschaffung und Betrieb (g.e.b.b.) mbH,
Head of Business Unit SCM,
Ferdinand-Porsche-Str. 1a,
51149 Cologne, Germany
Stephan.Klein-Schmeink@gebb.de
[2] Munich University of Applied Sciences,
Professor of Strategy,
Am Stadtpark 20,
81243 Munich, Germany
thomas.peisl@hm.edu

1 Innovation and Risk: Conflict or Opportunity?

"In the breast of one who wishes to do something new, the forces of habit rise up and bear witness against the embryonic project. A new and another kind of effort of will is therefore necessary in order to wrest, amidst the work and care of the daily round, scope and time for conceiving and working out the new combination and to bring oneself to look upon it as a real possibility and not merely as a day-dream." (Schumpeter 1934, p. 86).

In 1912 Schumpeter published the first edition of his theory of economic development, and explained fundamental findings about economic relations in general and driving forces of economic systems in particular. Concerning the latter, Schumpeter distinguishes between hedonistic (static) and entrepreneurial (dynamic) motivations of individuals and organisations. While the type of hedonistic host focuses on the satisfaction of wants, the energetic entrepreneur is driven by the joy of creating and the will to conquer (cf. Schumpeter 1934, p. 90-94). From Schumpeters point of view, innovation in economies is carried out by entrepreneurs, combining given resources within new processes. The underlying principle of 'creative destruction' since then represents the foundation of innovation theories (cf. Röpke/Stiller 2005, p. IX).

Innovation is a precondition for economic development, as it opens up opportunities for competitive advantages and long-term success [fig. 1]. But innovation also means risk, as it takes place under conditions of uncertainty (cf. Corsten et al. 2006, p. 1-11).

M. Essig et al. (Eds.): Supply Chain Safety Management, LNL, pp. 309–326.
springerlink.com

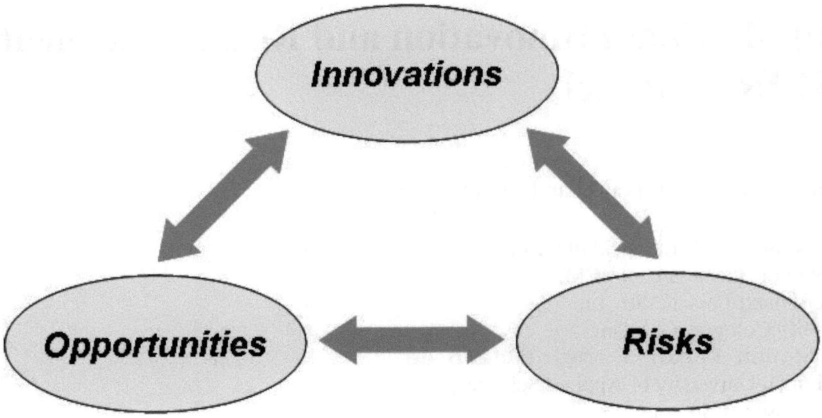

Fig. 1 Interrelation of innovations, opportunities, and risks (own illustration)

Pavitt (2006, p. 88) characterises this 'confusing mosaic' out of innovation, opportunities and risk as follows: "Innovation processes involve the exploration and exploitation of opportunities for new or improved products, processes or services (…). Innovation is inherently uncertain, given the impossibility of predicting accurately the cost and performance of a new artefact, and the reaction of users to it."

There is no innovation without risk, and there are no opportunities without innovation (cf. Gassmann 2006, p. 3-21). If an organisation disclaims innovation, on the one hand it eliminates possible causes of risk. On the other hand, it loses a vital source to generate opportunities, and thus the precondition for long-term success (which definitely means risk, too). How can this conflict of interest be managed?

The Supply Chain Innovation and Risk Assessment (SCIRA) model provides an approach to balance innovation and risk. Supply Chain Management (SCM) is the matching initial position of the model for three reasons: First, SCM itself represents a specific type of organisational innovation (cf. OECD 2005, p. 51). Second, the intention behind SCM is the generation of opportunities (cf. Powell/Grodal 2006, p. 59). Third, SCM causes a specific risk situation for value creating networks (cf. Götze/Mikus 2007b, p. 35).

Out of this context, a "new research area called Supply Chain Risk Management has developed." (Paulsson 2004, p. 80). The overall purpose of Supply Chain Risk Management (SCRM) is to identify risks affecting a supply chain, and to develop measures to reduce deviations from determined goals. Present publications of SCRM mainly focus on operational disturbances of material or information flow (cf. Kersten/Blecker 2006; Vahrenkamp/Siepermann 2007). Actually, SCRM is dominated by pure risks (i.e. negative deviations) and thus represents the 'traditional way' of specific risk management (cf. Mikus 2001, p. 10-13). Waters (2007) and Ziegenbein (2007) follow a structure that is quite common for a specific risk management process (identifying risks, analysing risks, and responding to risks). But these approaches have two main deficits: They disregard interrelations with opportunities and innovation, and the definition of objectives (which is a basic requirement to identify deviations) is missing. Insofar the suggested SCIRA model represents an advanced approach of SCRM, with focus on strategic issues.

2 Research Framework

This chapter provides a brief overview on the relevant management areas, which span the broad range from strategic management, risk management, and innovation management to supply chain management, business process management, and network management [fig. 2].

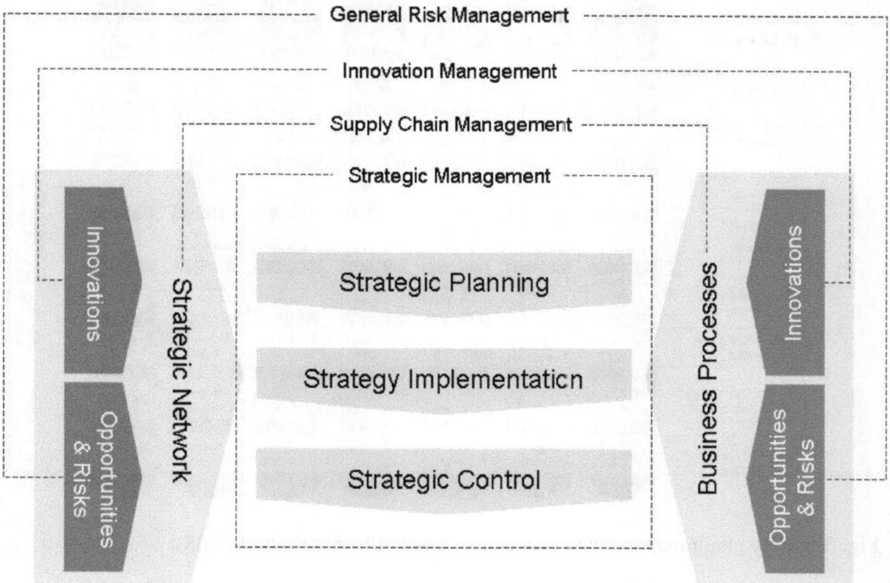

Fig. 2 Relevant areas of management (own illustration)

The intention is not to discuss these areas in detail, but to point out selected key aspects that determine the development of the SCIRA model.

2.1 SCM Is Business Process Management for Strategic Networks

The term 'supply chain management' was initially used in the early 1980s (cf. Erdmann 2003, p. 7-15; Poluha 2005, p. 29-35). Since then, SCM has been discussed in many publications with several different definitions and interpretations. It is obvious that the development of SCM is strongly driven by corporate practice (cf. Cooper et al. 1997, p. 1). Scientific research has delivered some more or less recognized explanations and models, but a generally acknowledged definition is missing so far (cf. Stewens 2005, p. 5-21). Especially the difference between logistics management and supply chain management is discussed quite controversially. Waters (2007, p. 38) points out that "the choice of terms is largely a matter of semantics". Attempts to classify SCM range from a minimum (advanced logistics) to a maximum (value creating networks) definition (cf. Konrad 2007, p. 29-59).

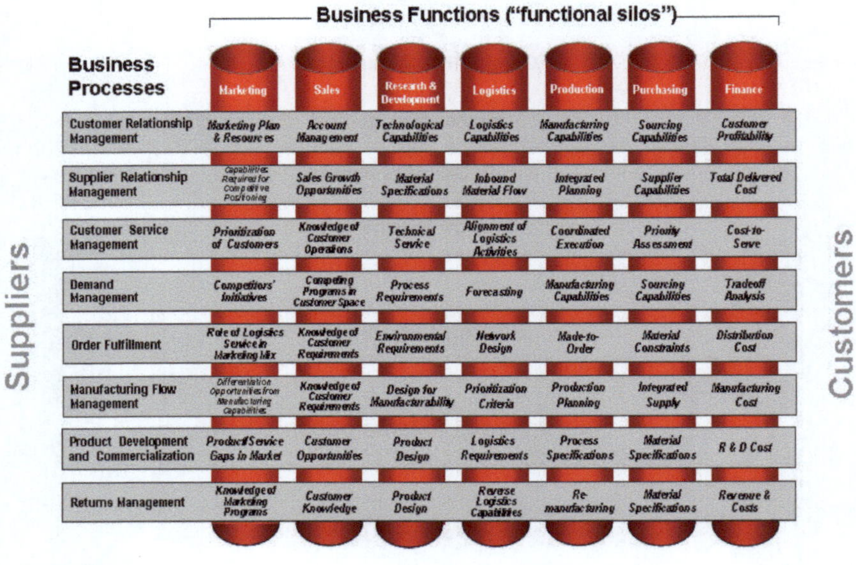

Fig. 3 Supply chain management processes (adapted from Lambert 2008)

Scientific literature offers a number of so-called reference models to explain or determine SCM (cf. Konrad 2005, p. 9-21; Corsten/Gössinger 2001, p. 124-151). One of these models is the conceptual framework by Lambert, composed of three elements: Supply chain management processes [fig. 3], supply chain network structure, and supply chain management components (cf. Lambert 2008, p. 1-24). Taking a broad view, the quintessence of SCM is to apply business process management to strategic networks.

Business process management contributes to the achievement of corporate (strategic) goals as it aligns the main value creating processes of an organisation with the demands of customers (or markets) and corporate strategy. The scope of business process management therefore is not only cross-functional, but also inter-organisational by integration of suppliers and customers (cf. Schmelzer/Sesselmann 2008, p. 4-12).

With regard to the conceptual framework of Lambert (2008, p. 9-15), eight SCM business processes have to be considered:

1. Customer relationship management,
2. Supplier relationship management,
3. Customer service management,
4. Demand management,
5. Order fulfilment,
6. Manufacturing flow management,
7. Product development and commercialisation, and
8. Returns management.

Strategic networks can be characterised by structure, by principles of coordination, and by relationships. A strategic network [fig. 4] owns a polycentric structure, but is managed by focal institutions (e.g. hub firms). The coordination of economic activities of a network takes place in an area between market (external) and hierarchy (internal). The relations between (autonomous) network partners are long-term, they require mutual trust, and they are based on cooperation rather than competition (cf. Sydow 2005, p. 78-83).

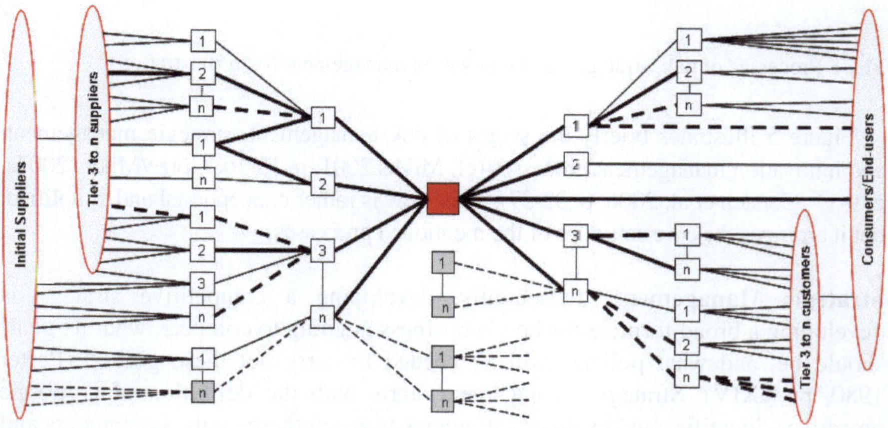

Fig. 4 Supply chain network structure (adapted from Lambert 2008)

SCM is considered as a specific type of a strategic network (cf. Corsten/Gössinger 2001, p. 81-94). “Strictly speaking, the supply chain is not a chain of businesses, but a network of businesses and relationships.” (Lambert 2008, p. 2).

Supply chain management components comprise methods by which processes are integrated and managed across the supply chain. They are classified into structural (e.g. control, workflow, knowledge) and behavioural (e.g. leadership, culture, trust) components. Compared with business processes and network structures, this part of the SCM framework appears as a broad mixture of tasks, methods, and instruments, which are deduced rather unsystematically (cf. Stölzle 1999, p. 166; Corsten/Gössinger 2001, p. 139).

2.2 Coherence of Strategy, Risk, and Innovation

Strategy, risk, and innovation represent distinctive disciplines of management science. Anyhow boundaries between strategic, risk and innovation management are overlapping. As a common feature, all of these management disciplines represent processes.

Fig. 5 Processes of risk, strategic, and innovation management (own illustration)

Figure 5 illustrates briefly the stages of risk management, strategic management, and innovation management processes (cf. Mikus 2001, p. 13-16; Götze/Mikus 2007a, p. 9-11; Corsten et al. 2006, p. 32-37). This view is rather conceptional and simplified, but it arranges the core activities of the mentioned processes.

Strategic Management. „Essentially, developing a competitive strategy is developing a broad formula for how a business is going to compete, what its goals should be, and what policies will be needed to carry out these goals.“ (Porter 1980, p. XXIV). Strategic management starts with the definition of corporate objectives, identifies and evaluates strategies to reach these goals, implements and enforces the chosen strategy, and controls the achievement of objectives (cf. Götze/Mikus 2007a; Welge/Al-Laham 2008; Bea/Haas 2005). The overall purpose of strategic management is to ensure the long-term success and existence of an organisation.

Success potentials are of significance for long-term existence, as they describe the core capabilities and competences that determine the competitive position of a corporation. From Gälweilers (2005, p. 26-35) point of view, success potentials represent the key parameters to ensure success on strategic level, because of their 'pre-regulating' function [fig. 6].

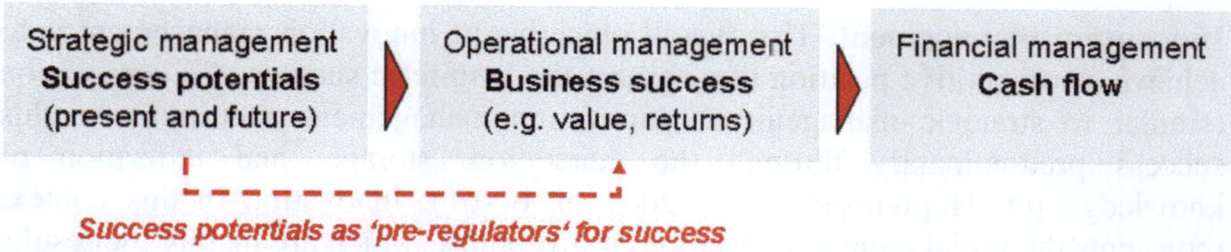

Fig. 6 Success potentials (own illustration based on Gälweiler 2005)

Effective management (i.e. arranging, maintaining, and developing) of success potentials is a vital requirement for long-term existence, and success potentials also build the foundation to determine corporate objectives within strategic management (cf. Götze/Mikus 2007a, p. 13-22; Welge/Al-Laham 2008, p. 213-220).

Risk Management. Risk is a consequence of decisions under uncertainty, and risk causes deviations (negative as well as positive) from defined objectives. Literature distinguishes between two dimensions of risk management.

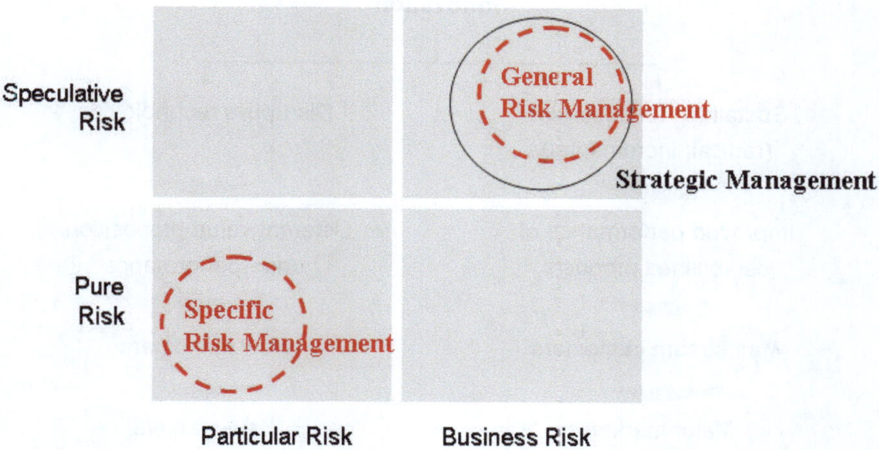

Fig. 7 Dimensions of risk management (adapted from Mikus 2001)

While specific risk management covers pure and particular risk, general risk management is concerned with speculative risk (which means opportunities, too) from an overall business perspective (cf. Mikus 2001, p. 9-13).

General risk management serves the same purpose as strategic management, and hence it is considered as an integral part of it [fig. 7]. The limitation of risk and the utilisation of opportunities both contribute to success and growth of corporations. With general risk management, the scope of risk management is expanded towards speculative risk and overall business risk. General risk management is a comprehensive managerial function that covers not only pure risk aspects, but corporate management as a whole (cf. Strohmeier 2007, p. 46-47).

Innovation management. The overall objective of innovation management is to achieve a competitive position that ensures the long-term success of a corporation (similar to strategic management). Innovation management contributes to this success predominantly through the generation, storage, and utilisation of knowledge (cf. Hopfenbeck et al. 2001, p. 60-61). Innovation in this context represents the initial commercialisation of inventions, which are mainly the results of applied knowledge combined with research and development (R&D) activities (cf. Corsten et al. 2006, p. 10-13).

Regarding types of innovation, literature offers a variety of categorisations. For example, OECD (2005, p. 47-52) distinguishes between four main types of innovation: product innovation, process innovation, marketing innovation, and organisational innovation. In this context, the introduction of SCM is considered as kind of organisational innovation. „However, many innovations may have characteristics that span more than one type of innovation." (OECD 2005, p. 53).

It has been mentioned before that innovation is inherently uncertain, and risks occur as a consequence of decisions under uncertainty (cf. Mikus 2001, p. 5; Waters 2007, p. 17). The extent of uncertainty and risk affecting innovation is closely related to the chosen path of innovation.

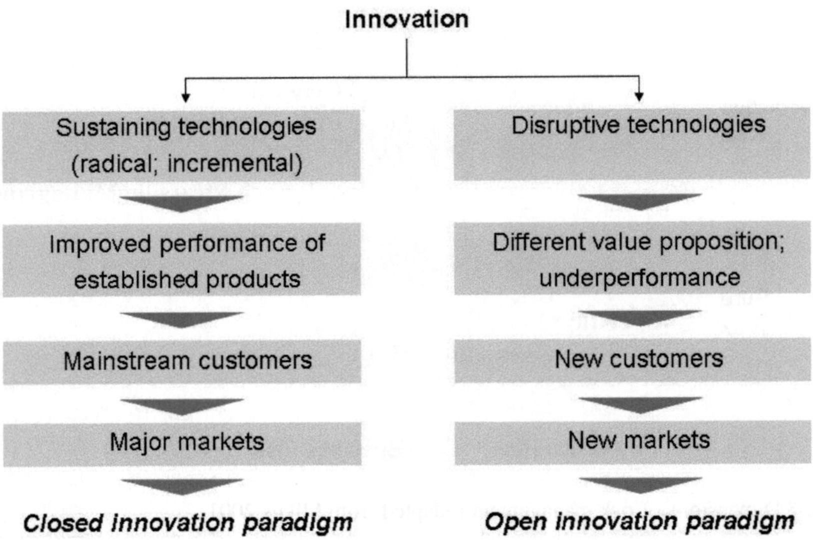

Fig. 8 Innovation paradigms (own illustration based on Chesbrough 2006)

Chesbrough explains the ongoing paradigm shift from closed to open innovation. "Open innovation is a paradigm that assumes that firms can and should use external ideas as well as internal ideas, and internal and external paths to market, as the firms look to advance their technology." (Chesbrough 2006, p. XXIV). Open innovation is characterised by a broad variety of knowledge sources, different value propositions, unknown demand, and new markets [fig. 8]. Consequently, incidence and impact of uncertainty and risk become importantly serious compared with the 'static' closed innovation path.

In summary, the mentioned management disciplines appear collectively somehow 'strategic', because their common purpose is the long-term success and existence of a corporation. They represent the theoretical background of the SCIRA model, which attempts to integrate and synchronise a selection of methods and instruments.

3 Blueprint of the SCIRA Model

This chapter outlines the Supply chain innovation and risk assessment (SCIRA) model. The (prescriptive) process of strategic management (composed of strategic planning, strategy implementation, and strategic control) represents the basic principle of the model [fig. 9].

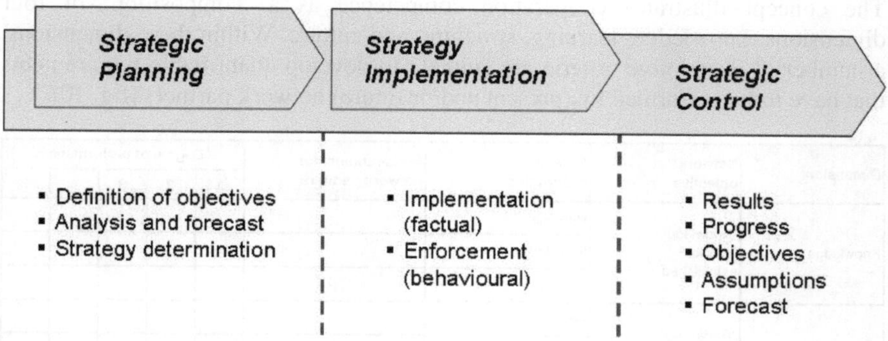

Fig. 9 Strategic management process (own illustration based on Götze/Mikus 2007a)

Current SCRM approaches do not consider business objectives and they disregard interrelations between risk, opportunities, and innovation. Exactly these issues are key functions of strategic planning, comprising the definition of objectives, the analysis of internal and environmental factors, and the derivation of strategies to achieve determined goals (cf. Götze/Mikus 2007a, p. 13-50). Accordingly, the study focuses on strategic planning activities at this point.

3.1 Identifying Success Potentials of SCM

Success potentials are pre-regulators for success in business, and thus they are appropriate factors to define business objectives (cf. Gälweiler 2005, p. 26-35; Götze/Mikus 2007a, p. 13-22; Welge/Al-Laham 2008, p. 213-220). If SCM represents business process management for strategic networks, it appears quite obvious that relevant success potentials have to be directly related to the quality of processes and the ability to cooperate within networks.

3.1.1 Cooperation Competences

SCM is rather cooperative than competitive, and thus the ability to cooperate is a key precondition for successful corporate networks. „Alliances between companies, whether they are from different parts of the world or different ends of the supply chain, are a fact of life in business today. (…) Whatever the duration and objectives of business alliances, being a good partner has become a key corporate asset." (Kanter 1995, p. 98). Cooperation competences are not hard facts, but soft skills including knowledge, relationships, behaviour, and many other aspects of interaction between humans and organisations. The challenge is to develop a method to translate these amorphous characteristics into a measureable structure. Suggestions to define and quantify cooperation competences in scientific literature are rare (cf. Ross 2006, p. 63-67). A comprehensive approach is represented by the meta-competences framework of Landt (2009, p. 160-180). The concept illustrates cooperation competence as a 'composition' of four dimensions: knowledge, learning, structure, and culture. Within these dimensions, a number of descriptive criteria are suitable to develop quantifiable requirements that have to be performed by (present and/or future) network partners [fig. 10].

Dimension	Network objective	Criteria of description	Requirements for network partners	Degree of performance				
				1	2	3	4	5
Knowledge	"Knowledge-logistics" established	Persons						
		Knowledge						
		Transfer						
Learning	"Radar" for change implemented	Determination						
		Receptivity						
		Transparency						
Structure	Independency and collaboration balanced	Direction						
		Objectivity						
		Rules						
		Relationships						
Culture	Different cultures integrated	Affectivity						
		Cognitivity						
		Behaviour						

Fig. 10 Criteria of cooperation competence (own illustration based on Landt 2009)

The application of this framework allows evaluating the degree of performance for any individual partner within a network, using selected criteria and decisive requirements. The range of performance indication is from 1 (initial) up to 5 (completely performed), similar to established approaches of capability levels. As a result of the evaluation, the present cooperation competence profile (CCP) of a network partner evolves, pointing out individual strength and weaknesses. To prepare the definition of objectives, a target CCP is developed, and consequently deviations between target and present profiles can be identified and analysed.

3.1.2 Process Capability and Maturity Levels

Porter (1985) introduced the value chain as a method to analyse business processes of a corporation with regard to identify their contribution to competitive advantage. „The value chain disaggregates a firm into its strategically relevant activities in order to understand the behavior of costs and the existing and potential sources of differentiation.“ (Porter 1985, p. 33). If business processes are strategically relevant for a corporation, SCM processes similarly are relevant for corporate networks. Quality, effectiveness and efficiency of business processes can be estimated by capability and/or maturity models. Literature offers a variety of methods and concepts (for example GPM, SPICE, PMMA; cf. Schmelzer/Sesselmann 2008, p. 315-340). For an application to service industries, the Capability Maturity Model Integration (CMMI-SVC) was developed by Software Engineering Institute (SEI). It suggests 24 areas to improve process maturity, from which five areas are originally dedicated to process management itself (process definition, process focus, training, process performance, innovation and deployment; cf. SEI 2009, p. 30-38). CMMI offers two alternative approaches to use the model, either with focus on capabilities (continuous), or with focus on maturity levels (staged). “Capability levels apply to an organization’s process improvement achievement in individual process areas. These levels are a means for incrementally improving the processes corresponding to a given process area.” (SEI 2009, p. 21). Maturity levels on the other hand apply to a predefined set of process areas and offer an extensive methodology to improve the overall performance and maturity of an organisation (cf. SEI 2009, p. 26).

Level	Capability levels (CL) continuous	Maturity levels (ML) staged
0	Incomplete	(not applicable)
1	Performed	Initial
2	Managed	Managed
3	Defined	Defined
4	Quantitatively Managed	Quantitatively Managed
5	Optimizing	Optimizing

Fig. 11 Capability and maturity levels (adapted from SEI 2009)

A common feature of both approaches is the definition of levels, which represent the degree of improvement in process quality [fig. 11]. They range from 0 (incomplete; only applicable for capabilities) to 5 (optimizing). “Reaching capability level 5 for a process area assumes that you have stabilized the selected subprocesses and that you want to reduce the common causes of variation in that process." (SEI 2009, p. 25). Both approaches (continuous and staged) provide a method to measure process quality and to improve processes. While capability levels (CL) are applied to a selection of individual process areas (for example process management for defined SCM processes), the achievement of maturity levels (ML) follows comprehensive regulations.

3.1.3 Objectives and Deviations

The given success potentials provide a sound basis for the derivation of objectives. In this sense, objectives represent observable and measurable results to be achieved within a determined timeframe. Hence, defined objectives are a necessary precondition to identify positive as well as negative deviations (cf. Götze/Mikus 2007b, p. 31-34).

In view of cooperation competence, the identified profile (CCP) is translated into a cooperation competence level (CCL). This figure represents the arithmetical average at a range from 1 up to 5, and consists of the preceding performance indications (cf. Ross 2006, p. 177-183). Having defined the CCL, the following activities are:

- Identification of relevant network partners,
- Specification of target CCL for individual network partners,
- Comparison of target CCL and present CCL, and
- Assessment of deviations, analysis of causes, and determination of responding measures.

For process capability/maturity, the continuous as well as the staged approach are suitable. The preferable application depends on the extent and complexity of the evaluated network, represented by the number of partners and the scope of processes to improve. Generally, both approaches enable to compare target levels of capability or maturity with present levels. As far as capability levels for selected SCM processes are concerned, target levels for individual processes have to be determined. Subsequently, the present achievement of capability levels is evaluated, deviations are analysed, and measures are determined.

Additionally, possible interdependences between cooperation competence and process capability/maturity may be cross-checked. If an individual network partner is the main driver (or owner) of a specific process, cooperation competence and process maturity can be closely related to each other. First, distinct connections between partners and processes have to be identified. That followed, target and present levels of cooperation competence and process maturity are compared, and correlations are identified. If a partner/process-connection for example features low levels of both, cooperation competence and process maturity, an in-depth analysis appears essential. If otherwise both levels are high, analysis should also take place in order to find out and describe the characteristics of successful practice.

3.2 Impact Assessment of Innovation

Taking general risk management into account, deviations from objectives may be positive as well as negative. How does innovation influence the success potentials of SCM?

3.2.1 Overlapping Innovation Processes

Literature offers a variety of suggestions to structure and describe the innovation process. A conceptional differentiation between stages of idea generation, idea acceptance, and idea realisation is generally accepted [fig. 12]. Some approaches suggest adding an initial stage and a diffusion stage (cf. Corsten et al. 2006, p. 32-37).

Fig. 12 Innovation process (own illustration based on Corsten et al. 2006)

The initiation of innovation can be characterised as intuitive without coordination and systematic planning. Based on a more or less random identification and analysis of opportunities, R&D projects are initiated to study these opportunities in more detail. The main task of idea generation is to research and to discover solutions for identified opportunities. This stage is dominated by creativity and draws conclusions from R&D projects. The acceptance of ideas shows certain similarities to risk management. At this stage, solutions are examined and possible consequences in case of realisation are estimated. Additionally, uncertainties are evaluated in order to select an appropriate solution. The stage of idea realisation focuses on the first commercial utilisation of a solution. The initial market positioning of a new product or service requires systematic coordination of activities (e.g. marketing, timing strategies). Finally, diffusion means the saturation of markets through penetration or imitation (cf. Corsten et al. 2006, p. 32-37).

Regardless of the applied structure, the process of innovation does not follow a linear continuum. Hence, the boundaries between stages are in motion. Pavitt (2006, p. 109) points out that "most innovation processes are overlapping and intertwined, terms like 'stages' or 'phases' impose an unrealistic linearity on the various innovation processes." Despite this, the conceptual differentiation provides a suitable indication to analyse the stages of an innovation process.

3.2.2 Interrelations with SCM Success Potentials

The impact of innovation on cooperation competence and process maturity is apparently variable, depending on the stages of the innovation process. Accordingly, an analysis of interrelations has to take the different characteristics of stages into account. Portfolio technique (cf. Bea/Haas 2004, p. 136-140) appears as an appropriate instrument to estimate the interdependences between success potentials and innovation. Portfolios have to be designed for both success potentials [fig. 13], one regarding the cooperation competence level (CCL), and the other related to the process maturity level (PML). As mentioned before, an alignment with process capabilities is alternatively possible.

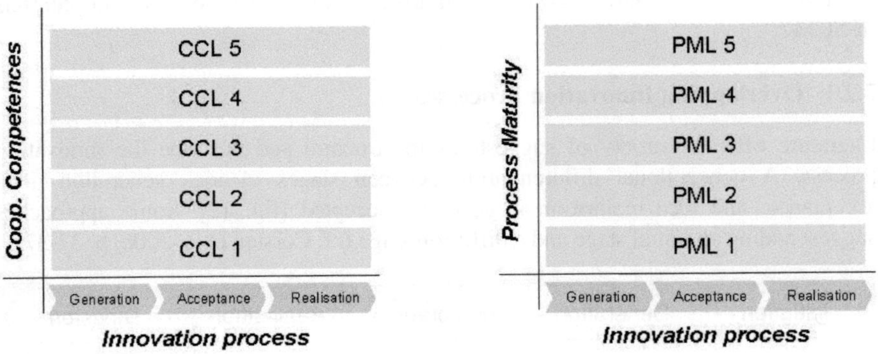

Fig. 13 Interrelations of innovation and success potentials (own illustration)

During the stage of idea generation, the impact of innovation activities on process maturity is low, because the actual generation of ideas does not influence the arrangement of established supply chain processes. Regarding cooperation competences, the impact of innovation is low, too. But vice versa, the capabilities of a network to research and discover opportunities by using its inherent knowledge is depending on cooperation competences to great extent.

The acceptance of ideas involves examination of possible solutions and estimation of consequences, in order to minimise risk and to utilise opportunities. The influence of innovation on process maturity at this stage is rather low. But the awareness about process arrangements within the network is a necessary precondition to estimate opportunities on process level as well as risks. If an innovation causes fundamental changes in business processes, the balance between opportunities and risks has to be evaluated carefully. Regarding cooperation competences, the actual acceptance of ideas does not cause significant impact. But cooperation competence is a relevant measure to estimate if and how network partners and the network as a whole will be able to manage opportunities and risks, caused by innovation.

At the stage of idea realisation, change takes place. On the one hand, established (and hence mature) business processes have to be re-arranged to implement an innovation. Consequently, the impact of idea realisation on process maturity is high. On the other hand, realisation of ideas influences the structure of a network, and thus the cooperation competence within the network changes. For example, the integration of new network partners (or the replacement of present partners) causes changes in knowledge, learning, structure, and culture.

The impact of innovation during idea generation and acceptance is rather low, but process maturity and cooperation competence are preconditions to generate and accept ideas in a successful and responsible way. At the stage of idea realisation, innovation means change, and therefore its influence on process maturity and cooperation competence is significantly high. And thus innovation can cause risks as well as opportunities.

3.3 (Re-)Configurating the Supply Chain

Cooperation competence and process maturity are considered as key success potentials for SCM. Effective management of these success potentials is a precondition for long-term existence of a value network (SCM). From a strategic point of view, achieving the best fit of network partners and business processes represents the way to lasting competitive advantages of a network. For a continuous (re-)configuration of the network, both dimensions of success potentials have to be evaluated. This can be done by analysing present network partners in view of their cooperation competence and with regard to the maturity of processes owned or mainly driven by network partners. For example, network partners providing high cooperation competence in combination with high process maturity are of best fit for extended business relationships. If cooperation competence is high, but process maturity low, a network partner should be actively supported to enhance process quality.

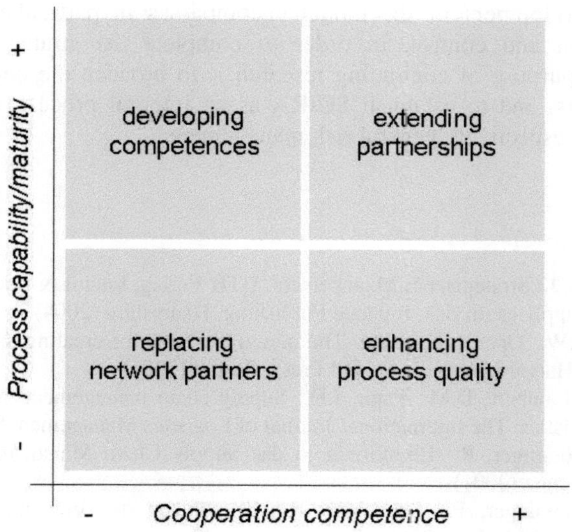

Fig. 14 Configuration portfolio (own illustration)

The configuration portfolio [fig. 14] provides the decisive background to achieve the aspired fit of partners and processes. An assessment of network partners should be arranged and carried out by the focal institution of the supply chain, in order to apply a consistent standard for evaluation.

4 Conclusion and Future Prospects

The conceptional findings of the SCIRA model can be summarised as follows:

- SCM represents the application of business process management to strategic networks.
- Related key success potentials of SCM are process maturity (or capability) and cooperation competence.
- Process maturity and cooperation competence are adequate variables to define objectives for SCM.
- Innovation influences both, process maturity as well as cooperation competence.
- Interdependence of innovation, process maturity, and cooperation competence causes opportunities as well as risk.
- Assessing the interrelation of innovations, opportunities, and risks is a vital issue of strategic risk management for corporate networks.

Based on these insights, SCIRA introduces a practice to define objectives for supply chains, and an approach to manage the tension between innovation, opportunities and risks. Accordingly, the model represents an advancement of supply chain risk management on a strategic level.

Further development of the model encompasses in particular the stages of implementation and control, in order to complete the strategic management process. The purpose of continuing research is to broaden the concept, to refine the instruments, and to establish SCIRA as an integral process within strategic management, respectively general risk management.

References

Bea, F.X., Haas, J.: Strategisches Management. UTB Verlag, Lucius & Lucius (2005)
Brindley, C.: Supply chain risk. Ashgate Publishing, Hampshire (2004)
Chesbrough, H.W.: Open Innovation. The new imperative for creating and profiting from technology. Harvard Business School Press, Boston (2006)
Cooper, M.C., Lambert, D.M., Pagh, J.D.: Supply chain management. More than a new name for logistics. The International Journal of Logistics Management 8(1), 1–14 (1997)
Corsten, H., Gössinger, R.: Einführung in das Supply Chain Management. Oldenbourg Verlag, München (2001)
Corsten, H., Gössinger, R., Schneider, H.: Grundlagen des Innovationsmanagements. Verlag Vahlen, München (2006)
Erdmann, M.K.: Supply chain performance measurement. Operative und strategische Management- und Controllingansätze. Josef Eul Verlag, Lohmar (2003)
Fagerberg, J., Mowery, D.C., Nelson, R.R.: The Oxford handbook of innovation. Oxford University Press (2006)
Gälweiler, A.: Strategische Unternehmensführung. Campus Verlag, Frankfurt (2005)
Gassmann, O., Kobe, C.: Management von Innovation und Risiko. Quantensprünge in der Entwicklung erfolgreich managen. Springer, Berlin (2006)

Gassmann, O.: Innovation und Risiko - zwei Seiten einer Medaille. In: Gassmann, O., Kobe, C. (eds.) Management von Innovation und Risiko, pp. 3–24. Springer, Berlin (2006)

Götze, U., Mikus, B.: Risikomanagement mit Instrumenten der strategischen Unternehmensführung. In: Götze, U., Henselmann, K., Mikus, B. (eds.) Risikomanagement, pp. 385–412. Physica Verlag, Heidelberg (2001)

Götze, U., Mikus, B.: Strategisches Management. Verlag der GUC, Chemnitz (2007a)

Götze, U., Mikus, B.: Der Prozess des Risikomanagements in Supply Chains. In: Vahrenkamp, R., Siepermann, C. (eds.) Risikomanagement in Supply Chains, pp. 29–58. Erich Schmidt Verlag, Berlin (2007b)

Hopfenbeck, W., Müller, M., Peisl, T.: Wissenbasiertes Management. Verlag Moderne Industrie, Landsberg (2001)

Kanter, R.M.: Collaborative advantage: The art of alliances. In: Harvard Business Review on Strategic Alliances (2002), pp. 97–128. Harvard Business School Press, Boston (1994)

Kersten, W., Blecker, T. (eds.): Managing risks in supply chains. Erich Schmidt Verlag, Berlin (2006)

Konrad, G.: Theorie, Anwendbarkeit und strategische Potenziale des Supply Chain Management. Deutscher Universitäts-Verlag, Wiesbaden (2005)

Lambert, D.M.: An executive summary of supply chain management. Processes, partnerships, performance. SCMI, Sarasota (2008)

Landt, N.: Kooperationskompetenz als Metakompetenz. Ein mehrdimensionaler Bezugsrahmen. Verlag der GUC, Chemnitz (2009)

Mikus, B.: Risiken und Risikomanagement - ein Überblick. In: Götze, U., Henselmann, K., Mikus, B. (eds.) Risikomanagement, pp. 3–28. Physica Verlag, Heidelberg (2001)

Organisation for Economic Cooperation and Development, OECD. Oslo Manual. Guidelines for collecting and interpreting innovation data. Joint publication of OECD and Eurostat (2005)

Paulsson, U.: Supply chain risk management. In: Brindley, C. (ed.) Supply Chain Risk, pp. 79–96. Ashgate Publishing, Hampshire (2004)

Pavitt, K.: Innovation processes. In: Fagerberg, J., Mowery, D.C., Nelson, R.R. (eds.) The Oxford Handbook of Innovation, pp. 86–114. Oxford University Press (2006)

Poluha, R.G.: Anwendung des SCOR-Modells zur Analyse der Supply Chain. Explorative Untersuchung von Unternehmen aus Europa, Nordamerika und Asien. Josef Eul Verlag, Lohmar (2005)

Porter, M.E.: Competitive strategy. Techniques for analysing industries and competitors. First free press export edition 2004. Free Press, New York (1980)

Porter, M.E.: Competitive Advantage. Creating and sustaining superior performance. First free press export edition 2004. Free Press, New York (1985)

Powell, W.W., Grodal, S.: Networks of innovators. In: Fagerberg, J., Mowery, D.C., Nelson, R.R. (eds.) The Oxford Handbook of Innovation, pp. 56–85. Oxford University Press (2006)

Röpke, J., Stiller, O.: Einführung zur Theorie der wirtschaftlichen Entwicklung. In: Schumpeter, J.A. (ed.) (printing of 1912 edition) Theorie der Wirtschaftlichen Entwicklung, p. V–XLIII. Verlag Duncker und Humblot, Berlin (2005)

Ross, A.: Wertsteigerung durch Netzwerkkompetenz. Konzeption und praktisches Vorgehen. Eul Verlag, Lohmar (2006)

Schmelzer, H.J., Sesselmann, W.: Geschäftsprozessmanagement in der Praxis. Carl Hanser Verlag, München (2008)

Software Engineering Institute, SEI. CMMI for Services, Version 1.2. Carnegie Mellon University (2009)

Schumpeter, J.A.: The theory of economic development. An inquiry into profits, capital, credit, interest, and the business cycle. 14th printing 2008. Transaction Publishers, New Brunswick (1934)

Stewens, M.: Gestaltung und Steuerung von Supply Chains. Josef Eul Verlag, Lohmar (2005)
Stölzle, W.: Industrial Relationships. Oldenbourg Verlag, München (1999)
Strohmeier, G.: Ganzheitliches Risikomanagement in Industriebetrieben. Grundlagen, Gestaltungsmodell und praktische Anwendung. Deutscher Universitäts-Verlag, Wiesbaden (2007)
Sydow, J.: Strategische Netzwerke. Evolution und Organisation. Gabler Verlag, Wiesbaden (2005)
Vahrenkamp, R., Siepermann, C. (eds.): Risikomanagement in Supply Chains. Erich Schmidt Verlag, Berlin (2007)
Waters, D.: Supply chain risk management. Kogan Page, London (2007)
Welge, M.K., Al-Laham, A.: Strategisches Management. Grundlagen, Prozess, Implementierung. Gabler Verlag, Wiesbaden (2008)
Ziegenbein, A.: Supply Chain Risiken. Identifikation, Bewertung und Steuerung. VDF Hochschulverlag, Zürich (2007)

Supply Chain Safety: A Diversification Model Based on Clustering

Andreas Brieden[1], Peter Gritzmann[2], and Michael Öllinger[3]

[1] Universität der Bundeswehr München,
Inhaber der Professur für Statistik, insbesondere Risikomanagement
(English: Statistics and Risk Management), Werner-Heisenberg-Weg 39,
85577 Neubiberg, Germany
Andreas.brieden@unibw.de

[2] Technische Universität München, Zentrum Mathematik, Lehrstuhl für Angewandte Geometrie und Diskrete Mathematik (English: Applied Geometry und Discrete Mathematics),
Boltzmannstr. 3, 85747 Garching, Germany
gritzman@ma.tum.de

[3] Universität der Bundeswehr München, Wissenschaftlicher Mitarbeiter an der Professur für Statistik, Insbesondere Risikomanagement, Werner-Heisenberg-Weg 39,
85577 Neubiberg, Germany
michael.oellinger@unibw.de

Abstract. The issue of supply chain safety has received broad attention which has led to a wide range of methodologically different approaches; for a survey see (Pfohl, Köhler & Thomas, 2010). The present paper introduces a novel quantitative algorithm that provides a multiple covering of the commodity graph via constrained clustering. In fact, we construct supply chain components in the overall supply network of a company, each being able to account for some percentage of the company's overall production. They are all isomorphic to and can hence be viewed as different realizations of the commodity graph which are most independent with respect to known hazards. Consequently, suppliers (of each level) are assigned to supply chain components so as to minimize the probability for a total (or severe enough) breakdown. The basic new model is given in detail, complemented by an outline of more involved ramifications that are able to deal with realistic scenarios. Also, we give computer simulations that indicate the favorable behavior already of our basic model in terms of risk reduction.

1 Introduction

The 9.0 earthquake followed by a tsunami that hit the east cost of Japan on March 11, 2011, and the subsequent nuclear fallout were a tremendous humanitarian disaster. Naturally, it also had severe direct regional and national economic consequences. There are, however, also substantial global implications due to the breakdown of global supply chains. A large German company estimated its related economic loss to more than 200 million Euros. While the total worldwide economic effect of Japan's recent natural catastrophe is still not fully accounted for, the effect of the much smaller '*Albuquerque accident*' on March 18, 2000, for Ericsson has been well analyzed; see

M. Essig et al. (Eds.): Supply Chain Safety Management, LNL, pp. 327–355.
springerlink.com

e.g. (Norrman & Jansson, 2004). A small fire at a sub-supplier's factory was made responsible for a loss of about 400 million US-Dollars and was at least partly decisive for the company's withdrawal from selling mobile phones. Nokia was also affected by the fire, but in contrast to Ericsson was able to obtain the component through alternative sources. Hence robustness of supply chains against external disruptions is not only of theoretical interest. As a matter of fact, total dependence on one supplier or sub-supplier may lead to enormous economic losses.

Consequently, various qualitative and quantitative methods for increasing supply chain safety have been developed, ranging from fuzzy set theory, see e.g. (Yang, Wang, Bonsall, Yang & Fang, 2005) to neural network approaches, see e.g. (Teuteberg, 2008). For a critical review see (Klibi, Martel & Guitouni, 2010).

There are various papers on the conceptual and normative level of the management of supply chain risks, see e.g. (Jüttner, Peck & Christopher, 2003) or (Peck, 2005), while others use qualitative research methods, see e.g. (Svensson, 2002), (Zsidisin & Ellram, 2003), (Jüttner, 2005) or (Zsidisin & Wagner, 2010). There is, however, need for quantitative methods. In particular, (Wagner & Neshat, 2010) emphasize that "in line with the frequently cited business wisdom 'You can't manage, what you don't measure' supply chain managers need support in quantifying and thus mitigating supply chain [risks]." Moreover, a look into the relevant (i.e., purchasing-, logistics-, and SCM-related) literature shows that – with regard to the use of quantitative research methods – only few publications can be identified which choose simulations and mathematical models as an alternative to large-scale surveys. In analyzing various quantitative models that deal with supply chain risks, (Tang, 2006a), for instance, points out that the existing quantitative models primarily focus on managing operational risks rather than on disruption risks. Consequently, he identifies the need for research on the demand and supply process, on the appropriate objective functions and on appropriate supply, demand, product, and information management strategies. (Deane, Craighead & Ragsdale, 2009) develop a model which allows organizations to mitigate two key global risks – environmental and density risks. The model is based on the idea that by adjusting the values of the underlying model parameters the procuring organization can limit its potential exposure to individual supplier failures and directly control the number of sources of supply. Thus, organizations will be able to employ and take advantage of sourcing strategies while simultaneously reducing the negative effects of environmental and proximity types of disruptions by increasing geographic dispersion of the supply base and selecting suppliers from different regions. (Ravindran, Bilsel, Wadhwa & Yang, 2010) develop two different types of multi-criteria supplier selection models incorporating supplier risk and apply them to a real organization. The risk-adjusted supplier selection problem is modeled as a multi-criteria optimization problem and is solved in two steps. Step 1 serves as a pre-qualification, where a large set of initial suppliers is reduced to a smaller set of manageable suppliers using various multi-objective ranking methods. Step 2 focuses on the allocation of order quantities among the short listed suppliers by using a multi-objective optimization model. Methods from stochastic optimization are also used in (Santoso, et al., 2005), (Goh, et al., 2005), and (Schütz, et al., 2009).

Of course, quantitative strategies require quantitative information on the structure and the (potential) dynamics of the underlying supply network but also on the kinds and probabilities of possible disturbances. As a diversification of the suppliers typically involves economic costs, and, in some sense, can be viewed as an insurance plan against a certain degree of failure with a certain probability, there is a trade-off between maximizing economic profit and minimizing economic risk. Naturally, in a risk-free environment, one would aim at lowest possible production cost, hence, in particular, at maximal quantity discounts, thus minimize the number of suppliers of items. On the other hand, companies with a positive degree of risk aversion may prefer supply chains with some amount of redundancies. In any case, supply chain risk and supply chain performance are not independent of each other, see (Wagner & Bode, 2008) for an ordinary least square regression to empirically quantify this dependence.

One way to improve the performance of a supply chain while taking redundancy into consideration is to enhance its resiliency to the threats and occurrence of supply chain disruptions, see e.g. (Christopher & Peck, 2004), (Sheffi & Rice, 2005) and (Tang, 2006b). Supply chain resiliency is a comparatively new concept emerging from events such as fuel protests in 2000 or foot-and-mouth-disease in 2001, see e.g. (Tandler & Eßig, 2011). It is targeted on enabling a supply chain which has been affected by a disruption to rapidly return to its normal performance levels, see e.g. (Christopher & Peck, 2004, p. 2, 7, 10), (Ponomarov & Holcomb, 2009, p. 124) and (Zsidisin & Wagner, 2010, p. 3). Supply chain resiliency can be achieved through redundancy (e.g., safety stock, diversification of suppliers, etc.) or through flexibility (e.g., distribution and customer-facing activities, control systems, etc.), see e.g. (Christopher & Peck, 2004) and (Sheffi & Rice, 2005). Whereas flexibility includes the creation of capabilities and infrastructure, which are used during normal operations, redundancy includes the creation of capacities that are needed in case a disruption occurs, see (Sheffi & Rice, 2005, p. 44).

In this article we regard supply chains as the multiple realization of the commodity graph of a company. Each supply chain component is a subgraph of the supply network that is isomorphic to the commodity graph. Hence the supply chain of a company falls into different components, each of which accounts for a certain fraction of the company's overall production. As different critical events like earthquakes, strikes or fires will typically affect different suppliers differently we are aiming at supply chains whose constituting components are as independent as possible with respect to the different risks.

We will (at a somewhat informal level) introduce a mathematical model for identifying supply chains with minimal probability for a given degree of failure based on constrained clustering. Of course, as is true for any quantitative model, our approach is based on exact or at least estimated data on suppliers and events. It allows, however, to adjust the degree of depth and granularity. In particular, the model may be restricted to major commodities or to a restricted number of supply levels. For the simplicity of the exposition we restrict ourselves here to a basic version of the new model which, in particular, is based on complete information.

Our paper is organized as follows. We will begin with a simple example in Section 2. In Section 3 we then give a basic version of our mathematical model (informally, but in some detail) and indicate ramifications that can be accommodated in a more general algorithm. Section 4 provides results of computer simulations for Bernoulli-distributed and independent risks that indicate the practical performance of the algorithm. Section 5 contains some final remarks.

2 An Example

We begin with a simple example to illustrate the underlying concept. Suppose a company manufactures an end product for which two major intermediate products are required (Commodity 1 and Commodity 2). While Commodity 1 relies on two resources (Commodity 3 and Commodity 4), only one ingredient (Commodity 5) is required for Commodity 2. Figure 1 depicts the underlying *commodity graph*.

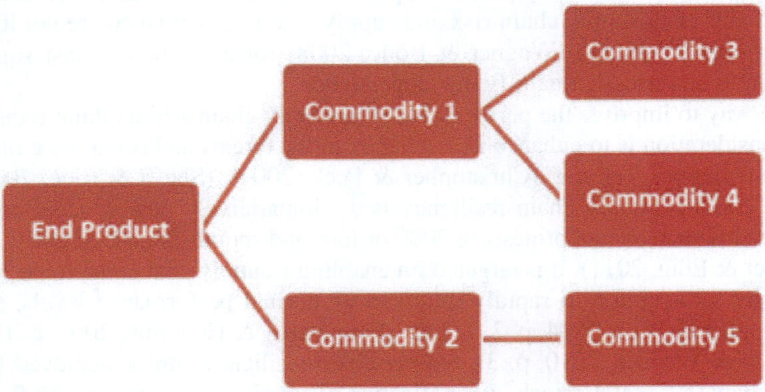

Fig. 1 Commodity graph

For each component there are different possible suppliers available; see Figure 2.

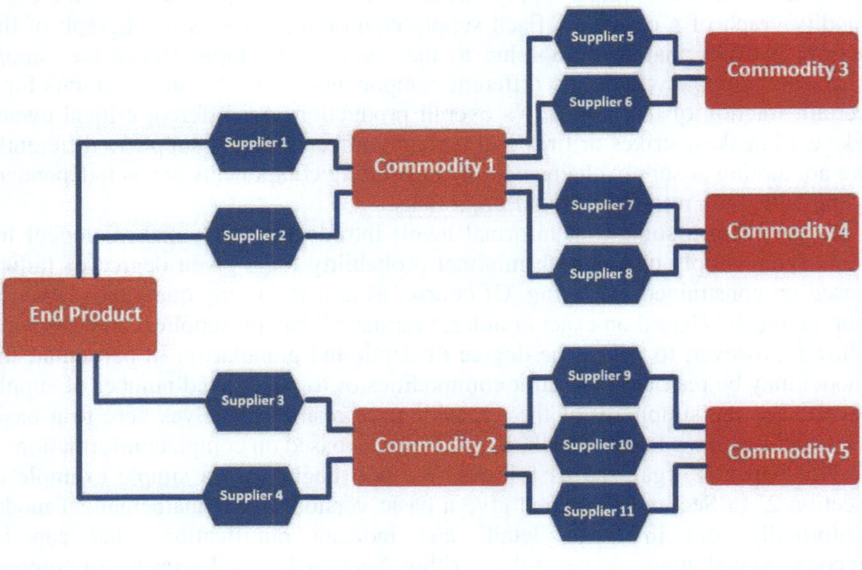

Fig. 2 Supply network

Hence the *supply network* contains the complete current supply structure, i.e., the dependencies on the level of commodities as well as all possible suppliers. Note that the suppliers correspond to the edges of the commodity graph. A possible *supply chain component* is now obtained by choosing one of the available suppliers for each edge of the commodity graph. Figure 3 shows one possible supply chain component.

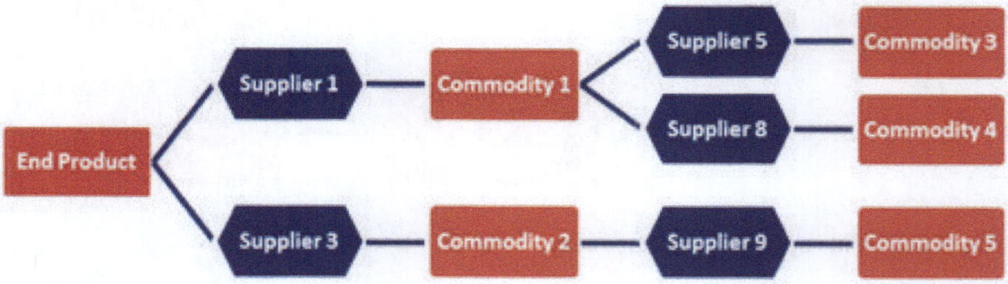

Fig. 3 Supply chain component

Clearly, this supply chain component is a realization of the commodity graph, and, as a matter of fact, could also be depicted as in Figure 4.

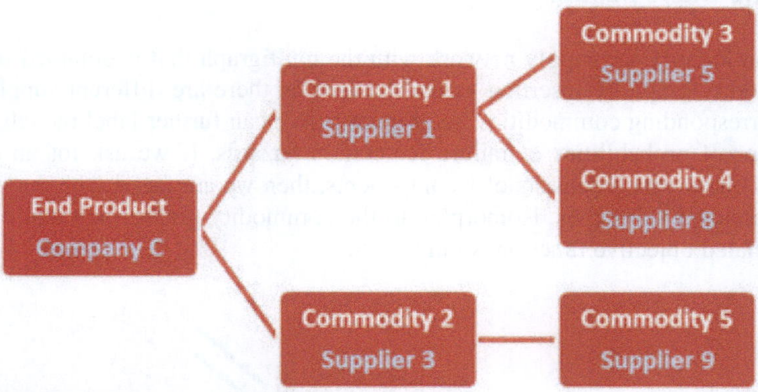

Fig. 4 Different depiction of a supply chain component

Now suppose that, in the case of an earthquake, Supplier 5 is completely shut down. Then the supply chain component of Figures 3 or 4 breaks down. If both Suppliers 5 and 8 are affected by the earthquake, resulting in a simultaneously total failure, the result is, of course, the same. Hence in terms of the risk posed by an earthquake, having Supplier 8 in the same supply chain component as Supplier 5 does not further increase the blackout probability. Now suppose that all other suppliers are earthquake-safe, but are possibly affected by other independent hazards. Then selecting a second supply chain component according to Figure 5 will result in a total supply chain (the union of both) that is only partially affected by an earthquake.

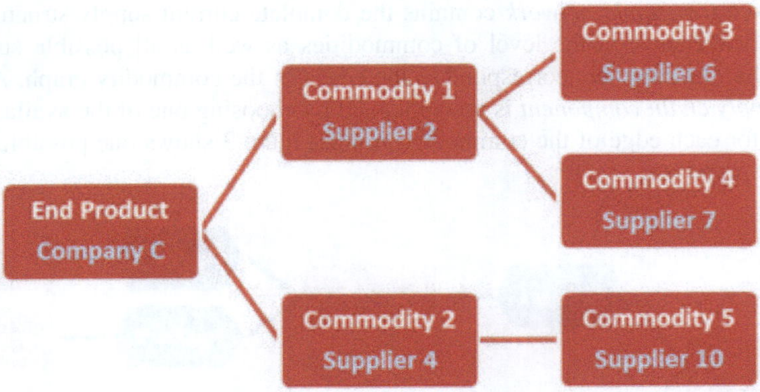

Fig. 5 Different supply chain component

The idea underlying our model is now to construct a supply chain that is composed of a suitable number of supply chain components that are as independent as possible with respect to the various hazards under consideration.

3 The Basic Model

We can identify the supply network with the multigraph that is obtained from the commodity graph by inserting as many edges as there are different suppliers for the corresponding commodities; see Figure 5. We can further label the edges with a vector of probabilities of failure for certain hazards. If we ask for an optimal supply chain which consists of *k* components, then we are faced with the problem of selecting *k* subgraphs, isomorphic to the commodity graph, such that a certain risk-related objective function is minimized.

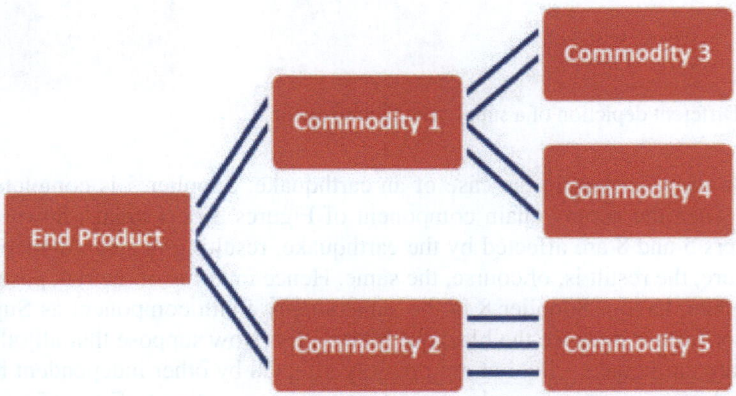

Fig. 6 Commodity-supplier multigraph

In the following we will outline a simplified basic packing model. The example of Section 2 will be used as a running illustration of the formal setting.

Let k denote the number of different supply chain components $C_1, \ldots, C_k$ a company is aiming for. (In the previous example k=2.) We assume for simplicity throughout this paper that, in the case of undisturbed operation, all supply chain components contribute the same fraction to the total production. The number of different commodities required for the production is denoted by l. (In our example l=5.) We assume that for each of these commodities exactly k suppliers are available. (In our example we have two suppliers for the end product and for Commodities 1 to 4. However, Commodity 5 can be provided by the three Suppliers 9 to 11. Here we assume that only two of them, say Suppliers 9 to 10 are actually relevant. It is not hard to adjust the final model to deal with the more general situation. We do not do it here in the basic model in order to better extract the main underlying paradigm.) Hence the total number m of different suppliers $S_1, \ldots, S_m$ involved in the production process is equal to $k \cdot l$. The suppliers are naturally partitioned into l subsets $P_1, \ldots, P_l$; each set P_i contains those k suppliers which contribute to commodity i. (There is no need to assume that the corresponding companies are different. In fact, in general the risks of failures may be highly correlated.) Of course, if a supplier belongs to two or more different supply chain components all of them break down if the supplier suffers from a complete failure. Hence such a situation may constitute a potentially dangerous bottleneck.

The suppliers may be threatened by a total number of n potential events $E_1, \ldots, E_n$. In our basic model we assume that these events occur independently Bernoulli-distributed with probabilities $p_1, \ldots, p_n$. (In our previous example we had n=3; an earthquake, a local fire or a strike in some subsector of the production line. Of course, there may be chances of severe accidental fires occurring at different suppliers. If the corresponding sites are sufficiently distant from each other it may be appropriate to introduce different and independent events 'fire' for each such supplier, possibly even with different probabilities according to their general production profile.) Each event is supposed to directly affect suppliers. While, for instance, an earthquake in Japan may create tremendous indirect damage to a supply chain, it does not cause direct devastation in Europe. We model the direct vulnerability of supplier S_i by the event E_j by means of a binary number e_{ij}. Accordingly, $e_{ij} = 1$ expresses the fact that, in the case that the event occurs, S_i is directly affected by E_j, while $e_{ij} = 0$ means that S_i will never be directly struck by E_j. For simplicity of exposition we assume in our basic model that the effect, if present at all, is a total breakdown. (It is, however, possible to associate with any pair (S_i, E_j) some arbitrary number or some random variable that expresses the quantity of the impairment; see Section 4.) For each supplier S_i, all such numbers are collected in an n-dimensional *event vector* $e_i = (e_{i1}, \ldots, e_{in})^T$.

As an example, consider four events E_1, E_2, E_3, E_4. For instance, event E_1 may stand for an earthquake in Japan, E_2 is a national strike in Greece, E_3 indicates that the quality of production of a high-precision component sinks below a certain level, and E_4 signifies the shutdown of a company due to a severe violation of environmental specifications. Now suppose, a supplier S_1 located on the island of Crete has the event vector $e_1 = (e_{11}, \dots, e_{14})^T = (0,1,1,0)^T$. This means S_1 will be affected by a national strike in Greece, and the quality of production is also at risk. However, an earthquake in Japan does not cause any direct damage to S_1, and the production of the related commodity is not regarded hazardous for the environment at all.

Now suppose that S_1 is contained in the supply chain component C_1. Clearly, under our simplifying assumption that failures are always total, the whole supply chain component C_1 breaks down if S_1 fails, i.e., if either E_2 (the national strike in Greece) or E_3 (shut down due to quality problems) occurs. (As mentioned before, this and other simplifying assumptions are made for the ease of exposition of the basic underlying idea of our approach. They can easily be lifted in a more involved model.) Of course, the supply chain component C_1 consists of l different suppliers for the l commodities. What would be a reasonable governing principle to select suppliers for the other $l - 1$ commodities in order to constitute C_1? To avoid additional risks it is natural to try to include in C_1 only suppliers for the remaining commodities whose event vectors have entries 0 in their first and fourth coordinate. This would ensure that the supply chain component C_1 can only be disrupted by the second or third event. Note that the entries in the second and third coordinate of the other suppliers' event vectors do not matter. In fact, there is no incentive to avoid entries 1 there since S_1 brings in these risks already, and there is no possibility to ever reduce these risks in C_1 by the subsequent choices of suppliers. This motivates the following aggregation. Given a supply chain component $C_i = (S_{i_1}, \dots, S_{i_l})$ we collect the relevant events in the *failure vector* $f_i = (f_{i1}, \dots, f_{in})^T$ that shows all events that threaten C_i; here $f_{ij} = \max\{e_{i_1,j}, \dots, e_{i_l,j}\}$, with event vectors $e_{i_j} = (e_{i_j,1}, \dots, e_{i_j,n})^T$.

Let us now consider Commodity 2, and suppose three suppliers S_2, S_3, and S_4 are capable of providing it, with event vectors $(1,0,0,1)^T, (0,1,1,0)^T$, and $(0,0,0,0)^T$, respectively. Adding S_2 to C_1would lead to the simultaneous dependence of C_1 on all four events; resulting in the failure vector $(1,1,1,1)^T$. However, adding S_3 or S_4 would not induce any additional risk for C_1; the failure vector would be $(0,1,1,0)^T$. But, of course, one would not really want to waste S_4 on C_1 since, S_4 is risk-free and could be used in any supply chain component without affecting the chances of failure. S_3, on the other hand, may very well increase the risk of any other supply chain component. Intuitively, the goal of the optimization might then be to construct supply chains with the property that the associated failure vectors $f_1, \dots, f_k$ differ in as many coefficients as possible, or, more precisely, such that $\sum \|f_i - f_j\|^2$ is maximized, where $\|\,.\,\|$ denotes the Euclidean norm, and the sum is taken for each pair (i, j) of indices. (Note that, for 0-1-vectors, we can omit the square without changing the count; for more general entries it does, however, make a difference.) Alternatively one might aim at minimizing the total sum of the coordinates of all failure vectors. As it turns out, the corresponding latter objective function is preferable to the former and is therefore used in Section 4.

Obviously, things become more complicated if more events and (as it is typically the case) many more suppliers are involved and more supply chain components are required. In fact, the structure of interdependencies of the available suppliers may become quite involved and the best choice will typically not be clear on inspection. However, the same basic idea can be cast into a mathematical model for designing supply chains. We model this idea by means of constrained clustering. The constraints have to guarantee that in each supply chain component each commodity is delivered by precisely one supplier and each S_i belongs to only one supply chain component. If we identify each supplier with its event vector, a supply chain component can be viewed as a collection of event vectors, exactly one for each of the sets $P_1, \ldots, P_l$. In order to refer to this interpretation of a supply chain component, we will speak of a *supply chain cluster*. Then, of course, a supply chain can be regarded as a *supply chain clustering*.

Let us consider a simple example (of scalable size) for illustration. Suppose that $n = l = k$ and that each event affects exactly one supplier of each of the sets $P_1, \ldots, P_l$. Further consider the following two clusterings:

$$C_1 = \left(\begin{pmatrix} 1 \\ 0 \\ 0 \\ \vdots \\ 0 \\ 0 \end{pmatrix}, \ldots, \begin{pmatrix} 1 \\ 0 \\ 0 \\ \vdots \\ 0 \\ 0 \end{pmatrix} \right), \; C_2 = \left(\begin{pmatrix} 0 \\ 1 \\ 0 \\ \vdots \\ 0 \\ 0 \end{pmatrix}, \ldots, \begin{pmatrix} 0 \\ 1 \\ 0 \\ \vdots \\ 0 \\ 0 \end{pmatrix} \right), \ldots, C_k = \left(\begin{pmatrix} 0 \\ 0 \\ 0 \\ \vdots \\ 0 \\ 1 \end{pmatrix}, \ldots, \begin{pmatrix} 0 \\ 0 \\ 0 \\ \vdots \\ 0 \\ 1 \end{pmatrix} \right);$$

$$\tilde{C}_1 = \left(\begin{pmatrix} 1 \\ 0 \\ 0 \\ \vdots \\ 0 \\ 0 \end{pmatrix}, \begin{pmatrix} 0 \\ 1 \\ 0 \\ \vdots \\ 0 \\ 0 \end{pmatrix}, \ldots, \begin{pmatrix} 0 \\ 0 \\ 0 \\ \vdots \\ 0 \\ 1 \end{pmatrix} \right), \ldots, \tilde{C}_k = \left(\begin{pmatrix} 1 \\ 0 \\ 0 \\ \vdots \\ 0 \\ 0 \end{pmatrix}, \begin{pmatrix} 0 \\ 1 \\ 0 \\ \vdots \\ 0 \\ 0 \end{pmatrix}, \ldots, \begin{pmatrix} 0 \\ 0 \\ 0 \\ \vdots \\ 0 \\ 1 \end{pmatrix} \right).$$

Both are supply chain clusterings, yet with dramatically different risk structure. Let us explore that in terms of the stochastics governing the events. Suppose that the random events $E_1, \ldots, E_n$ are independently and identically Bernoulli-distributed with probability p, i.e., $p = p_1 = \cdots = p_n$, and let X denote the random variable that counts the failing supply chain components (in the upcoming period).

We begin our analyses with the supply chain clustering $(C_1, \ldots, C_k)$. Then X is identical to the total number of events that occur, and is hence binomially distributed with expectation np. The probability for a simultaneous failure of all k supply chains is $\binom{n}{k} p^k (1-p)^{n-k}$, and since $k = n$, is equal to p^n.

The situation is quite different for $(\tilde{C}_1, \ldots, \tilde{C}_k)$. Since the occurrence of any single event affects all supply chains simultaneously, X is either equal to 0 or equal to k. The probability that no failure occurs is $(1-p)^n$ and hence all supply chains fail with probability $1-(1-p)^n$. Thus the expected value of X is $k(1-(1-p)^n)$.

Table 1 Failure expectation and probabilities for $n = 10$

Probability p		0.1	0.05	0.03	0.01
Expectation for number of failing chains	$C_1, \ldots, C_k$	1	0.5	0.3	0.1
	$\tilde{C}_1, \ldots, \tilde{C}_k$	6.513	4.013	2.626	0.956
Probability of total failure	$C_1, \ldots, C_k$	10^{-10}	$< 10^{-13}$	$< 10^{-15}$	$< 10^{-19}$
	$\tilde{C}_1, \ldots, \tilde{C}_k$	0.6513	0.4013	0.2626	0.0956

Table 1 depicts, for $n = 10$, the two expected values and the probability of total failure for some values of p. It shows how dramatically different the two supply chains behave in terms of their robustness under risk. In analogy to extending the concept of Value at Risk to that of Conditional Value at Risk in financial risk management, see e.g. (Hull, 2011), we can, of course, also ask for the worst-case behavior under the condition that at least one event happens, i.e., *What is the (conditional) expected relative number of failing chains? What is the (conditional) probability of a total failure?* The answers for our example are given in Table 2.

Table 2 Conditional failure expectation and probabilities

Probability p		0.1	0.05	0.03	0.01
Cond. Expectation of the number of failing chains	$C_1, \ldots, C_k$	1.5353	1.2461	1.1425	1.0458
	$\tilde{C}_1, \ldots, \tilde{C}_k$	10	10	10	10
Conditional probability of total failure	$C_1, \ldots, C_k$	$< 10^{-9}$	$< 10^{-12}$	$< 10^{-14}$	$< 10^{-18}$
	$\tilde{C}_1, \ldots, \tilde{C}_k$	1	1	1	1

In the above example we analyzed the situation that any failure led to a complete shut-down. It is not hard to adjust the model to accommodate partial reductions specified by numbers in the interval [0,1]. The interpretation of a value of, say, 0.4 is that in case the event occurs the company's output is reduced by 40%, i.e. the supplier is only able to maintain 60% of its regular production. (A next step of extending the model would be to regard this number as a random variable with a certain distribution.) Examples that model partial failure are given in the next section.

There are, of course, various other issues that are relevant in practice and that have to and can be incorporated in our model. First, there is the financial impact of a disrupted supply chain component. It depends on various factors including economical and legal conditions related to the end product and the cost of substitution. It determines the level of risk a company is willing to accept. Then there are production and commodity flow costs that may be quite different for different supply chain components. Further, one has to consider setup and other transaction costs that are related to any change of the present supply chain towards further diversification. Then, there are different issues related to relevant time periods including that of possibly different duration of effects of the same event on different suppliers. Since it would by far extend the scope of the present contribution we will not go into further modeling details here; see however (Brieden, Gritzmann & Öllinger, 2012).

Let us close this section by discussing some algorithmic issues related to supply chain safety. Given the power of modern computer technology one might be tempted to try to enumerate all possible supply chains and then select a best according to whatever optimality criterion one may want to apply. As simple estimates show this approach is bound to fail already for quite moderate sized supply networks. Suppose, our production involves 50 commodities, we have three suppliers for each commodity, and we aim for three supply chain components, i.e., $l = 50$ and $k = 3$. Then there are a total of 3^{50} (or more than 10^{23}) different possible combinations. If we could analyze one million possibilities per second, then we would still need more than 10^{16} seconds or more than 100 million years to execute our algorithm. This phenomenon of *combinatorial explosion* rules out simple enumerative algorithms.

As pointed out before, the problem can be viewed as a special cluster optimization task. Due to an abundance of practically highly relevant applications there are well-studied clustering algorithms; arguably the best known and most frequently used being the *k-means algorithm.* The situation here is, however, quite special since we have to guarantee additional constraints. Also, we are aiming at a global optimum. However, the optimization can be done quite efficiently; see (Brieden, Gritzmann & Öllinger, 2012). (See also (Borgwardt, Brieden & Gritzmann, 2011) for another application of *constraint clustering* and (Brieden & Gritzmann, 2009, 2011) for a deep analysis of properties of optimal clusterings.) Here we restrict ourselves to some proof-of-concept simulations.

4 Computer Simulation

In this section we illustrate the potential of the approach described above by means of computer simulations performed with the aid of *Microsoft Excel 2010* and *@Risk 5.5* for Excel. After explaining how the instances of event vectors were generated we describe different ways for determining supply chains. Then we give the results of four different simulations. While the number of events and their probabilities will be the same throughout, the numbers of commodities and supply chain components differ from Examples 1 to 2 to Examples 3 to 4. Examples 1 and 3 will assume that if struck by an event the supplier faces a total black out, while in Examples 2 and 4 only a certain fraction of the production is affected.

4.1 Generation of Data

To keep the examples accessible we choose $n = 10$ for the number of events. The event probabilities are defined by the vector $p = (p_1, \ldots, p_{10})^T = (0.005, 0.005, 0.01, 0.01, 0.02, 0.02, 0.01, 0.01, 0.005, 0.005)^T$, and we assume that all events are independent. The choice of the above values for the event probabilities implies that the probability of undisturbed supply, i.e., that none of the events occurs is about 90,4%.

Example 1

Let the number l of commodities be 16 and the number k of supply chain components be 4. Hence we consider $m = 64$ suppliers, 4 for each commodity. To specify the 640 binary coordinates of the 64 event vectors we use simulations, performed with *@Risk 5.5*, where the expected number of '$1's$' was chosen to be ten percent. The event vectors obtained that way are listed in Columns 3-12 of the Tables 9-10. (The total number of ones actually came out to be 68.) The first column gives the commodities by their number, Column 2 contains the numbers of the suppliers (or their event vectors) while the third block depicts the event vectors.

Example 2

The second example uses the data of Example 1. We model, however, that a supplier may only partially fail. Accordingly, for each entry 1 in an event vector we simulated a discrete random variable that uniformly attains the values $0.1, 0.2, \ldots, 1.0$. This means a drop of supply by $10\%, 20\%$, etc. respectively. The resulting (fractional) event vectors are listed in Columns 3-12 in Tables 11-12.

Example 3

As compared to the previous two examples, here the numbers of commodities and supply chain components have been interchanged, i.e., $l = 4$ and, $k = 16$. This time we study event vectors with a total number of 80 entries equal to 1. They are listed in Columns 3-12 in Tables 13-14.

Example 4

We use the data of Example 3 but model again partial failure. Here every coefficient 1 of an event vector of Example 3 is replaced by 0.2,0.4, 0.6, 0.8 or 1.0. The resulting (fractional) event vectors are listed in Columns 3-12 in Tables 15-16.

4.2 Constructing Supply Chains

Examples 1 and 2

We compare three different methods for constructing supply chains. The results of the constrained clustering algorithm whose basic ideas were outlined in Section 3 will be given in the columns marked *'cc'*. To illustrate its performance we compared it with two heuristic approaches, *h1* and *h2*. The first, *h1*, is a naive heuristic designed to produce a bad solution so as to give an impression of the possible total range of supply chain performance. *h2*, however, is a fair heuristic approach for generating a reasonably safe supply chain.

h1 allocates the four suppliers that produce the first commodity to supply chain components 1 to 4 in their numerical order. Then it proceeds inductively as follows. Suppose that the first j commodities have already been dealt with, i.e., the first $4j$ suppliers have already been assigned to the four supply chain components. Adding the event vectors assigned so far for each supply chain component, we obtain four integer vectors a_1, a_2, a_3, a_4. For each of the four suppliers that are in charge of commodity $j + 1$, we sum over all entries of a_1 which correspond to a coefficient 1 in the supplier's event vector. Then the (or, if there is a tie, a) supplier is assigned to the first supply chain component whose sum is the least. Similarly, we continue with the second supply chain component, now using a_2 and the remaining three suppliers, and so on.

h2 analogously assigns companies to supply chain components, except that in every iteration the company with the most conformities, i.e., the largest sum is assigned to the current supply chain component.

As, in the second example, a supplier may only partially fail we adapt the failure vectors. Whenever a supplier's output is obstructed by, say 70% , at least 70% of the corresponding supply chain component's production disappears. Therefore the event vectors of all suppliers of a supply chain component can be aggregated to its n-dimensional failure vector whose ith component is the maximal failure rate of the suppliers caused by the ith event. The failure vectors of the four supply chains of Examples 1 and 2 are listed in Tables 3-4, respectively.

Table 3 Failure vectors of Example 1

	h1				*h2*				cc			
Suppl. Chain Comp.	1	2	3	4	1	2	3	4	1	2	3	4
Events	1	1	1	1	1	1	1	1	1	1	0	0
	1	1	1	1	1	1	1	1	1	1	1	1
	1	1	1	1	1	1	1	1	1	1	1	0
	0	1	1	1	1	1	0	1	0	0	0	1
	0	0	0	1	1	0	0	0	0	0	0	1
	1	1	0	1	1	0	0	1	0	0	0	1
	1	0	1	1	0	1	0	1	0	0	0	1
	1	1	1	1	1	1	0	0	0	0	1	1
	1	1	1	1	1	1	1	1	0	1	0	1
	1	1	1	1	1	1	0	0	1	1	0	0

Table 4 Failure vectors of Example 2

	h1				*h2*				cc			
Chain	1	2	3	4	1	2	3	4	1	2	3	4
Events	0.3	0.5	0.6	0.9	0.6	0.9	0.3	0.2	0.1	0	0.3	0.9
	0.7	0.3	0.9	0.9	0.9	0.8	0.5	0.8	0.9	0.9	0	0.3
	1.0	0.5	0.9	0.8	1.0	0.8	0.5	0.9	0.5	1.0	0	0.8
	0	1.0	0.5	0.7	0.8	0.3	0	1.0	1.0	0	0	0
	0	0	0	1.0	1.0	0	0	0	1.0	0	0	0
	0.4	0.1	0	1.0	0.4	0	0	1.0	0.1	0	1.0	0
	0.4	0	0.1	0.4	0	0.3	0	0.4	0.3	0	0.4	0
	0.8	0.5	0.9	0.1	0.8	0.9	0	0	0.9	0.1	0	0
	0.7	0.3	0.6	0.9	0.6	0.9	0.1	0.6	0.9	0	0	0.7
	0.3	0.3	1.0	0.8	0.8	1.0	0	0	0	0	0.4	1.0

By our assumption that the (unaffected) output of all supply chain components is identical the average of their expected relative amount of production equals the overall expected relative amount of production.

Examples 3 and 4

We use two new heuristics, *h3* and *h4,* for comparison. *h3* simply assigns suppliers in their given order to the supply chain component. Since the event vectors are produced by means of randomization this heuristic is in fact a randomized assignment. In contrast, *h4* tries to exploit that the overall performance depends on the failure vectors. Hence it orders the suppliers of P_i, i.e., the companies responsible for Commodity i, according to the sum of their failures. Then, it successively assigns the (or a) supplier with the highest sum to the first chain, the (or a) company with second highest sum to the second, and so on. All assignments are taken for the binary and for the fractional case. The resulting failure vectors for the binary case (Example 3) are listed in Tables 13-14, the failure vectors for the fractional case (Example 4) are given in Tables 15-16.

4.3 Conditional Results

In this section we study conditional probabilities, i.e., we assume that at least one event occurs. These probabilities are produced by 100.000 iterations in a simulation done by *@Risk 5.5*.

Example 1

The simulation results for Example 1 are listed in Table 5. For instance, if we decide to use heuristic *h2* for the supply chain design, then, under the condition that at least one event occurs, the probability that three supply chain components fail is 10,9513%, while the expected number of failing supply chain components is 2.4326. For *cc* this number is 1.6349. This means that in case of at least one event, on the average the number of operating supply chain components is almost one higher in *cc* than in *h2*. So, in comparing the different ways to design the chains we see the effect of our clustering algorithm. It has the best average value and the conditional probability of a total black out is approximately one seventh of the conditional probability of *h1* and still one fourth of *h2*.

Table 5 Conditional Results for Example 1

Number of failing sup. chain comp.	*h1*	*h2*	*cc*
1	0.193290	0.193290	0.590497
2	0	0.438991	0.245493
3	0.393352	0.109513	0.102636
4	0.413358	0.258206	0.061373
Exp. no. of failing sup. chain comp.	3.0268	2.4326	1.6349

Example 2

The simulation results for Example 2 are listed in Table 6. For instance, under the condition that at least one event occurs, the probability that at most 50% of the production fails is 49.0647% for *h1* and 83.776% for *cc*.

While the worst case failure is 92.5% both for *h1* and *cc*, supply chains generated by the constrained clustering algorithm have a conditional average failure of 31.83%. The conditional probability of losing more than 75% of the production is less than 0.4% if the supply chains are constructed with *cc* but more than 10% if *h2* is applied.

Table 6 Results for Example 2

Conditional expected failure	*h1*	*h2*	*cc*
≤ 25%	0.289361	0.289361	0.482651
≤ 50%	0.490647	0.683963	0.837762
≤ 75%	0.887673	0.891841	0.996040
Maximal	0.925	0.85	0.925
Average	0.4726	0.4327	0.3183

Example 3

The simulation results for Example 3 are listed in Table 7. Under the condition that at least one event happens, for '*cc*-generated' chains the probability that at most three supply chain components fail is more than 72%; hence the probability that four or more supply chain components break down is less than 28%. For both heuristics the latter probability is 100%.

Example 4

The simulation results for Example 4 are listed in Table 8. For instance, under the condition that at least one event happens, for '*cc*-generated' chains the probability of a failure of at least 20% is less than 8%.

Table 7 Results for Example 3

No. of failures	*h3*	*h4*	*cc*
1			0.19494
2			0.18962
3			0.33907
4	0.28918	0.38687	0.00835
5	0.14329	0.19108	0.15049
6	0.14589	0.09570	0.10039
7	0.18962	0.00543	0.01211
8	0.19682	0.19776	0.00251
9	0.01252	0.05312	0.00198
10	0.01043	0.06001	0.00052
11	0.00459	0.00793	
12	0.00741	0.00115	
13	0.00010	0.00094	
14	0.00010		
15			
16			
Expected number	6.01	5.88	3.11

Table 8 Results for Example 4

Cond. Exp. failure	*h3*	*h4*	*cc*
≤ 10%	0.0942	0	0.4325
≤ 20%	0.4283	0.4785	0.9268
≤ 30%	0.7737	0.7345	0.9982
≤ 40%	0.9834	0.9811	1.0000
≤ 50%	0.9969	1.0000	
≤ 60%	1.0000		
Maximal	0.5875	0.5000	0.3375
Average	0.2275	0.2160	0.1135

5 Summary and Outlook

In this paper a quantitative model for supply chain safety is introduced. It can be used to measure the risk of the status quo of the supply chain of a production by calculating the (conditional) probability of failure. This risk can be judged in comparison to best assignments of suppliers to different supply chain components of the same size. Also, one can study the effect of allowing more or less supply chain components. As the simulation results presented in this paper indicate, the risk of failure can potentially be significantly reduced by using this quantitative approach.

Both, from a theoretical as well as from a practical point of view the results of the present paper are just the starting point. We mentioned already that various ramifications and extensions of the model can be accommodated; see (Brieden, Gritzmann & Öllinger, 2012). There are, of course, further issues that can be raised.

Let us close by addressing one important practical issue that has been briefly touched upon before. One might argue that it is not realistic to explicitly obtain all data that are necessary to run the algorithm. It is true that in order to fully analyze the risk inherent in the implemented supply chain one does need full knowledge of the supply chain and the potential risks of failures for each of its links. Without complete knowledge the identification of dangerous bottlenecks is impossible, and the threat of an 'Albuquerque-type incident' may always be present in the supply chain. Even more, if one wants to take full advantage of the potential of the method indicated in this paper one needs this information not only for one's business partners (at each level) but also for potential additional suppliers. The model allows, however, to work with incomplete information in various ways. In fact, one can concentrate on 'critical commodities' and reduce the supply chain structure dramatically. Alternatively, one can truncate the commodity graph at a specified level and work with estimates of the total risks of the cut off subgraphs (very much in the spirit of our failure vectors). Again, the complexity can be substantially reduced this way. Additionally, one can concentrate on the most likely events and neglect all others. Naturally, one has to handle the estimation and truncation errors. However, even working with partial and/or estimated data might already have substantial impact on the company's operation. Alas, implementing quantitative methods for supply chain risk management in practice might be seen as an own field of research; see (Pfohl, Köhler & Thomas, 2010) for a survey.

Acknowledgements. The authors are indebted to Michael Eßig for fruitful discussions on various economic issues of supply chain management.

References

Borgwardt, S., Brieden, A., Gritzmann, P.: Constrained minimum-k-star clustering and its application to the consolidation of farmland. Operational Research 11, 1–17 (2011)

Brieden, A., Gritzmann, P.: On clustering bodies: Geometry and polyhedral approximation. Discrete Comp. Geom. 44, 508–534 (2009)

Brieden, A., Gritzmann, P.: On optimal weighted balanced clusterings: gravity bodies and power diagrams. SIAM J. Discrete Math. (2011) (to appear)

Brieden, A., Gritzmann, P., Öllinger, M.: A new model for supply chain safety (2012) (in preparation)

Christopher, M., Peck, H.: Building the resilient supply chain. Intern. J. Logistics Manag. 15, 1–13 (2004)

Deane, J., Craighead, C., Ragsdale, C.: Mitigating environmental and density risk in global sourcing. Intern. J. Physical Distr. Logistics Manag. 39, 861–883 (2009)

Goh, M., Lim, J., Meng, F.: A stochastic model for risk management in global supply chain networks. European J. Operational Research 182(1), 164–173 (2007)

Hull, J.C.: Option, Futures, and other Derivatives, 8th edn. Pearson, New Jersey (2011)

Jüttner, U., Peck, H., Christopher, M.: Supply chain risk management. Outlining an agenda for future research. Intern. J. Logistics: Research & Applications 6, 197–210 (2003)

Jüttner, U.: Supply chain risk management. Understanding the Business Requirements from a Practitioner Perspective, Intern. J. Logistics Manag. 16, 120–141 (2005)

Klibi, W., Martel, A., Guitouni, A.: The design of robust value-creating supply chain networks: A critical review. Europ. J. Operational Research 203, 283–293 (2010)

Norrman, A., Jansson, U.: Ericsson's proactive supply chain risk management approach after a serious sub-supplier accident. Intern. J. Physical Distr. Logistics Manag. 34, 434–456 (2004)

Peck, H.: Drivers of supply chain vulnerability: an integrated framework. Intern. J. Physical Distr. Logistics Manag. 35, 210–232 (2005)

Pfohl, H.-C., Köhler, H., Thomas, D.: State of the art in supply chain risk management research: Empirical and conceptual findings and a roadmap for the implementation in practice. Logistics Research 2, 33–44 (2010)

Ponomarov, S., Holcomb, M.: Understanding the concept of supply chain resilience. Intern. J. Logistics Manag. 20, 124–143 (2009)

Ravindran, A.R., Bilsel, R.U., Wadhwa, V., Yang, T.: Risk adjusted multicriteria supplier selection models with applications. Intern. J. Production Research 48, 405–424 (2010)

Santoso, T., et al.: A stochastic programming approach for supply chain network design under uncertainty. European J. Operational Research 167(1), 96–115 (2005)

Schütz, P., Tomasgard, A., Ahmed, S.: Supply chain design under uncertainty using sample average approximation and dual decomposition. European J. Operational Research 199(2), 409–419 (2009)

Sheffi, Y., Rice, J.: A supply chain view of the resilient enterprise. MIT Sloan Management Review 47, 41–48 (2005)

Svensson, G.: A conceptual framework of vulnerability in firms' inbound and outbound logistics flows. Intern. J. Physical Distr. Logistics Manag. 32, 110–134 (2002)

Tandler, S., Eßig, M.: Supply Chain Safety Management: Konzeption und Gestaltungsempfehlungen. In: Bogaschewsky, R., Eßig, M., Lasch, R., Stölzle, W. (eds.) Supply Management Research: Aktuelle Forschungsergebnisse 2011, Gabler, Wiesbaden, pp. 57–92 (2011)

Tang, C.: Perspectives in supply chain risk management. Intern. J. Production Economics 103, 451–488 (2006a)

Tang, C.: Robust strategies for mitigating supply chain disruptions. Intern. J. Logistics: Research & Applications 9, 33–45 (2006b)

Teuteberg, F.: Supply chain risk management: A neural network approach. In: Ijioui, R., Emmerich, H., Ceyp, M. (eds.) Strategies and Tactics in Supply Chain Event Management, pp. 99–118. Springer, Berlin (2008)

Wagner, S., Bode, C.: An empirical examination of supply chain performance along several dimensions of risk. J. Business Logistics 29, 307–325 (2008)

Wagner, S., Neshat, N.: Assessing the vulnerability of supply chains using graph theory. Intern. J. Production Economics 126, 121–129 (2010)

Yang, Z.-L., Wang, J., Bonsall, S., Yang, J.-B., Fang, Q.-G.: A subjective risk analysis approach for container supply chains. Int. J. Automation Comput. 1, 85–92 (2005)

Zsidisin, G., Ellram, L.: An agency theory investigation of supply risk management. J. Supply Chain Management 39, 15–27 (2003)

Zsidisin, G., Wagner, S.: Do perceptions become reality? The moderating role of supply chain resiliency on disruption occurrence. J. Business Logistics 31, 1–20 (2010)

Appendix

Table 9 Event vectors and associated chains for Example 1 (Part 1)

		Events										Chain		
Commodity	Supplier	1	2	3	4	5	6	7	8	9	10	h1	h2	cc
1	1	0	0	0	0	0	0	0	1	0	0	1	1	3
	2	1	0	0	0	0	0	0	0	0	1	2	2	2
	3	0	0	0	0	0	0	0	0	0	0	3	3	1
	4	0	0	0	0	0	1	1	0	0	0	4	4	4
2	5	0	0	0	0	0	0	0	0	0	0	1	1	3
	6	0	1	0	0	0	0	0	0	0	0	2	3	1
	7	0	1	0	0	0	0	0	0	1	0	3	4	4
	8	0	0	0	0	0	0	0	0	0	1	4	2	2
3	9	0	0	0	0	0	0	0	0	0	1	1	2	1
	10	1	0	0	0	0	0	0	0	0	0	3	4	2
	11	0	0	0	1	0	1	0	1	1	0	2	1	4
	12	0	1	0	0	0	0	0	0	0	0	4	3	3
4	13	0	0	0	0	0	0	0	0	0	0	1	3	3
	14	0	0	1	0	0	0	0	0	0	1	3	2	1
	15	0	0	0	0	0	0	0	0	0	0	2	4	2
	16	0	1	0	1	1	0	0	1	0	0	4	1	4
5	17	0	0	0	0	0	0	1	0	0	0	1	4	4
	18	0	1	0	0	0	0	0	0	0	0	2	1	3
	19	1	1	0	0	0	0	0	0	0	0	3	2	2
	20	1	1	0	0	0	0	0	0	0	0	4	3	1
6	21	0	1	1	0	0	0	0	0	0	0	1	1	3
	22	0	0	1	0	0	0	0	0	1	0	2	3	2
	23	1	0	1	0	0	0	0	0	0	0	4	2	1
	24	0	0	0	0	0	0	0	0	0	0	3	4	4
7	25	0	0	1	0	0	0	0	0	0	1	3	1	1
	26	0	0	0	0	0	0	0	0	0	0	1	3	4
	27	0	0	0	0	0	0	0	0	0	0	2	4	3
	28	0	0	0	0	0	0	0	0	1	0	4	2	2
8	29	0	0	0	0	0	0	0	1	0	0	2	1	4
	30	1	0	0	0	0	0	0	0	0	0	1	3	1
	31	0	0	1	0	0	0	0	0	0	0	3	4	3
	32	0	0	0	0	0	0	0	0	0	1	4	2	2

Table 10 Event vectors and associated chains for Example 1 (Part 2)

		Events										Chain		
Commodity	Supplier	1	2	3	4	5	6	7	8	9	10	h1	h2	cc
9	33	0	0	0	0	0	0	0	0	0	0	1	3	3
	34	1	0	0	0	0	0	0	0	0	1	3	1	2
	35	0	0	0	0	0	0	0	0	0	0	2	4	4
	36	1	0	0	0	0	0	0	0	0	1	4	2	1
10	37	0	0	0	0	0	0	0	0	0	0	1	3	2
	38	0	0	0	0	0	0	0	0	0	0	2	4	1
	39	0	1	0	0	0	0	0	1	0	0	3	1	3
	40	0	1	0	1	0	0	1	0	1	0	4	2	4
11	41	0	0	0	0	0	0	0	0	0	0	1	3	2
	42	0	1	0	0	0	0	0	0	0	0	4	1	3
	43	0	0	0	0	0	0	0	0	0	0	2	4	1
	44	0	0	0	0	0	0	1	0	0	0	3	2	4
12	45	0	0	0	0	0	0	0	1	1	0	3	2	4
	46	0	1	0	0	0	0	0	0	1	1	4	1	2
	47	0	0	0	0	0	0	0	0	0	0	1	3	1
	48	0	0	0	0	0	0	0	0	0	0	2	4	3
13	49	0	0	0	0	0	1	0	0	0	0	1	1	4
	50	0	0	0	0	0	0	0	0	0	0	2	2	2
	51	0	0	0	0	0	0	0	0	0	0	3	3	3
	52	0	0	0	0	0	0	0	0	0	0	4	4	1
14	53	0	0	0	0	0	0	0	0	0	1	4	1	1
	54	0	0	0	0	0	0	0	0	1	0	1	2	2
	55	0	0	0	0	0	0	0	0	0	0	2	3	3
	56	0	0	0	1	0	0	0	0	0	0	3	4	4
15	57	0	0	0	0	0	0	0	0	0	1	3	2	1
	58	0	0	0	0	0	0	0	0	0	0	1	3	3
	59	0	1	1	0	0	0	0	0	0	0	4	1	2
	60	0	0	0	0	0	0	0	0	0	0	2	4	4
16	61	0	1	0	0	0	0	0	0	0	0	3	1	3
	62	0	0	0	0	0	0	0	0	0	0	1	3	1
	63	0	0	0	1	0	0	0	0	0	0	2	4	4
	64	0	0	0	0	0	0	0	0	0	1	4	2	2

Table 11 Event vectors and associated chains for Example 2 (Part 1)

		Events										Chain		
Commodity	Supplier	1	2	3	4	5	6	7	8	9	10	h1	h2	cc
1	1	0	0	0	0	0	0	0	0.8	0	0	1	1	1
	2	0.5	0	0	0	0	0	0	0	0	0.3	2	2	4
	3	0	0	0	0	0	0	0	0	0	0	3	3	2
	4	0	0	0	0	0	1.0	0.4	0	0	0	4	4	3
2	5	0	0	0	0	0	0	0	0	0	0	1	1	3
	6	0	0.3	0	0	0	0	0	0	0	0	2	3	2
	7	0	0.8	0	0	0	0	0	0	0.6	0	3	4	1
	8	0	0	0	0	0	0	0	0	0	0.7	4	2	4
3	9	0	0	0	0	0	0	0	0	0	0.3	1	2	3
	10	0.2	0	0	0	0	0	0	0	0	0	3	4	4
	11	0	0	0	0.8	0	0.1	0	0.5	0.3	0	2	1	1
	12	0	0.5	0	0	0	0	0	0	0	0	4	3	2
4	13	0	0	0	0	0	0	0	0	0	0	1	3	3
	14	0	0	0.5	0	0	0	0	0	0	0.10	3	2	4
	15	0	0	0	0	0	0	0	0	0	0	2	4	2
	16	0	0.9	0	0.7	1.0	0	0	0.1	0	0	4	1	1
5	17	0	0	0	0	0	0	0.4	0	0	0	1	4	3
	18	0	0.3	0	0	0	0	0	0	0	0	2	1	2
	19	0.2	0.2	0	0	0	0	0	0	0	0	3	2	4
	20	0.1	0.2	0	0	0	0	0	0	0	0	4	3	1
6	21	0	0.7	1.0	0	0	0	0	0	0	0	1	1	2
	22	0	0	0.5	0	0	0	0	0	0.1	0	2	3	1
	23	0.9	0	0.8	0	0	0	0	0	0	0	4	2	4
	24	0	0	0	0	0	0	0	0	0	0	3	4	3
7	25	0	0	0.1	0	0	0	0	0	0	0.6	3	1	4
	26	0	0	0	0	0	0	0	0	0	0	1	3	2
	27	0	0	0	0	0	0	0	0	0	0	2	4	3
	28	0	0	0	0	0	0	0	0	0.9	0	4	2	1
8	29	0	0	0	0	0	0	0	0.2	0	0	2	1	1
	30	0.3	0	0	0	0	0	0	0	0	0	1	3	3
	31	0	0	0.9	0	0	0	0	0	0	0	3	4	2
	32	0	0	0	0	0	0	0	0	0	0.8	4	2	4

Table 12 Event vectors and associated chains for Example 2 (Part 2)

		Events										Chain		
Commodity	Supplier	1	2	3	4	5	6	7	8	9	10	h1	h2	cc
9	33	0	0	0	0	0	0	0	0	0	0	1	3	1
	34	0.6	0	0	0	0	0	0	0	0	0.8	3	1	4
	35	0	0	0	0	0	0	0	0	0	0	2	4	2
	36	0.2	0	0	0	0	0	0	0	0	0.4	4	2	3
10	37	0	0	0	0	0	0	0	0	0	0	1	3	3
	38	0	0	0	0	0	0	0	0	0	0	2	4	4
	39	0	0.5	0	0	0	0	0	0.1	0	0	3	1	2
	40	0	0.8	0	0.3	0	0	0.3	0	0.8	0	4	2	1
11	41	0	0	0	0	0	0	0	0	0	0	1	3	2
	42	0	0.9	0	0	0	0	0	0	0	0	4	1	1
	43	0	0	0	0	0	0	0	0	0	0	2	4	4
	44	0	0	0	0	0	0	0.1	0	0	0	3	2	3
12	45	0	0	0	0	0	0	0	0.9	0.6	0	3	2	1
	46	0	0.3	0	0	0	0	0	0	0.6	0.8	4	1	4
	47	0	0	0	0	0	0	0	0	0	0	1	3	3
	48	0	0	0	0	0	0	0	0	0	0	2	4	2
13	49	0	0	0	0	0	0.4	0	0	0	0	1	1	3
	50	0	0	0	0	0	0	0	0	0	0	2	2	2
	51	0	0	0	0	0	0	0	0	0	0	3	3	4
	52	0	0	0	0	0	0	0	0	0	0	4	4	1
14	53	0	0	0	0	0	0	0	0	0	0.2	4	1	3
	54	0	0	0	0	0	0	0	0	0.7	0	1	2	4
	55	0	0	0	0	0	0	0	0	0	0	2	3	2
	56	0	0	0	0.5	0	0	0	0	0	0	3	4	1
15	57	0	0	0	0	0	0	0	0	0	0.5	3	2	4
	58	0	0	0	0	0	0	0	0	0	0	1	3	3
	59	0	0.4	0.3	0	0	0	0	0	0	0	4	1	2
	60	0	0	0	0	0	0	0	0	0	0	2	4	1
16	61	0	0.9	0	0	0	0	0	0	0	0	3	1	2
	62	0	0	0	0	0	0	0	0	0	0	1	3	3
	63	0	0	0	1.0	0	0	0	0	0	0	2	4	1
	64	0	0	0	0	0	0	0	0	0	0.2	4	2	4

Table 13 Event vectors and associated chains for Example 3 (Part 1)

		Events										Chain		
Commodity	Supplier	1	2	3	4	5	6	7	8	9	10	h3	h4	cc
1	1	1	0	0	0	0	0	0	0	0	1	1	1	3
	2	0	1	0	0	0	1	0	0	0	0	2	2	13
	3	0	0	1	0	0	0	1	0	0	0	3	3	14
	4	0	0	0	1	0	0	0	1	0	0	4	4	16
	5	1	0	0	0	0	0	0	0	0	0	5	6	4
	6	0	0	1	0	0	0	0	0	1	0	6	5	5
	7	0	0	0	0	1	0	0	0	0	0	7	7	12
	8	0	0	0	0	0	0	1	0	0	0	8	8	1
	9	1	0	0	0	0	0	0	0	0	0	9	9	2
	10	0	0	0	1	0	0	0	0	0	0	10	10	10
	11	0	0	0	0	0	0	1	0	0	0	11	11	8
	12	0	0	0	0	0	0	0	0	0	1	12	12	15
	13	1	0	0	0	0	0	0	0	0	0	13	13	11
	14	0	0	0	0	0	0	0	0	0	0	14	15	6
	15	0	0	0	0	0	0	0	0	0	0	15	16	7
	16	0	0	0	0	1	0	0	0	0	0	16	14	9
2	17	1	0	0	0	0	0	0	0	0	0	1	5	2
	18	0	0	0	0	0	0	0	0	0	1	2	6	15
	19	0	0	0	0	0	0	0	0	0	0	3	15	6
	20	0	0	0	0	0	0	0	0	0	0	4	16	7
	21	1	0	0	1	0	0	0	0	0	0	5	3	16
	22	0	1	0	0	1	0	0	0	0	0	6	4	11
	23	0	0	1	0	0	1	0	0	0	1	7	1	13
	24	0	1	0	0	1	0	0	0	1	0	8	2	9
	25	1	0	0	0	0	0	0	0	0	0	9	7	4
	26	0	1	0	0	0	0	0	0	0	0	10	8	3
	27	0	0	1	0	0	0	0	0	0	0	11	9	5
	28	0	0	0	1	0	0	0	0	0	0	12	10	10
	29	0	0	0	0	0	0	0	0	0	1	13	11	1
	30	0	0	0	0	0	0	0	0	1	0	14	12	8
	31	0	0	0	0	0	0	0	1	0	0	15	13	12
	32	0	0	0	0	0	0	1	0	0	0	16	14	14

Table 14 Event vectors and associated chains for Example 3 (Part 2)

		Events										Chain		
Commodity	Supplier	1	2	3	4	5	6	7	8	9	10	h3	h4	cc
3	33	0	0	0	1	0	0	0	0	0	1	1	2	10
	34	0	0	0	0	1	0	0	0	0	0	2	7	11
	35	0	0	0	0	0	1	0	0	0	0	3	8	13
	36	0	0	0	0	0	0	1	0	0	1	4	3	1
	37	1	1	0	0	0	0	0	0	0	0	5	4	3
	38	0	0	0	0	0	0	0	0	0	0	6	14	6
	39	0	0	1	0	1	0	0	0	0	0	7	5	9
	40	0	0	0	0	0	0	0	0	0	0	8	15	8
	41	1	1	1	0	0	0	0	0	0	0	9	1	2
	42	0	1	1	0	0	0	0	0	0	0	10	6	5
	43	0	0	1	0	0	0	0	0	0	0	11	9	14
	44	0	0	0	0	0	0	0	0	0	0	12	16	7
	45	1	0	0	0	0	0	0	0	0	0	13	10	4
	46	0	0	0	0	1	0	0	0	0	0	14	11	12
	47	0	0	0	0	0	0	0	1	0	0	15	12	16
	48	0	0	0	0	0	0	0	0	0	1	16	13	15
4	49	0	0	0	0	0	0	0	0	0	0	1	15	6
	50	1	0	0	0	0	0	0	0	0	0	2	7	4
	51	1	0	1	0	0	0	0	0	0	0	3	2	2
	52	1	0	1	0	1	0	0	0	0	0	4	1	9
	53	0	0	0	0	0	0	1	0	1	0	5	3	8
	54	0	0	0	0	0	0	0	1	0	0	6	8	16
	55	0	0	0	0	0	0	0	0	0	1	7	9	15
	56	0	0	0	0	0	0	0	0	0	0	8	16	7
	57	1	0	0	0	0	0	0	0	0	0	9	10	3
	58	0	1	0	0	0	1	0	0	0	0	10	4	13
	59	0	0	0	0	0	0	1	0	0	0	11	11	14
	60	0	0	0	0	0	0	0	0	0	1	12	12	1
	61	1	0	0	0	1	0	0	0	0	0	13	5	11
	62	0	0	0	0	1	0	0	1	0	0	14	6	12
	63	0	0	0	0	0	0	0	0	1	0	15	13	5
	64	0	0	0	0	0	0	0	0	0	1	16	14	10

Table 15 Event vectors and associated chains for Example 4 (Part 1)

		Events										Chain		
Commodity	Supplier	1	2	3	4	5	6	7	8	9	10	h3	h4	cc
1	1	0.8	0	0	0	0	0	0	0	0	0.8	1	1	7
	2	0	0.2	0	0	0	1.0	0	0	0	0	2	2	12
	3	0	0	0.4	0	0	0	0.6	0	0	0	3	3	5
	4	0	0	0	0.6	0	0	0	0.4	0	0	4	4	4
	5	0.8	0	0	0	0	0	0	0	0	0	5	6	13
	6	0	0	0.6	0	0	0	0	0	0.4	0	6	5	15
	7	0	0	0	0	0.8	0	0	0	0	0	7	7	2
	8	0	0	0	0	0	0	0.6	0	0	0	8	8	8
	9	0.2	0	0	0	0	0	0	0	0	0	9	9	10
	10	0	0	0	0.6	0	0	0	0	0	0	10	10	9
	11	0	0	0	0	0	0	0.4	0	0	0	11	11	6
	12	0	0	0	0	0	0	0	0	0	0.4	12	12	3
	13	0.6	0	0	0	0	0	0	0	0	0	13	13	11
	14	0	0	0	0	0	0	0	0	0	0	14	15	14
	15	0	0	0	0	0	0	0	0	0	0	15	16	1
	16	0	0	0	0	0.8	0	0	0	0	0	16	14	16
2	17	0.6	0	0	0	0	0	0	0	0	0	1	5	13
	18	0	0	0	0	0	0	0	0	0	0.4	2	6	3
	19	0	0	0	0	0	0	0	0	0	0	3	15	6
	20	0	0	0	0	0	0	0	0	0	0	4	16	1
	21	0.6	0	0	0.6	0	0	0	0	0	0	5	3	9
	22	0	0.4	0	0	0.8	0	0	0	0	0	6	4	16
	23	0	0	0.4	0	0	0.8	0	0	0	0.6	7	1	12
	24	0	0.2	0	0	0.8	0	0	0	0.2	0	8	2	2
	25	0.2	0	0	0	0	0	0	0	0	0	9	7	10
	26	0	0.6	0	0	0	0	0	0	0	0	10	8	11
	27	0	0	0.4	0	0	0	0	0	0	0	11	9	15
	28	0	0	0	0.6	0	0	0	0	0	0	12	10	4
	29	0	0	0	0	0	0	0	0	0	0.4	13	11	7
	30	0	0	0	0	0	0	0	0	0.4	0	14	12	8
	31	0	0	0	0	0	0	0	0.4	0	0	15	13	14
	32	0	0	0	0	0	0	0.4	0	0	0	16	14	5

Table 16 Event vectors and associated chains for Example 4 (Part 2)

		Events										Chain		
Commodity	Supplier	1	2	3	4	5	6	7	8	9	10	h3	h4	cc
	33	0	0	0	0.6	0	0	0	0	0	0.4	1	2	4
	34	0	0	0	0	0.6	0	0	0	0	0	2	7	2
	35	0	0	0	0	0	0.8	0	0	0	0	3	8	12
	36	0	0	0	0	0	0	0.8	0	0	0.2	4	3	8
	37	0.4	0.6	0	0	0	0	0	0	0	0	5	4	11
	38	0	0	0	0	0	0	0	0	0	0	6	14	3
	39	0	0	0.2	0	0.6	0	0	0	0	0	7	5	16
3	40	0	0	0	0	0	0	0	0	0	0	8	15	6
	41	0.2	0.2	0.4	0	0	0	0	0	0	0	9	1	9
	42	0	0.4	0.4	0	0	0	0	0	0	0	10	6	5
	43	0	0	0.6	0	0	0	0	0	0	0	11	9	15
	44	0	0	0	0	0	0	0	0	0	0	12	16	1
	45	0.6	0	0	0	0	0	0	0	0	0	13	10	13
	46	0	0	0	0	0.4	0	0	0	0	0	14	11	7
	47	0	0	0	0	0	0	0	0.4	0	0	15	12	14
	48	0	0	0	0	0	0	0	0	0	0.2	16	13	10
	49	0	0	0	0	0	0	0	0	0	0	1	15	5
	50	0.6	0	0	0	0	0	0	0	0	0	2	7	13
	51	0.6	0	0.2	0	0	0	0	0	0	0	3	2	9
	52	0.4	0	0.4	0	1.0	0	0	0	0	0	4	1	16
	53	0	0	0	0	0	0	0.8	0	0.4	0	5	3	8
	54	0	0	0	0	0	0	0	0.4	0	0	6	8	14
	55	0	0	0	0	0	0	0	0	0	0.4	7	9	4
4	56	0	0	0	0	0	0	0	0	0	0	8	16	1
	57	0.4	0	0	0	0	0	0	0	0	0	9	10	11
	58	0	0.6	0	0	0	0.8	0	0	0	0	10	4	12
	59	0	0	0	0	0	0	0.4	0	0	0	11	11	6
	60	0	0	0	0	0	0	0	0	0	0.4	12	12	3
	61	0.2	0	0	0	0.6	0	0	0	0	0	13	5	2
	62	0	0	0	0	0.4	0	0	0.2	0	0	14	6	7
	63	0	0	0	0	0	0	0	0	0.4	0	15	13	15
	64	0	0	0	0	0	0	0	0	0	0.4	16	14	10

Table 17 Failure vectors of Example 3 (*h3*)

	Chains (*h3*)															
	1	2	3	4	5	6	7	8	9	10	11	12	13	14	15	16
Events	1	1	1	1	1	0	0	0	1	0	0	0	1	0	0	0
	0	1	0	0	1	1	0	1	1	1	0	0	0	0	0	0
	0	0	1	1	0	1	1	0	1	1	1	0	0	0	0	0
	1	0	0	1	1	0	0	0	0	1	0	1	0	0	0	0
	0	1	0	1	0	1	1	1	0	0	0	0	1	1	0	1
	0	1	1	0	0	0	1	0	0	1	0	0	0	0	0	0
	0	0	1	1	1	0	0	1	0	0	1	0	0	0	0	1
	0	0	0	1	0	1	0	0	0	0	0	0	0	1	1	0
	0	0	0	0	1	1	0	1	0	0	0	0	0	1	1	0
	1	1	0	1	0	0	1	0	0	0	0	1	1	0	0	1

Table 18 Failure vectors of Example 3 (*h4*)

	Chains (h4)															
	1	2	3	4	5	6	7	8	9	10	11	12	13	14	15	16
Events	1	1	1	1	1	1	1	0	1	1	0	0	1	0	0	0
	1	1	0	1	0	1	0	1	0	0	0	0	0	0	0	0
	1	1	1	0	1	1	0	0	1	0	0	0	0	0	0	0
	0	1	1	1	0	0	0	0	0	1	0	0	0	0	0	0
	1	1	0	1	1	1	1	0	0	0	1	0	0	1	0	0
	1	1	0	1	0	0	0	1	0	0	0	0	0	0	0	0
	0	0	1	0	0	0	0	1	0	0	1	0	0	1	0	0
	0	0	0	1	0	1	0	1	0	0	0	1	1	0	0	0
	0	1	1	0	1	0	0	0	0	0	0	1	1	0	0	0
	1	1	1	0	0	1	0	0	1	0	1	1	1	1	0	0

Table 19 Failure vectors of Example 3 (*cc*)

	Chains (cc)															
	1	2	3	4	5	6	7	8	9	10	11	12	13	14	15	16
Events	0	1	1	1	0	0	0	0	1	0	1	0	0	0	0	1
	0	1	1	0	1	0	0	0	1	0	1	0	1	0	0	0
	0	1	0	0	1	0	0	0	1	0	0	0	1	1	0	1
	0	0	0	0	0	0	0	0	0	1	0	0	0	0	0	0
	0	0	0	0	0	0	0	0	1	0	1	1	0	0	0	0
	0	0	0	0	0	0	0	0	0	0	0	0	1	0	0	0
	1	0	0	0	0	0	0	1	0	0	0	0	0	1	0	0
	0	0	0	0	0	0	0	0	0	0	0	1	0	0	0	0
	0	0	0	0	1	0	0	1	1	0	0	0	0	0	0	0
	1	0	1	0	0	0	0	0	0	1	0	0	1	0	1	1

Table 20 Failure vectors of Example 4 (*h3*)

	Chains (h3)															
	1	2	3	4	5	6	7	8	9	10	11	12	13	14	15	16
Events	0.8	0.6	0.6	0.4	0.8	0	0	0	0.4	0	0	0	0.6	0	0	0
	0	0.2	0	0	0.6	0.4	0	0.2	0.2	0.6	0	0	0	0	0	0
	0	0	0.4	0.4	0	0.6	0.4	0	0.4	0.4	0.6	0	0	0	0	0
	0.6	0	0	0.6	0.6	0	0	0	0	0.6	0	0.6	0	0	0	0
	0	0.6	0	1.0	0	0.8	0.8	0.8	0	0	0	0	0.6	0.4	0	0.8
	0	1.0	0.8	0	0	0	0.8	0	0	0.8	0	0	0	0	0	0
	0	0	0.6	0.8	0.8	0	0	0.6	0	0	0.4	0	0	0	0	0.4
	0	0	0	0.4	0	0.4	0	0	0	0	0	0	0	0.2	0.4	0
	0	0	0	0	0.4	0.4	0	0.2	0	0	0	0	0	0.4	0.4	0
	0.8	0.4	0	0.2	0	0	0.6	0	0	0	0	0.4	0.4	0	0	0.4

Table 21 Failure vectors of Example 4 (*h4*)

	Chains (h4)															
	1	2	3	4	5	6	7	8	9	10	11	12	13	14	15	16
Events	0.8	0.6	0.6	0.4	0.6	0.8	0.6	0	0.2	0.6	0	0	0.6	0	0	0
	0.2	0.2	0	0.6	0	0.4	0	0.6	0	0	0	0	0	0	0	0
	0.4	0.2	0.4	0	0.6	0.4	0	0	0.6	0	0	0	0	0	0	0
	0	0.6	0.6	60	0	0	0	0	0	0.6	0	0	0	0	0	0
	1.0	0.8	0	0.8	0.6	0.4	0.8	0	0	0	0.4	0	0	0.8	0	0
	0.8	1.0	0	0.8	0	0	0	0.8	0	0	0	0	0	0	0	0
	0	0	0.8	0	0	0	0	0.6	0	0	0.4	0	0	0.4	0	0
	0	0	0	0.4	0	0.2	0	0.4	0	0	0	0.4	0.4	0	0	0
	0	0.2	0.4	0	0.4	0	0	0	0	0	0	0.4	0.4	0	0	0
	0.8	0.4	0.2	0	0	0.4	0	0	0.4	0	0.4	0.4	0.2	0.4	0	0

Table 22 Failure vectors of Example 4 (*cc*)

	Chains (cc)															
	1	2	3	4	5	6	7	8	9	10	11	12	13	14	15	16
Events	0	0.2	0	0	0	0	0.8	0	0.6	0.2	0.6	0	0.8	0	0	0.4
	0	0.2	0	0	0.4	0	0	0	0.2	0	0.6	0.6	0	0	0	0.4
	0	0	0	0	0.4	0	0	0	0.4	0	0	0.4	0	0	0.6	0.4
	0	0	0	0.6	0	0	0	0	0.5	0	0	0	0	0	0	0
	0	0.8	0	0	0	0	0.4	0	0	0	0	0	0	0	0	1.0
	0	0	0	0	0	0	0	0	0	0	0	1.0	0	0	0	0
	0	0	0	0	0.6	0.4	0	0.8	0	0	0	0	0	0	0	0
	0	0	0	0.4	0	0	0.2	0	0	0	0	0	0	0.4	0	0
	0	0.2	0	0	0	0	0	0.4	0	0	0	0	0	0	0.4	0
	0	0	0.4	0.4	0	0	0.8	0.2	0	0.4	0	0.6	0	0	0	0

Risk Management through Flexible Capacity Allocation and Price Control – Auctions in the New Car Sales Process

Thomas Ruhnau[1] and Thomas Peisl[2]

[1] BMW AG, Petuelring 130, 80788 München, Germany
thomas.ruhnau@bmw.com
[2] Munich University of Applied Sciences, Professor of Strategy, Am Stadtpark 20, 81243 München, Germany
thomas.peisl@hm.edu

1 Introduction

Auctions are already used in the automotive industry on a regular basis in two areas. They are applied in used car sales (cf. Yaron 2008, p. 325 ff.; Berger 2009, p. 10 ff.) and in procurement as reverse auctions, in which suppliers bid to win the order (cf. Talluri/van Ryzin 2004, p. 243). In sales of new automobiles there is no comprehensive application of auctions in place so far. This research proposes a model for a forward auction with one supplier and a large number of bidders for the indirect sales of new vehicles aiming at risk reduction in a manufacturer's sales channel. Based on characteristics of indirect car sales, the auction model is defined and explained by a process flow. Based on the proposed auction model, challenges are discussed and solutions are suggested.

1.1 Contribution Margin as the Control Variable for the Vehicle Allocation

Car manufacturers use a multi-channel approach for the distribution of new vehicles. In the context of this research topic, a distinction needs to be made between direct and indirect sales. The most common method of distribution in the automotive industry is the indirect sales via distributors and takes the central role in the distribution channels of the international automotive sales (cf. Diez 2006, p. 274 ff.). Dealers buy vehicles from the original equipment manufacturer (OEM) based on certain conditions and then sell these cars in their own name and on their own account to customers. Thus, this attribute is of particular importance because it means that the OEM generates revenue from sales to dealers.

M. Essig et al. (Eds.): Supply Chain Safety Management, LNL, pp. 357–372.
springerlink.com

Direct sales of vehicles are carried out either via the central sales departments of the OEM, which are specialized in selling to specific customer groups (e.g. employees, VIPs, fleet customers, authorities) or OEM-owned sales subsidiaries (cf. Diez 2006, S. 271). These subsidiaries are organized like a dealer and operate using a business design of a profit center. The proposed auction model will be validated for OEM indirect sales as the first step of our research.

The manufacturer's contribution margin (CM) results in general from the difference between ex-factory selling price and production costs. For vehicles and optional equipment, a certain ex-factory price is defined on which the retailer may order a vehicle from the OEM (from the dealer's point of view, the ex-factory price is consistent with the acquisition price or dealer cost). Initially, it is not important whether the dealer specifies a vehicle on the basis of a specific customer order or for inventory. These ex-factory prices are based on the vehicle's manufacturer's suggested retail price (MSRP) which grants the dealer a certain margin. Due to different price levels, different currencies and exchange rate fluctuations, international ex-factory prices are a subject to variation which results in different revenues at the OEM.

As certain market conditions (e.g. demand fluctuations, the level of discounts) can lead to a situation in which the regular dealer margin is not sufficient to ensure vehicle sales, OEMs often support the dealers with various forms of sales promotions (cf. Homburg/Krohmer 2009, S. 792 f.; Kotler et al 2007, S. 763 ff.). These promotions are used consistently within one market and include premiums (trade-in premiums, conquest premiums, loyalization premiums, registration premiums, etc.), bonuses (for example, to achieve certain sales targets), or temporary reductions of dealer purchase prices. All instruments have in common that they increase the dealer margin and, thus, allow a higher discount level. The OEM aims at a short-term adjustment of the transaction price in order to influence demand. However, the OEM has no direct control over the transaction price.

The following figure shows the price hierarchy and the contribution sources in vehicle sales as explained above.

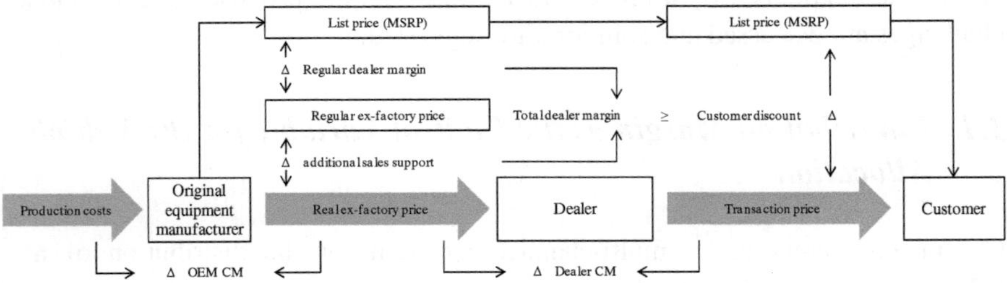

Fig. 1 Price hierarchy and contribution sources in the automotive sales (own illustration)

1.2 The OEM's (Re-)Active Price and Allocation Control

Different price levels, fluctuating exchange rates and the dynamically changing market developments lead to different contribution margins for the same products on the OEM side. Therefore, the assignment of identical capacity units in the OEM's production facility results in different yields, depending on the final market. During periods of demand backlog, OEMs will try to allocate their production capacity in the short-term with the objective to maximize the contribution margin. This approach may result in a situation in which it is better to offer vehicles in market *A* using additional sales promotions because the contribution margin (despite higher costs of retail) is still higher than in market *B* which is characterized by a lower price level.

Therefore, the OEMs try to respond flexibly to different market developments in order to control the utilization of capacity to gain an optimum yield.

The following aspects describe the OEM's core challenges:

Short term price control: In order to influence demand in each market, OEMs are limited to an indirect and time-delayed influence on demand through price control. For example, dealer margins may be determined in the dealer contracts. Changes in sales promotion can only be made as a reaction to market development. Due to competitive situations, list price adjustments are also feasible only within a certain range (cf. Diez 2006, p. 242). The OEM cannot directly control the transaction price due to the indirect sales structure.\

International production allocation: The OEM can only respond to short-term macroeconomic changes through flexible production allocation. The allocation optimization which is specified in the production planning is based either on historical data and suffers, in consequence, from a time lag; or it is based on assumptions about the future and is a subject to uncertainty (cf. Klug 2010, S. 259 ff.). For instance, if a market develops better than expected, capacity can be re-allocated with only a delay.

Intra-national production allocation: The OEM publishes one list price per market only. Within this market, there will inevitably be dealers who need to use more of their margin as sales support than others (e.g., due to local conditions, such as differences in income levels). For the OEM, this causes ex-factory prices for a market to be either too low or too high, because there cannot be the right price due to intra-national differences. Therefore, the allocation within a market to single dealerships is an additional challenge.

Conflict between control costs and error costs: The control efforts for market or dealer specific margin levels and necessary sales promotions are significant. OEMs define standardized margin levels or sales promotions for specific markets or regions. The disadvantage is that this leads automatically to the previously described errors in the control of ex-factory prices or sales promotions and production allocation. If the OEM attempts to reduce the faultiness, the control efforts will inevitably increase.

In summary, OEMs' current price and sales control do not allow for a balance of supply and demand between OEMs and dealers. The OEM always affects supply and demand by (re-)active pricing, setting of sales promotions and (re-)active capacity allocation. A free competition for the existing, ultimately always limited production capacity does not exist, which therefore leads to a non-optimal utilization of capacity.

2 Auctions in the New Car Sales Process

2.1 Auctions as a Tool to Manage Risks

The different price levels in different markets are the result of spatial price discrimination, which is applied in automobile sales (cf. Homburg/Krohmer 2009, p. 702).[1] The basic idea of price discrimination is to skim potential customers' willingness to pay as much as possible (cf. Simon/Butscher 2001, p. 110 f.; Nagle/Hogan 2007, p. 95.). A price effect can be achieved through price discrimination (skimming of the consumer surplus) and, in comparison to the flat price positioning, a volume effect can be caused by additional demand (cf. Klein/Steinhardt 2008, p. 49 f).

Ideal price discrimination is obtained when a price is set at each customer's maximum willingness to pay, equivalent to first-degree price discrimination (cf. Stole 2007, p. 2229; Varian 1989, p. 524). The predominant forms of price discrimination used in car sales by the OEMs are second and third degree price discrimination (cf. Diez 2006, p. 243 ff.; Homburg/Krohmer 2009, p. 701 ff.). Even if the optimal first-degree price discrimination is awarded to have mainly theoretical benefits (cf. Diller 2008, p. 228), there are price-policy practices, which tend to have the goal to achieve perfect price discrimination. Individual price negotiation with the customer is one of these practices. However, it is not implemented by the OEM, but between dealers and customers. It should be noted that in the individual price negotiation, the problem occurs that the sales person has limited information regarding the buyer's willingness to pay. Auctioning is a potential approach to determine the maximum willingness to pay (cf. Talluri/van Ryzin 2004, p. 242).

In the context of our research, an auction process is proposed in which the OEM auctions limited capacity to dealers and determines the market's maximum willingness to pay. Hence, the OEM maximizes the contribution margin per unit. In addition to price determination, an auction also provides allocation by identifying one or more auction winners (cf. Kalagnanam/Parkes 2004, p. 145). The special attribute of the auction process outlined here is that not a single unit of a product is to be sold, but a greater quantity of identical products, in contrast to, for example, the auction of art objects. The central goal of the auction is not only to determine the maximum willingness to pay, but also to obtain a self-regulating optimal allocation of production capacity. The auction represents a particular kind of marketplace, where supply meets demand. Classically, auctions are always used in a situation of monopoly (cf. McAfee/McMillan 1987, p. 703). There is no monopoly in the car market; rather an oligopoly is assumed. However, each OEM is a monopolist for specific brands being the sole supplier of his authorized dealers.

---

[1] Next to spatial price discrimination, the automotive industry also applies product-related price differentiation (cf. Wübker 1998, S. 18). For further information about the concept of price discrimination please refer to Tacke 1989, Diller 2008 und Simon/Faßnacht 2009.

Through the auction process the OEM's (re-)active control of ex-factory prices, sales promotion and production allocation can be eliminated. The auction, in theory, builds up a self-regulatory approach that solves these tasks in an optimal way because it puts dealers internationally and intra-nationally in a competitive situation for a scarce commodity. Ideal pricing and capacity allocation may be achieved in a contribution-optimal way without extensive investigation of buyer behavior and willingness to pay (cf. Talluri/van Ryzin 2004, p. 242). The auction system in our research is limited to transactions between OEMs and dealers (B2B). For final customers, the process remains the same. The OEM publishes a list price and customers buy their cars from the dealer using an individual negotiation process. Conflicts with regional pricing strategies or adverse effects to the brand do not occur.

2.2 Classification and Design of Auctions

McAfee/McMillan (1987, p. 701) define an auction "[…] as a market institution with an explicit set of rules determining resource allocation and prices on the basis of bids from the market participants." The term *auction* is defined as both, an offer to sale as well as an offer to purchase (cf. McAfee/McMillan 1987, p. 701). In order to classify an auction, Kalagnanam/Parkes (2004, p. 144 ff.) propose a number of key characteristics:

Resources: It is necessary to define exactly which resource is auctioned. The resource can just consist of one single item (single auctions) or several items (combinatorial auctions) and the resource can be offered individually (single unit) or in a majority (multi-unit). Additionally, it needs to be determined whether the resource is a standard commodity or if it needs to be specified by certain attributes.

Market structure: An auction is an instrument for negotiation between buyer and seller. One-sided auctions can appear as *forward auctions* and *reverse auctions*. In forward auctions one single seller meets a variety of buyers. In reverse auctions, one buyer faces multiple vendors (typical for procurement auctions). A market structure with multiple sellers and multiple buyers is called *double auction*.

Preference structure: In an auction, the bidder's preference structure is of special importance as it affects other factors. Preferences define the bidder's benefit from different results of the auction. In a multi-unit auction, for example, the marginal utility of a bidder might decrease when offering additional units.

Bid structure: The bid structure defines the flexibility by which a bidder can express his needs. In a single-item single-unit auction, merely a statement of willingness to pay (auction with rising prices = English auction) or price acceptance (auction with falling prices = Dutch auction) is necessary. In a multi-unit auction, the quantity needs to be specified by the bidder, as well.

Matching supply and demand: Matching supply and demand is the core of an auction, which means it must fixed how to determine the winner of an auction. In a single-unit auction just one seller and one buyer are defined. In a multi-unit auction, multiple buyers might win an auction facing one single seller.

Information feedback: In a direct auction mechanism, bids are placed without providing any information feedback from the auction to the bidders. In an indirect mechanism, the bidders receive information within the framework of the auction and they may adjust their bids as it is common in art auctions. The information feedback typically consists of price information and a statement on the provisional allocation.

2.3 Basic Attributes of an Auction Model for Indirect Car Sales

Resources: Auction resources are units of capacity in the production at the OEM. A capacity unit is perceived as a single item, because it is not offered in combination with other objects. Our research proposes to auction not only one single unit but a plurality of identical units of capacity in a given period as larger manufacturers may produce a couple of million vehicles a year. Since cars need to be specified in more detail, such as engine power, features or color, the attributes need to be assigned to the resources by the bidders.

Market Structure: In the case of this auction model, a forward-auction is to be assumed, since only one seller (OEM) faces of a large number of bidders (dealers).

Preference structure: The bidders' preference structure is characterized by decreasing benefits, and, in accordance, declining willingness to pay per additional unit. Here, the willingness to pay is based on the price elasticity of each dealer's local market environment. If a dealer buys a car at a high price, he will find a correspondingly low number of customers who are ready to purchase this car at this price plus an additional surcharge which reflects the dealer's contribution. The more vehicles a dealer aims to sell, the lower the price needs to be, because this is the only way the dealer can lower the transaction price, while ensuring a stable margin. The dealer's margin is subject to variation but the preference structure is oriented on the price-demand function of the appropriate sales market. The price-demand function is defined ranging from OEM production cost to the MSRP, as the OEM cannot offer the vehicles below production costs[2] and customers will not be prepared to pay a price above the MSRP.[3]

The following figure illustrates the simplified preference structure assuming a linear price-demand function.

---

[2] Special cases, e.g. vehicles designed as brand shapers with low volume targets or models regarded as marketing investments, are not considered.

[3] Special cases, e.g. highly emotional cars with a strict limitation to a very few units, are not considered.

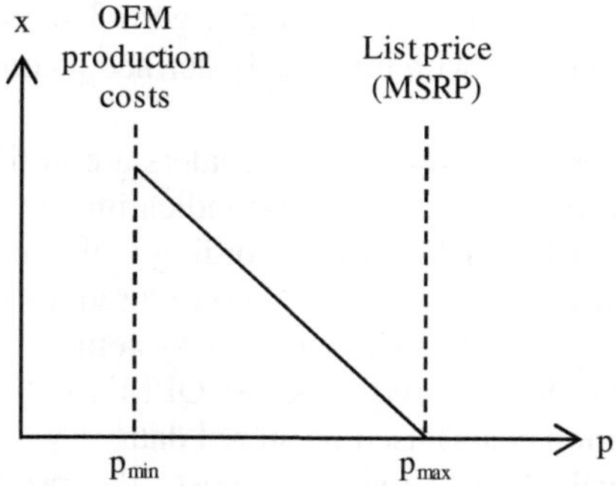

Fig. 2 Basic preference structure of automobile dealers (own illustration)

Bid structure: Since, as defined above, a multi-unit auction with attributes is applied, the dealer has to specify both, the quantity and the attributes. We propose a Dutch auction which only requires acceptance of the called price. In this case, each vehicle can be confirmed separately. For larger dealers, there may be demand for several hundred cars per week. The Dutch auction simplifies operations at the dealer. It is possible that – in case of remaining unused capacity – the dealer places additional orders over time in the knowledge that prices fall. Thus, a competitive environment for the remaining capacity can be ensured and bidding for additional vehicles along the preference curves can be carried out. Additional details of the bid structure are discussed in chapter 2.4.

Matching supply and demand: During a Dutch auction the winner of the auction is determined by acceptance of the price. In this case, the order of the pre-specified vehicle is initiated by the dealer. In a multi-unit auction bidders can simultaneously accept a plurality of certain prices and complete the auction. However, in case of remaining capacities, the dealers must have the possibility to demand additional vehicles, as mentioned above. Prices fall within the Dutch auction until supply and demand are balanced, i.e. no more capacity is available or the auction will be cancelled, because the contribution has reached the required minimum.

Information feedback: This auction model presumes a direct auction mechanism. The dealers accept round by round their placed orders with no information about the exact remaining capacity or the number of current supplements. Within one bidding round, no information feedback is provided. But in case of residual capacities, the dealer is informed, in order to allow the configuration of additional vehicles. If a dealer has already triggered vehicle orders in an earlier auction round by accepting the called price, the dealer now knows that those vehicles (or at least part of them) could have been bought at a lower price. In consequence, the auction mechanism is described as direct, but there is an information feedback, which may affect the retailer for future auctions.

2.4 Process Flow in the Auction Model

Within the process flow, the production capacity of a given period is sold. Usually, the total production capacity is divided into production weeks. After explaining the proposed process step by step, figure 3 shows the sequence of the new car auction in accordance with previously defined characteristics.

1) Via the OEM's ordering systems the dealers are notified and the auction is opened for a certain production period indicating the expected delivery date which might differ for each dealer according to the required transport time.
2) The dealers enter the required vehicle configurations in the ordering system. The corresponding MSRP is shown in the system.
3) After sending the delivery requests, the OEM's ERP systems calculate the accruing production costs based on stored data.
4) Based on the calculated production costs the price for each vehicle is generated by adding a required contribution margin.
5) Prices are transferred to the dealers. Each vehicle now has an individual price (shown in local currency). However, for the OEM, all vehicles provide the same contribution, independent from vehicle type or requested specification.
6) Every dealer is able to inform himself about his margin (difference between MSRP and called price). If the margin is sufficient to sell the car to the market and if an acceptable profit is provided, the dealer decides to accept the called price.
7a) It needs to be tested if all accepted orders can be covered by the available capacity.
8) In the most likely case that the number of accepted prices is not exactly equal to the offered capacity, the OEM needs to conduct an order selection, e.g. by quota or strategic considerations.
9) Negatively selected orders are refused.
10) Positively selected orders are confirmed and the order execution (production) is initiated. Afterwards, an auction can be started for a new production period.
7b) Caused by different price levels, the dealer margin is negative in certain markets or it is too low to ensure successful sales at the beginning of the auction process. In this case, the dealer does not accept the called prices. Instead, he waits for the next round of bidding at lower prices. This next bidding opportunity will occur only if there is residual capacity left.
11) If there is residual capacity for the appropriate period, the dealer which already accepted prices, receive an order cconfirmation and the order execution is started.
12) All dealers are informed that further capacity is available.
13) The dealers get the opportunity to set up additional delivery requests. For example, it is possible that a dealer's requests are completely confirmed. Knowing that prices will fall, he might be prepared to buy further vehicles. Thus, it can be ensured that open delivery requests (e.g. coming from markets with a low price level) remain in a competitive situation for the limited capacity.
14) Based on the new delivery requests, the production costs are calculated, the CM-requirement is lowered and a new price for all delivery requests is generated.

15a) The OEM's CM-level is requested to remain positive respectively to stay above a certain minimum requirement which is needed to be tested. If the contribution fulfills the desired level respectively is positive, the price are transferred to the dealers.

15b) If this is not the case, there is no further offer of capacity in this certain production period and a new auction process for a new production period is started.

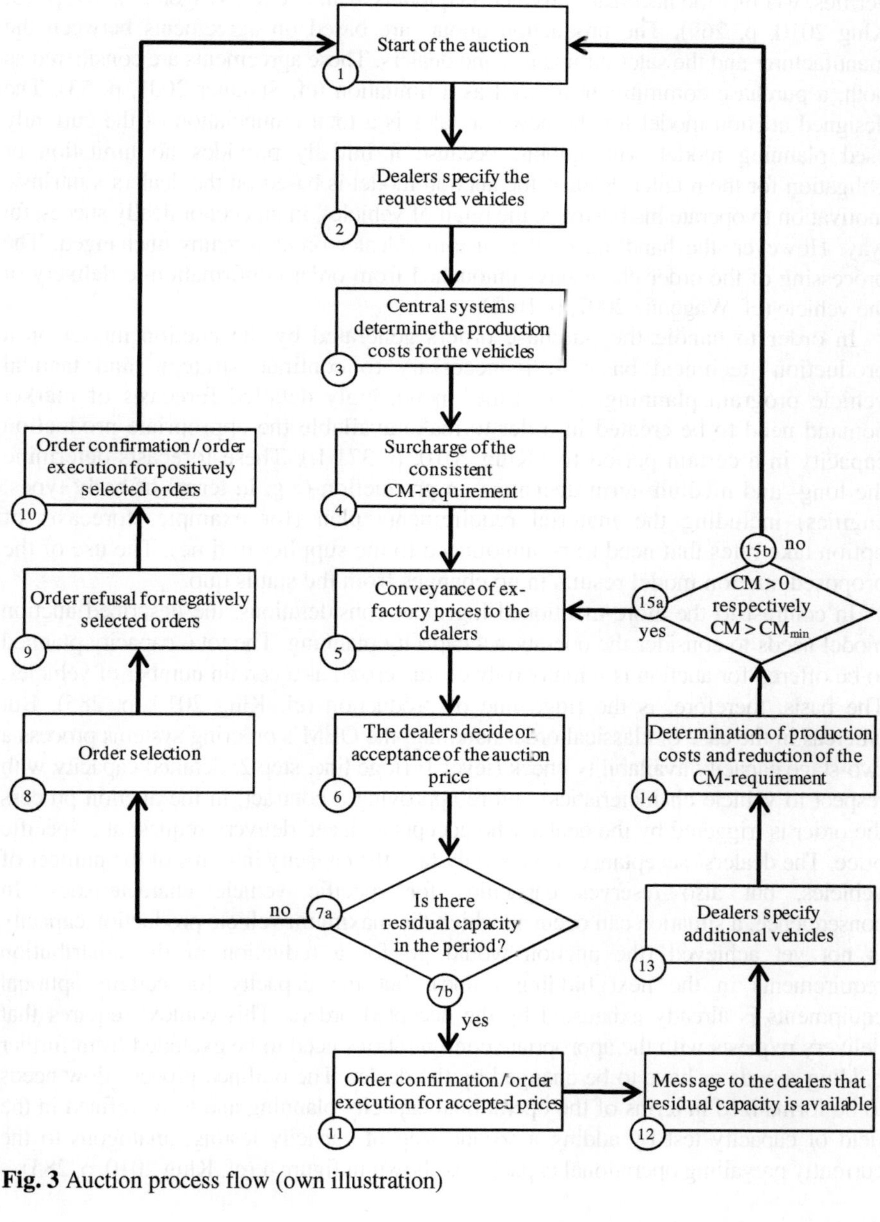

Fig. 3 Auction process flow (own illustration)

2.5 *Classification of the Auction Process Model to the OEM's Capacity and Production Control*

The implementation of the auction model requires the consideration of prevailing logistics processes in the OEM's capacity and production control. The common practice of the customer order process provides that the dealer places customer orders or orders for his stock in the order systems. Afterwards the OEM's sales organization verifies, whether the necessary production quota is available (cf. Wagenitz 2007, p. 15; Klug 2010, p. 269). The production quotas are based on agreements between the manufacturer and the sales companies and dealers. These agreements are considered as both, a purchase commitment as well as a limitation (cf. Stautner 2001, p. 53). The designed auction model for the new car sales is a total renunciation of the currently used planning model with quotas, because it initially provides no limitation or obligation for the retailer. Rather, the auction model is based on the dealers's intrinsic motivation to operate his business, the retail of vehicles, in an economically successful way. However, the handling of the customer/dealer order remains unchanged. The processing of the order chain stays untouched from order confirmation to delivery of the vehicle (cf. Wagenitz 2007, p. 16 ff).

In order to handle the purchase orders generated by the auction model on a production- technical basis, it is necessary to continue strategic and tactical vehicle program planning. Over time, increasingly detailed forecasts of market demand need to be created in order to make available the appropriate production capacity in a certain period (cf. Klug 2010, p. 371 f.). These forecasts determine the long- and medium-term decisions on production (e.g. in terms of body types, engines) including the material requirements plan (for example, forecasts of option take rates that need to be announced to the supplier in time). The use of the proposed auction model results in no changes from the status quo.

In contrast to the afore-mentioned logistical considerations, the described auction model needs to consider the operational capacity planning. The total capacity planned to be offered for auction is initially only characterized as a certain number of vehicles. The basis, therefore, is the ridge line of production (cf. Klug 2010, p. 285). But whereas in the case of classical order handling, the OEM's ordering systems process a two-stage capacity availability check (level 1: ridge line, step 2: detailed capacity with respect to vehicle characteristics) before approving a contract, in the auction process the order is triggered by the dealer who accepts a placed delivery request at a specific price. The dealers' acceptances do not only face the capacity in terms of the number of vehicles, but also reserves capacities for specific vehicle characteristics. In consequence, a situation can occur in which the maximum vehicle production capacity is not yet achieved (the auction would go for a reduction of the contribution requirements in the next bidding round), but the capacity for certain optional equipments is already exhausted by the accepted orders. This context requires that delivery requests with the appropriate configurations need to be excluded from further bidding rounds or have to be changed by the dealer. The outlined process flow needs to be formalized in terms of the operational capacity planning and to be refined in the field of capacity test by adding a second step of capacity testing, analogous to the currently prevailing operational capacity as shown in figure 4 (cf. Klug 2010, p. 285).

7c) After testing if there is residual capacity in terms of vehicles, the same test needs to be executed in terms of potentially ordered specifications. If there is sufficient capacity available, the order confirmation can be given and the auction can go for a new bidding round, respectively is finished if the contribution requirements are not fulfilled, as shown above.

7d) If the capacity for certain options or specifications is already exhausted, at first, an order selection needs to be carried out.

7e) Negatively selected orders are refused.

7f) All open delivery requests that haven't been accepted by the dealers have not been tested in terms of capacity availability. If these delivery requests contain specifications that would face capacity limitations, they are excluded from the next round of bidding. Thus, it can be ensured that the appropriate delivery requests can be changed by the dealers in order to participate in the next auction round.

7g) The OEM's ordering systems are adjusted for the next auction lap in order to avoid the ordering of options/specification that already ran out of capacity. After this second step of capacity availability testing, the auction process can continue as described above.

2.6 Implementation Challenges and Initial Solutions

Implementation challenges are discussed on the basis of the involved parties' perspectives. OEM-related, dealer-related and customer-related issues are addressed.

OEM-related challenges: The OEM is responsible for the systemic implementation of the auction as a part of the ordering and distribution process. In order to ensure a successful implementation of the auctioning process, the prerequisites include a standard online-ordering system, consideration of country specific legislation, and sufficient computational power to ensure quick response times.

This ordering system must have an interface to the OEM's ERP system. The cost information for each vehicle specification needs to be allocated and production systems must reflect the capacity. An accurate cost and capacity planning is a prerequisite for the application of the auction model. This means, for example, that not only the production costs need to be integrated, but also country-specific tax rules must be captured, as these may influence costs. Capacity planning must reach a level of detail that also covers all attributes, such as drive train, options and colors.

OEMs do not always subordinate long-term goals to short-term optimization potential. For example, if a manufacturer aims to reach a specific market share target in order to achieve a strategic competitive position, it is possible that short-term profit optimizing may be subordinate to strategic objectives. In the proposed auction model, this may be simply carried out by applying a subsidy value. For the respective markets, the OEM can simply lower the contribution claim (Step 4 and Step 14 in the process flow). Automatically, dealers gain more competitiveness and will be able to bid on a larger number of vehicles. The desired strategic quantity effect can occur without negatively influencing the optimization effect.

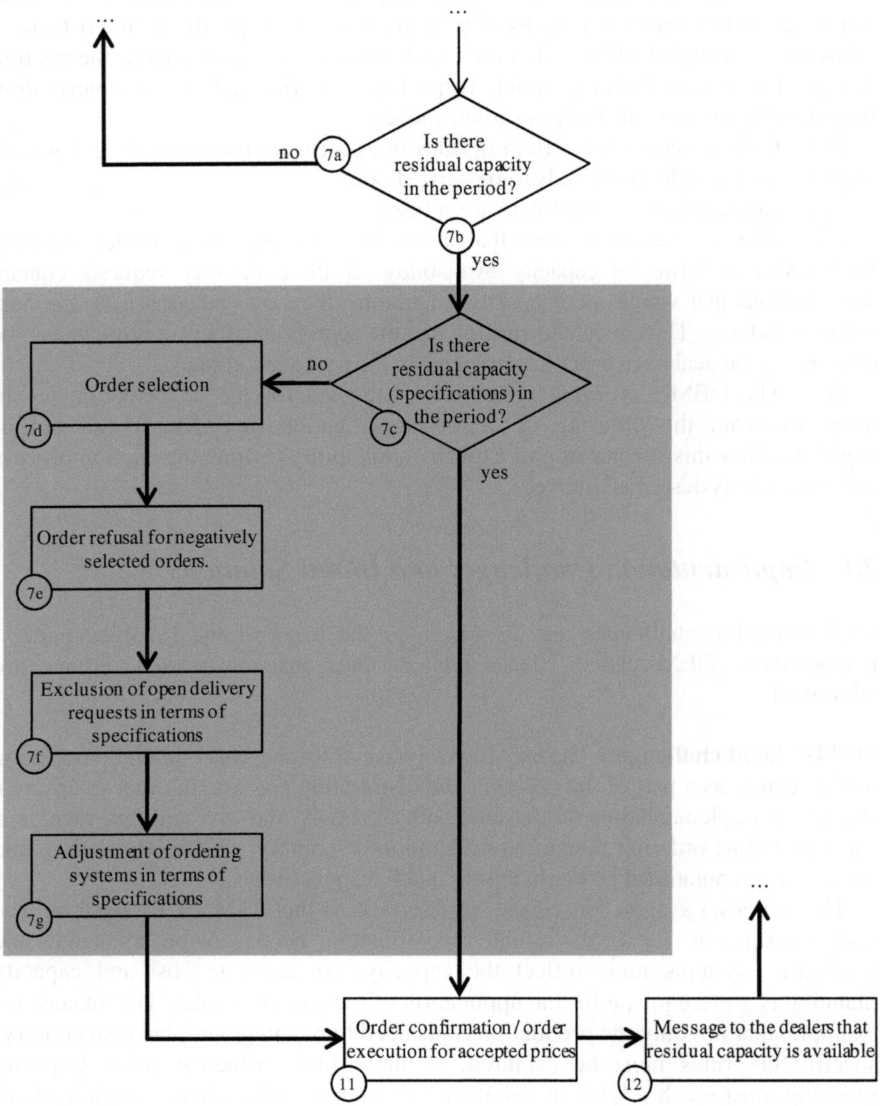

Fig. 4 Integration of a second capacity test into the process flow (own illustration)

Dealer-related challenges: The dealer's objective is to maximize the margin per vehicle sold to the customer. This objective should be encouraged by the OEM in order to realize the highest possible ex-factory price. For the dealer, the intention to achieve high margins is equivalent to the goal of providing customers the lowest possible discount on list prices. In order to enable the dealer to employ available margins as effectively as possible, the OEM needs to support and train the dealers in identifying the optimal discount type for various customer segments.

The proposed auction model assumes a myopic behavior (cf. Talluri/van Ryzin 2004, S. 223). It is assumed that the dealer directly accepts an auction price on the basis of relative advantage and realizes the maximum willingness to pay. Dealers located in markets with a relatively high price level have a structural advantage over dealers in markets with lower prices. With time, such a dealer may realize that there is still capacity available, although all orders have been confirmed. In order to increase surplus, the dealer converts behavior from myopic to strategic. The dealer waits until prices fall and generates a higher margin due to lower acquisition prices. From a manufacturer's point of view, there is a need to encourage the dealers to act myopically. A potential solution to reach this objective can be provided by stopping capacity offers to certain markets on an irregular basis, even if there is residual capacity available. As it is to be assumed that the dealer tries to maintain an access supply situation in his own interest, such an action might lead the dealer to exhibit myopic order acceptance behavior. The disadvantage of this method is that the auction no longer provides the optimum price and capacity allocation decision for the specific period. However, the artificially shortened capacity can, at least partly, be covered by direct sales orders. Fluctuating capacity quantity for periods of equal length can be achieved by the manufacturer in this way. For the bidding dealers it becomes more difficult to discern regularity in the available capacity.

Another option to encourage the dealers to adopt myopic behavior is in the adjustment of the price reduction steps. Depending on how big the next price reduction step is, a dealer competes with a certain number of other dealers who are also in the area of relative advantages and potentially raise demand for capacity. In order to reduce risk that dealers become able to estimate this context over time, the OEM may vary the reduction steps. Therefore, dealers cannot know the characteristics of the competitive situation after the next reduction step. They will tend to bid on a vehicle immediately, as soon as the called price generates a relative advantage and will neglect waiting and observing.

Customer-related challenges: The auction model presumes that the dealer composes the specification of the vehicle in the ordering systems of the OEM, which is in line with current situation. In many markets, this is also the only way, as customers are used to buy vehicles *off the shelf* from the show room or stock. Even long distances between sales market and county of origin can lead to a long delivery time which may be a reason for a distribution based on a built-to-stock approach. Depending on the sales model, however, some of the orders are made based on an actual customer request. The dealer can conduct a sales negotiation with the customer, but information on delivery time and dealer purchase price will be ultimately only available after the auction process. Thus, there is an uncertainty on the dealer side, resulting in inaccurate information about delivery time and unclear revenue situation. This challenge can be solved either by simply accepting and managing this weakness or by excluding customer orders from the auction. In the first case, a dealer negotiates with reserved terms of delivery time and then tries to place the customer's order in one of the next auctions. If he succeeds, he can confirm to the customer on the contract and if not, he has to offer a different car to the customer or - as a last resort - to reject the customer. In particular, premium OEMs will consider this solution as not appropriate. The possibility remains to apply a standard margin for customer order. Depending on the desired adherence to delivery dates, the OEM is flexible in the field of capacity commitment. Hence, capacity can be blocked for customer orders or order information is included in the auction process. For example, an order which is marked in the system as customer occupied can always be included in a production period, when the contribution from this customer order is greater than or equal to the contribution level which is achieved through the auction. If this is not the case, the production of the customer order is postponed until a desired latest production date.

3 Summary and Outlook

The proposed auction model for the indirect sales process of new vehicles provides an innovative and comprehensive approach to the key challenges of short-term price optimization and production allocation control at OEMs. In auctioning, ex-factory prices can be optimized and the maximum willingness to pay can be skimmed. Simultaneously, the production capacity is allocated in exactly these markets (internationally) and to those dealers (intra-nationally) that can reach these maximum prices with their customers. The control effort is minimized as the optimal distribution is defined as a self-adjusting mechanism and sales channel risks are lowered. The OEM may dispense with the costly collection of short-term preference and demand information that is otherwise required to be able to control prices and production capacities.

In order to achieve the maximum impact of the proposed auction model, in theory, the distribution needs to focus on a pull strategy. In contrast to a push strategy, which attempts to reach targets via price variation and building up inventory pressure on the market, the pull strategy joins the real market demand (cf. Diez 2006, p. 357). It is assumed that the dealer has a vested interest to turn as many vehicles as possible. In this case, pressure by the manufacturer is not necessary as the auction balances supply and demand in an ideal way. The price acts regulatory and provides the dealer with order incentives through the instrument of automatic price reduction as a part of the rounds of bidding which supports the pull approach. Thus, price discrimination close to the actual price-demand function can be achieved.

Future research will show whether the outlined challenges that arise in connection with the auction model can be overcome by the approaches discussed above. We will model the proposed auction process using quantitative methods. In addition, the extension of the process model to direct car sales will be further step of our future research.

It should also be noted that such an auction sales system cannot be implemented without adequate testing. Prior to implementation, simulations need to be set up analyzing the behavior of auction attendees involving game theory approaches. Also other auction procedures need be illuminated, such as the use of *second-price auctions* (cf. Vickrey 1961, p. 8 ff). If tests run successfully, it is advisable to roll out the auction process to a few pilot markets. This also allows comparing the auction model with traditional price and allocation control before the system is launched globally.

Finally, the presented research opens a field of research, which leaves space for an extensive scope of investigations in fields such as risk management, logistics, process management, information technology, and operations research.

References

Baker, T., Murthy, N.N.: Viability of Auction-Based Revenue Management in Sequential Markets. Decision Sciences, 259–286 (2005)

Berger, M.: Gradmesser für die Branche. In: Gebrauchtwagen Praxis, pp. 10–15 (2009)

Caldentey, R., Vulcano, G.: Online Auction and List Price Revenue Management. Management Science, 795–813 (2007)

Carroll, K., Coates, D.: Teaching Price Discrimination: Some Clarification. Southern Economic Journal, 466–480 (1999)

Diez, W.: Automobilmarketing - Navigationssystem für neue Absatzstrategien. Landsberg am Lech, mi-Fachverlag (2006)

Diller, H.: Preispolitik. Stuttgart, Kohlhammer (2008)

Homburg, C., Krohmer, H.: Marketingmanagement: Strategie, Instrumente, Umsetzung, Unternehmensführung. Wiesbaden, Gabler (2009)

Kalagnanam, J., Parkes, D.C.: Auctions, Bidding and Exchange Design. In: Simchi-Levi, D., Wu, S.D., Shen, Z.J. (eds.) Handbook of Quantitative Supply Chain Analysis - Modeling in the E-Business Era, pp. 143–212. Kluwer Academic Publishers, Boston (2004)

Klein, R., Steinhardt, C.: Revenue Management - Grundlagen und mathematische Methoden. Springer, Berlin (2008)

Klemperer, P.: Auction Theory: A Guide to the Literature. Journal of Economic Surveys, 227–286 (1999)

Klug, F.: Logistikmanagement in der Automobilindustrie. Springer, Berlin (2010)

Kotler, P., Keller, K.L., Bliemel, F.: Marketing Management - Strategien für wertschaffendes Handeln. Pearson Studium, München (2007)

McAfee, R.P., McMillan, J.: Auctions and Bidding. Journal of Economic Literature, 699–738 (1987)

Milgrom, P.: Auctions and Bidding: A Primer. Journal of Economic Perspectives, 3–22 (1989)

Nagle, T.T., Hogan, J.E.: Strategie und Taktik in der Preispolitik. Pearson, München (2007)

Raviv, Y.: The Role of the Bidding Processes in Price Determination: Jump Bidding in Sequential English Auctions. Economic Inquiry, 325–341 (2008)

Shen, Z.J.M., Su, X.: Customer Behavior Modeling in Revenue Management and Auctions: A Review and New Research Opportunities. Production & Operations Management, 713–728 (2007)

Simon, H., Butscher, S.A.: Individualised Pricing: Boosting Profitability with the Higher Art of Power Pricing. European Management Journal, 109–114 (2001)

Simon, H., Faßnacht, M.: Preismanagement - Strategie, Analyse, Entscheidung. Gabler, Wiesbaden (2009)

Stautner, U.: Kundenorientierte Lagerfertigung im Automobilvertrieb. Deutscher Universitäts-Verlag, Wiesbaden (2001)

Stole, L.A.: Price Discrimination and Competition. In: Armstrong, M., Porter, R.H. (eds.) Handbook of Industrial Organization, vol. 3, pp. 2221–2300. Elsevier Science & Technology, Amsterdam (2007)

Tacke, G.: Nichtlineare Preisbildung: Höhere Gewinne durch Differenzierung. Gabler, Wiesbaden (1989)

Talluri, K.T., van Ryzin, G.J.: The Theory and Practice of Revenue Management. Springer, New York (2004)

Varian, H.R.: Price Discrimination. In: Schmalensee, R., Willing, R.D. (eds.) Handbook of Industrial Organization, pp. 597–654. Elsevier Science & Technology, Cambridge (1989)

Vickrey, W.: Counterspeculation, Auctions and Competitive Sealed Tenders. Journal of Finance, 8–37 (1961)

Vulcano, G., van Ryzin, G., Maglaras, C.: Optimal Dynamic Auctions for Revenue Management. Management Science, 1388–1407 (2002)

Wübker, G.: Preisbündelung - Formen, Theorie, Messung und Umsetzung. Gabler, Wiesbaden (1998)

Author Index

MIX
Papier aus verantwortungsvollen Quellen
Paper from responsible sources
FSC® C105338

If you have any concerns about our products,
you can contact us on
ProductSafety@springernature.com

In case Publisher is established outside the EU,
the EU authorized representative is:
Springer Nature Customer Service Center GmbH
Europaplatz 3, 69115 Heidelberg, Germany

Printed by Libri Plureos GmbH
in Hamburg, Germany